# Lecture Notes in Computer Science    13776

More information about this series at https://link.springer.com/bookseries/558

Anisur Rahaman Molla · Gokarna Sharma ·
Pradeep Kumar · Sanjay Rawat (Eds.)

# Distributed Computing and Intelligent Technology

19th International Conference, ICDCIT 2023
Bhubaneswar, India, January 18–22, 2023
Proceedings

 Springer

*Editors*
Anisur Rahaman Molla (iD)
Indian Statistical Institute
Kolkata, India

Gokarna Sharma (iD)
Kent State University
Kent, OH, USA

Pradeep Kumar (iD)
Indian Institute of Management
Lucknow, India

Sanjay Rawat (iD)
Technology Innovation Institute
Abu Dhabi, United Arab Emirates

ISSN 0302-9743 ISSN 1611-3349 (electronic)
Lecture Notes in Computer Science
ISBN 978-3-031-24847-4 ISBN 978-3-031-24848-1 (eBook)
https://doi.org/10.1007/978-3-031-24848-1

This Springer imprint is published by the registered company Springer Nature Switzerland AG
The registered company address is: Gewerbestrasse 11, 6330 Cham, Switzerland

# Preface

This volume contains the papers selected for presentation at the 19th International Conference on Distributed Computing and Intelligent Technology (ICDCIT 2023) held during January 18–22, 2023, in Bhubaneswar, India.

Starting from its first occurrence in 2004, the ICDCIT conference series has grown to an annual conference of international repute and has become a global platform for computer science researchers to exchange research results and ideas on the foundations and applications of distributed computing and intelligent technology. ICDCIT strives to provide an opportunity for students and young researchers to be exposed to topical research directions of distributed computing and intelligent technology.

ICDCIT is broadly organized into two tracks: Distributed Computing (DC) and Intelligent Technology (IT). The DC track solicits original research papers contributing to the foundations and applications of distributed computing, whereas the IT track solicits original research papers contributing to the foundations and applications of Intelligent Technology. Each track has a separate Program Committee (PC) including PC chairs, who evaluate the papers submitted to that track.

This year we received 55 full paper submissions–18 papers in the DC track and 37 papers in the IT track. Each submission considered for publication was reviewed by three (IT track) or four (DC track) PC members, with the help of reviewers outside of the PC. Based on the reviews, the PC decided to accept 29 papers–20 regular papers and nine short papers - for presentation at the conference, with an acceptance rate of 36.36% for full papers. The DC track PC accepted six regular papers and one short paper, with an acceptance rate of 33.33% for full papers. The IT track PC accepted 14 regular papers and eight short papers, with an acceptance rate of 37.83% for full papers. ICDCIT 2023 adopted a double blind review process to help PC members and external reviewers come to a judgment about each submitted paper without bias. Additionally, each paper that was in conflict with a chair/PC member was handled/reviewed by another chair/PC member who had no conflict of interest with the paper.

We would like to express our gratitude to all the researchers who submitted their work to the conference. Our special thanks go to all colleagues who served on the PC, as well as the external reviewers, who generously offered their expertise and time to help us select the papers and prepare the strong conference program.

We were fortunate to have seven distinguished invited speakers – Maurice Herlihy (Brown University, USA), Neeraj Mittal (University of Texas at Dallas, USA), Rajkumar Kettimuthu (Argonne National Laboratory, USA), John Augustine (Indian Institute of Technology Madras, India), Sanjay Madria (Missouri University of Science and Technology, USA), Mukesh Mohania (IIIT Delhi, India), and P. Radha Krishna (National Institute of Technology Warangal, India). Their talks provided us with the unique opportunity to hear research advances in various fields of DC and IT from the leaders in those respective fields. The abstracts related to the invited talks were also included in this volume.

We wish to express our thanks to the local organizing committee who worked hard to make this conference a success, especially our organizing chair Krishna Chakravarty. We also wish to thank the organizers of the satellite events as well as the many student volunteers. The School of Computer Engineering, Kalinga Institute of Industrial Technology (KIIT), the host of the conference, provided various support and facilities for organizing the conference and its associated events.

Finally, we enjoyed institutional and financial support from KIIT for which we are indebted. We express our appreciation to all the Steering Committee members, and in particular Subhasis Das and Samaresh Mishra, whose counsel we frequently relied on. Thanks are also due to the faculty members and staff of the School of Computer Engineering, KIIT, for their timely support.

January 2023

Anisur Rahaman Molla
Gokarna Sharma
Pradeep Kumar
Sanjay Rawat

# Organization

## General Chairs

Sathya Peri            IIT Hyderabad, India
Sandeep Kulkarni      Michigan State University, USA

## Program Committee Chairs

Anisur Rahaman Molla    ISI Kolkota, India
Gokarna Sharma        Kent State University, USA
Pradeep Kumar         IIM Lucknow, India
Sanjay Rawat          Technology Innovation Institute (TII), UAE

## Conference Management Chair

Samaresh Mishra      KIIT, India

## Organizing Chair

Krishna Chakravarty    KIIT, India

## Finance Chairs

Jagannath Singh       KIIT, India
Amiya Kumar Dash     KIIT, India

## Publicity Chairs

Roshni Pradhan        KIIT, India
Saurabh Bilgaiyan     KIIT, India

## Registration Chairs

Abhaya Kumar Sahoo    KIIT, India
Pratyusa Mukherjee    KIIT, India

## Session Management Chairs

| | |
|---|---|
| Manjusha Pandey | KIIT, India |
| Bhaswati Sahoo | KIIT, India |

## Publications Chairs

| | |
|---|---|
| Santwana Sagnika | KIIT, India |
| Satarupa Mohanty | KIIT, India |

## Student Symposium Chairs

| | |
|---|---|
| Hrudaya Kumar Tripathy | KIIT, India |
| Minakhi Rout | KIIT, India |
| Sushruta Mishra | KIIT, India |

## Industry Symposium Chairs

| | |
|---|---|
| Subhasis Dash | KIIT, India |
| Manas Ranjan Lenka | KIIT, India |

## Project Innovation Contest Chairs/Innovation Chairs

| | |
|---|---|
| Bindu Agarwalla | KIIT, India |
| Kunal Anand | KIIT, India |

## Workshop Chairs

| | |
|---|---|
| Chittaranjan Pradhan | KIIT, India |
| Rajat Kumar Behera | KIIT, India |

## Ph.D. Symposium Chairs

| | |
|---|---|
| Satya Ranjan Dash | KIIT, India |
| Himansu Das | KIIT, India |

## Steering Committee

| | |
|---|---|
| Raj Bhatnagar | University of Cincinnati, USA |
| Rajkumar Buyya | University of Melbourne, Australia |
| Diganta Goswami | IIT Guwahati, India |
| Sandeep Kulkarni | Michigan State University, USA |

# Program Committee

## Distributed Computing Track

| | |
|---|---|
| John Augustine | IIT Madras, India |
| Sruti Gan Choudhury | Jadavpur University, India |
| Arghya Kusum Das | University of Alaska Fairbanks, USA |
| Ayan Dutta | University of North Florida, USA |
| Dianne Foreback | Case Western Reserve University, USA |
| Barun Gorain | IIT Bhilai, India |
| Diksha Gupta | IBM Research, Singapore |
| Rory Hector | US Navy Research Laboratory, USA |
| Maleq Khan | Texas A&M University–Kingsville, USA |
| Partha Sarathi | Mandal IIT Guwahati, India |
| Yannic Maus | TU Graz, Austria |
| Kaushik Mondal | IIT Ropar, India |
| William K. Moses Jr. | University of Houston, USA |
| Mikhail Nesterenko | Kent State University, USA |
| Shreyas Pai | Aalto University, Finland |
| Gopal Pandurangan | University of Houston, USA |
| Debasish Pattanayak | LUISS University, Italy |
| Matthieu Perrin | Nantes Université, France |
| Pavan Poudel | Augusta University, USA |
| Peter Robinson | Augusta University, USA |
| Sushmita Ruj | University of New South Wales, Australia |
| Dibakar Saha | NIT Raipur, India |
| Himadri Sekhar | Paul TCS Research and Innovation, India |
| Suman Sourav | National University of Singapore, Singapore |
| Jerry Trahan | Louisiana State University, USA |
| R. Vaidyanathan | Louisiana State University, USA |
| Maxwell Young | Mississippi State University, USA |

## Intelligent Technology Track

| | |
|---|---|
| Debajyoty Banik | KIIT, India |
| Arani Bhattacharya | IIIT Delhi, India |
| Prachet Bhuyan | KIIT, India |
| Saurabh Bilgaiyan | KIIT, India |
| Leena Das | KIIT, India |
| Debasmita Dey | FORE School of Management, India |
| Vatharam Ganivada | University of Hyderabad, India |
| Puneet Goyal | IIT Ropar, India |
| Samrat Gupta | IIM Ahmedabad, India |

| | |
|---|---|
| Vivek Gupta | IIM Lucknow, India |
| Mukul Gupta | IIM Indore, India |
| Chittaranjan Hota | BITS Pilani, India |
| Manoj Jain | IBM, India |
| Sumit Jha | University of Texas at San Antonio, USA |
| Anjeneya Swami Kare | University of Hyderabad, India |
| Saurabh Kumar | IIM Indore, India |
| Nagamani M. | University of Hyderabad, India |
| Hitesh Mahapatra | KIIT, India |
| Pradeep Kumar Mallick | KIIT, India |
| Rajhans Mishra | IIM Indore, India |
| Bhabani S. P. Mishra | KIIT, India |
| Atul Negi | University of Hyderabad, India |
| Naveen Nekuri | University of Hyderabad, India |
| Radhakrishna P. | NIT Warangal, India |
| Nibedan Panda | KIIT, India |
| Mohit Ranjan Panda | KIIT, India |
| Manjusha Pandey | KIIT, India |
| Roshni Pradhan | KIIT, India |
| P. S. V. S. Sai Prasad | University of Hyderabad, India |
| Krishna Prasad | IIT Gandhinagar, India |
| Udai Pratap Rao | SVNIT Surat, India |
| Subba Reddy Oota | Inria, France |
| Minakhi Rout | KIIT, India |
| Gaurav Sarin | Delhi School of Business, India |
| Manoranjan Satpathy | IIT Bhubaneswar, India |
| Vishnu Kumar Saxena | Vikram University, India |
| Mayank Sharma | IIM Kashipur, India |
| Pradeep Kumar Singh | NIT Raipur, India |
| Sobha Rani T. | University of Hyderabad, India |
| Manas Tripathi | IIM Rohtak, India |
| Chiranjeevi Yarra | IIIT Hyderabad, India |

## Additional Reviewers

| | |
|---|---|
| Amit Chaudhary | Swarupa Rani K. |
| Amit Kumar Dhar | Thakare Kamalakar Vijay |
| Bibhas Chandra Dhara | Uttam Kumar Roy |
| Steven Fernandes | Niladri Sett |
| Antriksh Goswami | Sumit Tetarave |
| Aarushi Jain | Peng Wang |
| Santhosh K. K. | |

# Abstracts of Invited Talks

# Correctness Conditions for Cross-Chain Transactions

Maurice Herlihy ⓘ

Computer Science Department, Brown University, Providence, RI, USA
mph@cs.brown.edu

**Abstract.** Modern distributed data management systems face a new challenge: how can autonomous, mutually-distrusting parties cooperate safely and effectively? Addressing this challenge brings up many questions familiar from classical distributed systems. Nevertheless, many of these questions requires subtle rethinking when participants are autonomous and potentially adversarial.

We propose the notion of a *cross-chain deal*, a new way to structure complex distributed computations that manage assets in an adversarial setting. Deals are inspired by classical atomic transactions, but differ in important ways to accommodate the decentralized and untrusting nature of the exchange.

This talk is intended for a general audience.

Joint work with Barbara Liskov and Liuba Shrira.

# Harnessing Concurrency in Multicore Systems

Neeraj Mittal

Department of Computer Science, The University of Texas at Dallas,
Richardson, TX 75080, USA
neerajm@utdallas.edu

**Abstract.** Until two decades ago, general-purpose processor manufacturers were able to achieve regular improvements in CPU performance by using traditional approaches such as increasing the clock speed of the CPU, increasing the length of the instruction pipeline or increasing the size of the cache and/or the number of cache levels. These steady improvements in CPU performance, and to a lesser extent in memory and disk performances, enabled building of ever-faster mainstream computer systems. As a result, most classes of software applications enjoyed regular (and free) performance gains for several decades without even releasing new versions or doing anything special. Many of the traditional approaches for boosting CPU performance have now hit a Brick Wall, a term often used to describe the inherent physical limitations faced by hardware designers in boosting CPU performance further. The transistor count, which is the number of transistors in an integrated circuit chip, continues to increase as per the Moore's Law. To make use of these large number of additional transistors available on a chip and due to traditional approaches offering only limited gains, major general-purpose processor manufacturers (Intel, AMD and PowerPC) have turned to hyper-threading and multi-core architectures to improve hardware performance. A consequence of this trend is that the free ride that software programs have enjoyed for around four decades is finally over, and most current software applications will not benefit from this enormous parallel processing power offered by a modern computing device unless they are rewritten in a way that enables a program to distribute its tasks across several cores. Even a program written for a multi-core system may fail to scale well with the number of cores if poorly designed and coded.

Even though concurrency has been around for many decades, writing a concurrent program that runs correctly on a multi-core system is still known to be very hard, let alone writing a concurrent program that scales well with the number of cores. Not surprisingly, concurrent programming is largely the skill set of elite programmers often with doctoral degree in concurrent computing or related area. In this talk, I will present the current research on designing high-performance concurrent programs suitable for multi-core systems. I will also talk about the current research on using the new memory technology, called persistent memory (Pmem), that combines the low latency of main memory and the persistence of hard disk to design fault-tolerant concurrent programs.

# From File Transfers to Streaming: Enabling Distributed Science in the Exascale Era

Rajkumar Kettimuthu (ID)

Argonne National Laboratory Lemont, IL 60439, USA
`kettimut@mcs.anl.gov`

**Abstract.** Extreme-scale simulations and experiments can generate large amounts of data, whose volume can exceed the compute and/or storage capacity at the simulation or experimental facility. Moreover, as scientific instruments are optimized for specific objectives, both the computational infrastructure and the codes are becoming more specialized with the proliferation of AI workloads and accelerators. Distributed science is now a norm rather than an exception and it requires the rapid and automated movement of large quantities of data between federated scientific facilities. Traditionally, file-based data movement formed the backbone of distributed science and is still the predominant mode of data exchange across facilities. Near real-time analysis of streaming data (from scientific instruments) at remote facilities is emerging as a key requirement with recent technological advances that allow scientific instruments to generate data at rates that can exceed tens of gigabytes per second. In this talk, I will discuss our work in high-speed data movement for enabling distributed science in the Exascale era–ranging from moving a petabyte (large number of files) between two scientific facilities in a day to memory-to-memory data streaming between federated scientific instruments at 100 gigabits per second.

# Three Vignettes from the Distributed Trust Paradigm of Computing

John Augustine 🆔

Institute of Technology Madras, Chennai, India
augustine@cse.iitm.ac.in

**Abstract.** Byzantine fault tolerance has been studied extensively for over four decades. Much of the work is centered on Byzantine Agreement and related problems like State Machine Replication. These works have established mechanisms for trustworthy computation in distributed environments despite the presence of malicious nodes. Does this distributed trust paradigm make sense in broader contexts? In this talk, we will look at three vignettes from disparate domains that provide us with affirmative evidence.

We will begin with classical distributed graph computing in the congested clique model and show how we can approach the problem of connectivity despite the presence of Byzantine nodes. We will then present a fully decentralized mechanism to build sparse overlay networks that are resilient to Byzantine failures. We will conclude with techniques for rank aggregation wherein we obtain a global ranking by aggregating pair-wise comparisons of voters over a set of objects. Our approach provides reliable ranking as long as the proportion of Byzantine voters is strictly less than a half.

These are joint works with Soumyottam Chatterjee, Arnhav Datar, Anisur Rahaman Molla, Gopal Pandurangan, Arun Rajkumar and Yadu Vasudev. They have appeared in DISC, SPAA, and NeurIPS–all recently in 2022.

# Machine Learning for Emotion Prediction, Ideology Detection and Polarization Analysis Using COVID-19 Tweets

Sanjay K. Madria🆔

Department of Computer Science, Missouri University of Science
and Technology, Rolla, MO 65409, USA
madrias@mst.edu

**Abstract.** The adversarial impact of the Covid-19 pandemic has created a health crisis globally all over the world. This unprecedented crisis forced people to lockdown and changed almost every aspect of the regular activities of the people. Thus, the pandemic is also impacting everyone physically, mentally, and economically, and it, therefore, is paramount to analyze and understand emotional responses during the crisis affecting mental health. Negative emotional responses at fine-grained labels like anger and fear during the crisis might also lead to irreversible socio-economic damages. In this talk, I will discuss a neural network model trained using manually labeled data to detect various emotions at fine-grained labels in the Covid-19 tweets automatically. I will discuss about a manually labeled tweets dataset on COVID-19 emotional responses along with regular tweets data. A custom Q&A roBERTa model to extract phrases from the tweets that are primarily responsible for the corresponding emotions has been designed. None of the existing datasets and work currently provide the selected words or phrases denoting the reason for the corresponding emotions. Further, we propose a deep learning model leveraging the pre-trained BERT-base to detect the political ideology from the tweets for political polarization analysis. The experimental results show a considerable improvement in the accuracy of ideology detection when we use emotion as a feature.

# AI for Personalized Education

Mukesh K. Mohania

Indraprastha Institute of Information Technology Delhi,
New Delhi 110020, India
mukesh@iiitd.ac.in

**Abstract.** Online courses and learning systems have gained tremendous popularity over the last few years. While their ease of access and availability make them a very useful medium for knowledge sharing and learning, they do not keep the learners and their learning abilities in mind. The "one size fits all" approach to learning content and the question paper does not work in a large virtual classroom consisting of diverse students with different skill profiles, learning styles, aptitudes and capabilities. In a traditional classroom, teachers who interact closely with students are in a position to evaluate the pace and depth of the curriculum being taught and can also suggest learning content to students not being able to cope with the general classroom teaching. Such suggestions and guidance are absent in current online learning systems. In this talk, we aim to address how AI can help in (1) making content smarter through learning content analytics and automatic content tagging, (2) generating diverse but semantically related questions for evaluating the student's knowledge, (3) assisting in short answers evaluation, and finally (4) understanding the student's learning style/capacity through learning data analytics, thus enabling the adaptive and personalized education on Big Data platform.

# Attention-Based Representational Learning for Social Network Analysis

P. Radha Krishna [ORCID]

National Institute of Technology Warangal, Telangana 506004, India
prkrishna@nitw.ac.in

**Abstract.** Social networks carry high-level complex structural and semantic information in the form of nodes and edges. Network representations help in improving the analytical tasks such as community detection, link prediction and information propagation. Low dimensional feature representations generate features automatically and reduces the human's manual efforts in feature extraction. In this talk, heterogeneous network representation learning models for influence propagation are discussed and some thoughts on research questions will be presented. A focus will be laid on the aggregating various types of semantic information based on their importance and weightage to avoid semantic confusion and also employing an attention-based mechanisms though meta-path learning.

# Contents

**Distributed Computing**

# Filling MIS Vertices of a Graph by Myopic Luminous Robots

Subhajit Pramanick[1], Sai Vamshi Samala[1], Debasish Pattanayak[2]◉,
and Partha Sarathi Mandal[1(✉)]◉

[1] Department of Mathematics Indian Institute of Technology Guwahati,
Guwahati, India
{subhajit.pramanick,sai170123042,psm}@iitg.ac.in
[2] LUISS University, Rome, Italy
dpattanayak@luiss.it

**Abstract.** We present the problem of finding a maximal independent set (MIS) (named as *MIS Filling problem*) of an arbitrary connected graph with luminous myopic mobile robots. The robots enter the graph one after another from a particular vertex called the *Door* and move along the edges of the graph without collision to occupy vertices such that the set of occupied vertices forms a maximal independent set.

This paper explores two versions of the *MIS filling problem*. For the *MIS Filling with Single Door* case, our IND algorithm forms an MIS of size $m$ in $O(m^2)$ epochs under an asynchronous scheduler, where an epoch is the smallest time interval in which each participating robot gets activated and executes the algorithm at least once. The robots have three hops of visibility range, $\Delta + 8$ number of colors, and $O(\log \Delta)$ bits of persistent storage, where $\Delta$ is the maximum degree of the graph. For the *MIS Filling with Multiple Doors* case, our MULTIND algorithm forms an MIS in $O(m^2)$ epochs under a semi-synchronous scheduler using robots with five hops of visibility range, $\Delta + k + 7$ number of colors, and $O(\log(\Delta + k))$ bits of persistent storage, where $k$ is the number of doors.

**Keywords:** Distributed algorithms · Multi-agent systems · Mobile robots · MIS · Filling problem · Luminous robots

## 1 Introduction

### 1.1 Motivation

The coordination among large number of autonomous mobile robots or agents has gained significant interest in recent years. Under the framework of "Look-Compute-Move" cycles, the robots can perform various tasks such as exploration [1], gathering [6,7,12], pattern formation [5,17], dispersion [2,14], scattering

P.S.Mandal—Partially supported by SERB, Govt. of India, Grant Number: MTR/ 2019/001528

A. R. Molla et al. (Eds.): ICDCIT 2023, LNCS 13776, pp. 3–19, 2023.
https://doi.org/10.1007/978-3-031-24848-1_1

[10,15] and others. In this work, we consider the underlying environment as a graph, and the robots can stay at the nodes and move along the edges.

In general, maximal independent set (MIS) of a network graph play a significant role in decomposing the network into clusters of low diameter, which is often very useful in designing and implementing distributed divide and conquer algorithms. MIS vertices can also be used as a network backbone for deploying communication infrastructure. For example, information dissemination in a low latency system where all robots form a network should be located at MIS vertices so that all other vertices are just one hop away.

The Filling problem, introduced by Hsiang et al. [11], considers the robots enter via particular vertices and fill an environment (graph) composed of pixels (vertices) and robots occupy every pixel (vertex). Later Hideg et al. [9] presented the Filling problem for an arbitrary connected graph. It is of interest to cover the entire graph but using a smaller number of robots. Thus forming an MIS by the robots that enter the graph becomes a natural extension. We call this problem the *MIS Filling problem.*

In this paper, we consider *luminous robots*, that are mobile robots possessing externally visible persistent memory (or *lights*). Each vertex can contain at most one robot at a time. We say a collision happens when two or more robots move to the same vertex. Only one robot can travel along one edge at a time. In this problem, the robots enter the graph one by one through a specific vertex called the *Door* and move in the graph along the edges from one vertex to another while avoiding a collision. The objective is only to occupy vertices that form an MIS. We solve two flavors of the problem: graphs with a single Door under an asynchronous (ASYNC) scheduler and graphs with multiple Doors under a semi-synchronous (SSYNC) scheduler. We use *epochs* to denote the time complexity, where an epoch is the smallest amount of time required for all the participating robots to activate once. On each activation a robot executes a *Look-Compute-Move (LCM)* cycle. In ASYNC, the cycles are independently executed within finite but unpredictable time. In SSYNC, time is discretely separated into rounds, and a subset of the robots are activated in each round and finish the execution of a cycle in the same round. Having multiple Doors instead of just one offers redundancy in situations where a Door can be blocked.

## 1.2  Related Works

Kamei and Tixeuil [13] solve two variations of the maximum independent set (MAX_IS) placement problem for grid networks. The first one assumes knowledge of port-numbering for each node. It uses three colors of light and a visibility range of two. The other one removes the assumption of port-numbering and uses seven colors of light and a visibility range of three. Barrameda et al. [3] proposed algorithms for uniform dispersal or filling problem on any simply connected orthogonal space using identical asynchronous sensors. They present two algorithms; one for the single door, where sensors have one unit of visibility range and two-bit of persistent memory, and the other for multiple doors, where sensors have two units of visibility and a constant amount of persistent memory.

They also prove that oblivious sensors cannot solve the problem deterministically even if they have unlimited visibility. For multiple doors, they show that the problem is unsolvable if the visibility range is less than two, even if sensors have unbounded memory. Further, even with unbounded visibility and memory, they show that the problem is unsolvable if the sensors are identical. Barrameda et al. [4] extended the problem of uniform dispersal for orthogonal domains with holes. They solve the problem when robots have a visibility range of six without any direct communication among themselves. Later, they solve the problem using direct communication among robots to reduce the visibility radius without increasing the memory requirement. Hideg and Lukovszki [8] solve the filling problem in orthogonal regions, where the robots enter the region through entry points, called doors. They propose two algorithms with run-time $O(n)$, one for single door and the other for multiple door case. Later Hideg and Lukovszki [9] presented the Filling problem for an arbitrary connected graphs in asynchronous setting where the goal is to fill the entire graph using myopic luminous robots.

The algorithm proposed by Hideg and Lukovszki [9] cannot be directly applied to *MIS Filling problem*, as the communication and movement of robots in PACK algorithm are limited to one hop. For starters, one needs to maintain a two hop gap between the chain of robots, while simultaneously ensuring that the chain never crosses itself. The chain crossing problem does not arise if the chain is closely packed at one hop distance, and it becomes challenging in the presence of multiple chains. Also, at no point we can allow more than the size of MIS of robots to enter the graph, since that would render the problem unsolvable. We show that our algorithms handle these additional requirements and correctly form an MIS. The state-of-the-art results and ours' are presented in Table 1, where $m$ is the number of robots that form MIS. The steps represent the total movement of robots throughout the execution of the algorithm. The number of nodes and the grid dimensions are represented by $n$, $L$, and $l$, respectively.

**Table 1.** The state of the art of previous results PACK [9], BLOCK [9] and MIS placement on grid [13] with our proposed algorithms IND and MULTIND.

| Algorithm | PACK [9] | BLOCK [9] | Algo 1 [13] | Algo 2 [13] | IND | MULTIND |
|---|---|---|---|---|---|---|
| Scheduler | ASYNC | ASYNC | ASYNC | ASYNC | ASYNC | SSYNC |
| Problem | Filling Problem | Filling Problem | MAX_IS Placement | MAX_IS Placement | *MIS Filling* | *MIS Filling* |
| Topology | Connected graph | Connected graph | Grid network | Grid network | Connected graph | Connected graph |
| Number of doors | Single | Multiple ($k$) | Single | Single | Single | Multiple ($k$) |
| Visibility range | 1 hop | 2 hops | 2 hops | 3 hops | 3 hops | 5 hops |
| Memory (in bits) | $O(\log \Delta)$ | $O(\log \Delta)$ | $O(1)$ | $O(1)$ | $O(\log \Delta)$ | $O(\log(\Delta + k))$ |
| Number of colors | $\Delta + 4$ | $\Delta + k + 4$ | 3 | 7 | $\Delta + 8$ | $\Delta + k + 7$ |
| Time complexity | $O(n^2)$ epochs | $O(n)$ epochs | $O(n(L+l))$ steps[3] | $O(n(L+l))$ steps | $O(m^2)$ epochs | $O(m^2)$ epochs |

## 1.3   Contributions

In this paper, we propose two algorithms, IND and MULTIND, corresponding to single and multiple doors.

- Algorithm IND solves the *MIS Filling problem* in graphs with a single Door under an ASYNC scheduler using robots with a visibility range of 3, $\Delta + 8$ number of colors, $O(\log \Delta)$ bits of persistent storage in $O(m^2)$ epochs, where $\Delta$ is the maximum degree of the graph, and $m$ is the number of robots that form an MIS.
- Algorithm MULTIND solves the *MIS Filling problem* in graphs with multiple Doors under an SSYNC scheduler using robots with a visibility range of 5, $\Delta + k + 7$ number of colors, $O(\log(\Delta + k))$ bits of persistent storage in $O(m^2)$ epochs, where $m$, $\Delta$, and $k$ are the number of robots, the maximum degree of the graph and number of Doors, respectively.

Rest of the paper is organized as follows. Section 2 discusses the necessary definition and the underline model. Section 3 describes the IND algorithm for single Door case. Section 4 describes the MULTIND algorithm for multiple Doors case. Section 5 is the discussion section and finally concluded in Sect. 6. The full version of the paper is available in the following reference [16].

## 2   Model

In this paper, we model the environment as a graph. We say that the graph contains a set of vertices that are connected to Doors from where robots can enter. The number of Doors in the graph is unknown to the robots, but they are equipped with colors to distinguish themselves if they enter the graph from different Doors. We assume a maximum of $k$ Doors are attached to the graph.

**Graph:** We consider an anonymous graph, i.e., the nodes of the graph are indistinguishable from each other. Each vertex $v$ of the graph contains port numbers corresponding to the incident edges from $[1, 2, \ldots, \delta(v)]$, where $\delta(v)$ is the degree of the vertex $v$. Given an anonymous connected port labeled graph $H = (V', E')$, we construct a graph $G = (V, E)$ with $k$ Doors, that adds two auxiliary vertices $\{d_i, d_i'\}$ corresponding to each Door that is connected to distinct vertices $\{v_1, v_2, \ldots, v_k\} \subset V'$. We have a path $d_i \rightarrow d_i' \rightarrow v_i$ corresponding to each Door $d_i$. The robots enter the graph through the Door, and a new robot appears at the Door immediately after it becomes empty. We say a vertex is *free* if none of the vertices adjacent to it are occupied by any robot. Since we add a buffer vertex corresponding to each Door, all the vertices in the original graph $H$ are *free* vertices in the beginning.

**Robots:** The focus is on completing the task using robots with minimal capabilities operating under certain adversarial conditions. The robots are *autonomous* (no central or external control), *myopic* (they have limited visibility range), *anonymous* (without distinguishable features or identification) and *homogeneous* (they all have the same capabilities/execute the same program). Additionally, the robots are *luminous*, i.e., they have a light attached to them that can display various *colors* that represent the value of a state variable. The light works as a mode of communication between the robots.

**Time Cycle:** Each robot operates in the *Look-Compute-Move* (LCM) model, in which the actions of the robots are divided into three phases.

- *Look:* The robot takes a snapshot of its surroundings, i.e., the vertices within the visibility range and the colors of the robots occupying them.
- *Compute:* The robot runs the algorithm using the snapshot as the input and determines a target vertex or chooses to remain in place.
- *Move:* The robot moves to the target vertex if needed. A robot moves two hops in a single move phase.

**Assumptions:** We have the following assumptions regarding the knowledge of a robot and the properties of the underlying graph.

- The robots have no knowledge of the graph but an upper bound of $\Delta$, the maximum degree of the graph.
- For a robot placed at $v$ with a visibility range of $z$, the port numbers of all the vertices in its visibility range are visible.
- Movement of robots is non-instantaneous.
- A robot knows all the colors, but it can only display one color corresponding to the Door via which it enters the graph. Note that this color corresponding to the Door is only used by one robot at a time, but all robots that come from the same Door can display it when the need arises. (We use unique colors for each Door to determine a hierarchy among them.)

Note that we use directions and port numbers interchangeably throughout the paper. Each port number corresponds to a DIR color, also the direction towards Successor or Predecessor.

**Problem:** We define the MIS filling problem formally as follows:

*Problem 1* (MIS filling problem). Given an anonymous connected port labeled graph $G = (V, E)$ with $k$ Doors, the objective is to relocate robots that appear at Doors such that at termination, the robots occupy a set of vertices $V_1$ ($V_1 \subset V$) that forms a maximal independent set of $G$.

## 3   Algorithm for MIS Filling with Single Door

We now describe the IND algorithm inspired by the PACK algorithm [9] and use the concept of the Virtual Chain Method [10]. The robots move throughout the graph like the depth-first search (DFS). We assume that the robots operate under an asynchronous (ASYNC) scheduler. An epoch is the shortest time in which each robot present in the graph not in the Finished state is activated at least once and performs an LCM cycle. Note that, an epoch may contain multiple cycles for some robots, but not for all. The length of an epoch may vary depending on the activation schedule of the robots. Each robot requires a visibility range of 3 hops, $O(\log \Delta)$ bits of persistent memory, and $\Delta + 8$ colors.

### 3.1   Preliminaries

**Colors:** The colors used by the robots are described here.

- ON - Used initially when a robot arrives at the door.
- DIR - $\Delta$ colors corresponding to a port number in $[1, \Delta]$.
- CONF - Used to confirm that the first DIR color has been seen and received.
- CONFC - Used to confirm that CONF color has been seen and received.
- CONF2 - Used to confirm that the second DIR color is seen and received.
- CONF3 - Used to confirm that the Packed state is achieved.
- WAIT - Used by a Leader while waiting for the Packed State.
- MOV - Used when a robot is in motion.
- OFF - Used by a robot in the Finished state.

Note that a DIR color pointing toward a successor is the special color to indicate the leadership transfer.

**Definition 1** *(k hops Neighborhood of a vertex v). For a vertex v, we define k-hops Neighborhood of v to be the set of all vertices that are k hops away from v and denote it by $N_v^k$.*

**Definition 2** *(k hops Visibility Set of a robot r). For a robot r placed at a vertex of the graph, we define the k hops Visibility Set of r to be the set of all vertices within k hops from the current location of r. We denote it by $V_r^k$.*

### 3.2   IND Algorithm

In this section, we present the rules for the robots that they follow to form a chain with robots occupying alternative vertices successfully. We say a vertex is *free* if none of its neighbors contain a robot. The first robot that enters the graph is called the Leader robot. Any robot that enters the graph after the Leader, is called follower robot. We define *Packed state* as the state of a chain where all the alternative vertices are occupied by robots. We define it formally as follows:

**Definition 3** *(Packed state).   Let L be a positive odd integer and P = $\{v_1, v_2, \ldots, v_L\}$ be a path starting from the Door at $v_1$ and the leader at $v_L$ such that every second vertex of P, i.e., $v_1, v_3, \ldots, v_{L-2}$ is visited by the Leader. A chain of robots is in a Packed state if the vertices $v_1, v_3, \ldots, v_{L-2}$ are occupied by follower robots.*

Table 2 lists down the variables used in our algorithm. Only the Leader is allowed to move in the Packed state. WAIT color is used by the Leader as soon as it reaches its target vertex to indicate that it is waiting for the chain to reach the Packed state. The packed state ensures that the Leader can choose a target such that no other robot will occupy neither the target nor any of the adjacent vertices. Secondly, the Leader (Predecessor) $r_1$ needs to communicate the directions it will take to its follower (Successor) $r_2$ so that $r_2$ can know in which direction $r_1$ has moved. The *Color* variable represents the color displayed

**Table 2.** List of variables used in algorithms IND and MULTIND.

| Variable | Description |
|---|---|
| State | State of the robot - None/Follower/Leader/Finished |
| Target | Directions to the target vertex |
| NextTarget | Directions to the vertex to which the robot has to move after reaching the Target vertex (Used by Followers) |
| Successor | Successor robot |
| Predecessor | Predecessor robot |
| Color | Color displayed by robot's light |
| Entry | Previous location of the robot/location of the follower stored in terms of directions |

by the robot's light, and the *Target* variable represents directions to the target vertex. The *NextTarget* variable represents directions to the next target after reaching the target vertex. *Entry* variable represents the direction of the two hops a robot moved so that the robot knows the location of its follower.

We explain communication between Predecessor and Successor and the restoration of the Packed state after movement with an example as shown in Fig. 1.

**Communicating the Movement Directions:** The robots establish the Predecessor and Successor relationship by their order of arrival at the Door. A Predecessor communicates its destination to a Successor by showing the colors corresponding to the port numbers at the vertex. Suppose $r_1$, $r_2$ and $r_3$ are located at $e$, $c$ and $a$ as shown in Fig. 1.

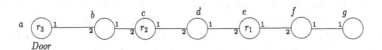

**Fig. 1.** Communication of color from $r_1$ to $r_2$

$r_1$ determines that it will move to vertex $g$. Now, it has to communicate that to $r_2$, so that $r_2$ can follow $r_1$. First, $r_1$ shows the DIR color corresponding to port 1 as it wants to take the path $ef - fg$, and $r_2$ responds by showing the CONF color. Then $r_1$ sets its color to CONFC to confirm that it has seen the CONF color. Now, $r_2$ also sets its color to CONFC to inform $r_1$ that it is ready to receive the second color (CONFC color is used to distinguish between two DIR colors that can be the same). Next, $r_1$ shows the DIR color of port 1 corresponding to edge $fg$. $r_2$ confirms that it has seen the color by setting its color to CONF2. Once $r_1$ sees CONF2, it can move to $g$ after setting its color to MOV. Now, $r_2$ does the same process to communicate DIR colors to $r_3$.

**Restoration of Packed State After Movement:** We describe this module with the Fig. 1. After the leader $r_1$ moves to $G$ with color MOV, $E$ becomes empty, so the Packed state is now distorted. So $r_1$ changes its color to WAIT. Now, $r_2$ moves to $E$ with color MOV after communicating DIR colors to $r_3$. So, $C$ becomes empty. After that, $r_3$ moves to $C$ with color MOV without communicating DIR colors as it does not have any successor at this movement. As soon as $r_3$ leaves the Door, a robot $r_4$ is placed at the Door with color ON. After $r_3$ reaches $C$, it sees $r_4$ with color ON and $r_3.Color =$ MOV. So, it changes its color to CONF3. Now, $r_2$ also changes its color to CONF3 after seeing $r_3.Color =$ CONF3 and $r_2.Color =$ MOV. The leader $r_1$ sees $r_2$ with color CONF3 and $r_1.Color =$ WAIT, so it understands that the Packed state is achieved. Now, $r_1$ looks for new target to move to.

**Transferring the Leadership:** The current leader $r_1$ transfers its leadership if either of the two scenarios occurs. First, $r_1$ gets stuck, i.e., there are no other free vertices left to move to from the current vertex of the leader. Secondly, for each free vertex $v \in V_{r_1}^2$, $r_1$ finds at least one vertex $v' \in N_v^2 \cap V_{r_1}^3$ with a robot not in Finished state. In both cases above, $r_1$ transfers the leadership to its successor $r_2$ by communicating the direction pointing to $r_2$. Since $r_2$ knows the directions to $r_1.Target$, it can realize that the DIR color corresponds to the reverse direction of the virtual chain.

**Detailed Description:** When a robot first appears at the Door, it initializes to the None state and sets color ON. Let $r_1$ be the first robot that appears at the Door. $r_1$ does not find a robot on any adjacent vertices, so it changes its state to Leader. Now $r_1$ is the Leader and chooses a target vertex two hops away and moves to it with color MOV. As soon as the Leader $r_1$ leaves the Door, the next robot $r_2$ appears at the Door. At this time, $r_1$ is still in motion and is nearest to $r_2$. $r_2$ sees this and becomes the follower of $r_1$ and sets $r_2.Color =$ ON, where $r_1.Successor = r_2$ and $r_2.Predecessor = r_1$. After $r_1$ reaches its target, it sets $r_1.Color =$ WAIT to indicate that it is waiting for Packed state. When $r_2$ sees $r_1.Color =$ WAIT and $r_2.Color =$ ON, $r_2$ changes its color to CONF3. $r_1$ now chooses a new free vertex (if any) as $r_1.Target$ and communicate its directions to $r_2$ as described above. When $r_1$ gets confirmation from $r_2$ ($r_2$ sets its color to CONF2), $r_1$ moves to $r_1.Target$ with color MOV and again changes its color to WAIT to indicate that it is waiting for the chain to be in Packed state.

When the chain is in Packed state, the leader $r_1$ chooses a free vertex $v$ from $V_{r_1}^2$ as target, if all $v' \in N_{r_1}^2 \cap V_{r_1}^3$ are either unoccupied or having robots at Finished state. Then, $r_1$ communicates the directions of $r_1.Target$ to $r_2$ as follows. $r_1.Target$ contains two DIR colors (*One* and *Two*), for two hop movement. $r_1$ displays DIR color corresponding to $r_1.Target.One$ by setting $r_1.Color$. $r_2$ sees this and stores the direction in $r_2.NextTarget.One$. $r_2$ confirms that it has seen the first DIR color by displaying CONF color. $r_1$ confirms that it has seen CONF on $r_2$ by setting $r_1.Color$ to CONFC (confirmation of confirmation). $r_2$ sees CONFC on $r_1$ and in turn, sets $r_2.Color$ to CONFC to show that it is ready to receive the second DIR color. The second DIR color $r_1.Target.Two$ is displayed by $r_1$ and is seen by $r_2$. $r_2$ stores this second direction in $r_2.NextTarget.Two$ and sets $r_2.Color$ to CONF2 to send confirmation that the second direction of $r_1$ is received. Once $r_1$ has seen CONF2 in $r_2$, it moves two hops to $r_1.Target$ in Move phase by setting $r_1.Color$ = MOV. After reaching to $r_1.Target$, $r_1$ sets $r_1.Color$ =WAIT. Now $r_2$ sets $r_2.Target$ based on the information stored in $r_2.NextTarget$. $r_2$ needs to reach the old position of $r_1$.

Finally, when the Leader $r_1$ can no longer find free vertices to move to, it communicates this information to its follower $r_2$ using the DIR color that points towards $r_2$. Then, $r_1$ changes its color to OFF and goes into the Finished state, and $r_2$ becomes the Leader and continues exploring the graph.

When a robot moves to its target vertex, it sets $r.Entry$ to the directions of the two hops it moved under $Entry$ variable (so that the robot knows the location of its follower).

### 3.3 Analysis of IND Algorithm

Here, we present a few lemmas that establish the robots' behavior and the correctness of the IND algorithm.

**Lemma 1.** *There can be at most one Leader robot, and the Leader robot $r_1$ moves to a free vertex $v \in V_{r_1}^2$ such that every $v' \in N_v^2 \cap V_r^3$ is either unoccupied or occupied by robots in Finished state.*

*Proof.* In the rules for *Transferring the Leadership*, when a Leader $r_1$ signals to its successor $r_2$ that it is stuck, $r_2$ can become the Leader only after the previous Leader $r_1$ has switched to Finished state (recognized by OFF color on $r_1$). The first robot placed becomes the Leader, and the robots appearing next can become a Leader only after the previous one reaches the Finished state. Therefore, there can be at most one Leader at any time during the dispersion.

A free vertex has none of its adjacent vertices occupied by a robot. As the current Leader $r_1$ can only move when the chain is in a Packed state, it chooses a vertex in $V_{r_1}^2$ that is free by checking all the vertices adjacent to potential target vertex are not occupied by any robot. This can be done since the robots have a visibility range of 3. Further, there is no possibility of this target vertex or any vertex adjacent to it being occupied by any other robot as the Leader is allowed to move only when the chain is in the Packed state. If there is a free vertex $v \in V_{r_1}^2$ as a potential target, then every 2 hops neighbor $v'$ of $v$ ($v' \in N_v^2 \cap V_{r_1}^3$), that is visible to $r_1$ needs to be either free or occupied by robots in Finished state. Due to this condition, a chain does not cross itself as shown in Fig. 2. If no such free vertex $v$ is found, then $r_1$ transfers the leadership to $r_2$ and switches to Finished state. □

**Corollary 1.** *The Leader $r$ moves to a free vertex $v$ such that every vertex $v' \in N_v^2 \cap V_r^3$ is either unoccupied or occupied by a robot in the Finished state. Consequently, the chain never crosses itself.*

**Lemma 2.** *The Robots do not collide.*

**Lemma 3.** *No two robots in the Finished state occupy adjacent vertices.*

From Lemma 3, we can say that when a robot enters the Finished state, none of the vertices adjacent to it are occupied by a robot which means that eventually, all the vertices occupied by robots form an independent set.

*Remark 1.* If a vertex is occupied by a robot in the packed state, it remains occupied thenceforth.

**Theorem 1.** *Algorithm IND fills a maximal independent set.*

**Lemma 4.** *Algorithm IND fills an MIS of G with luminous robots having a visibility range of 3 hops, $O(\log \Delta)$ bits of memory, and $\Delta + 8$ colors.*

Now we analyze the time complexity of the algorithm in terms of epochs. To find the total time required by the algorithm, we first individually establish the time-bound on the movement of robots. Consider a chain containing $i$ robots $\{r_1, r_2, \ldots, r_i\}$, where $r_1$ is the Leader and the foremost robot in the chain. The robots $r_1, r_2, \ldots, r_i$ are on alternative vertices on the path from the vertex occupied by the Leader to the Door vertex. Suppose the chain is in the Packed state. We determine the time required for two consecutive movements of the Leader. We first determine the time required for all the robots in the chain to occupy the position of their Predecessor. As a result, a new robot appears at the Door, increasing the size of the chain. Next, we find the time required for the robots in the chain to set their colors to CONF2, indicating that they have received the movement direction of their Predecessor. We also find the time required for the chain to reach in the Packed state again after the leader of the chain moves.

**Lemma 5.** *Algorithm IND takes at most $i$ epochs for all the robots in a chain of length $i$ to perform one MOVE operation.*

*Proof.* Once the chain is in the Packed state, the leader robot $r_1$ moves to its target vertex in one epoch and $r_1$ sets its color to WAIT. By the next epoch, $r_2$ observes that $r_1$ has left its previous vertex, so it moves to $v$. In a cascading manner all Successors move to their Predecessors location. Thus in the worst case, it takes $i$ epochs for all the robots in the chain to move.                    □

**Lemma 6.** *Algorithm IND takes at most $7i$ epochs for the robots in a chain of length $i$ to reach the Packed state after movement.*

*Proof.* As per the algorithm description, the communication between a Predecessor $r_p$ and its Successor $r_q$ by showing a series of seven colors: first DIR color at $r_p$, CONF at $r_q$, CONFC at $r_p$, CONFC at $r_q$, second DIR color at $r_p$, CONF2 at $r_q$ and CONF3 at $r_q$. A leader can only move when the chain is in the Packed state. For a chain to be in the Packed state, the Successor of the leader robot has to show CONF3 color. The process of communication of colors starts from the new robot $r_{i+1}$ that appears at the Door. It takes at most seven epochs for the communication between $r_i$ and $r_{i+1}$. Similar communication happens between all $i$ pairs of consecutive robots on the chain. So it takes at most $7i$ epochs for the chain to reach Packed state. □

**Lemma 7.** *Algorithm IND takes at most 4 epochs to transfer leadership from a leader to its Successor.*

*Proof.* Now consider the situation where the Leader $r_1$ cannot find any free vertices two hops away. If $r_1$ is at the Door, it sets its color to OFF and switches to Finished state; the maximal independent set is filled. Otherwise, $r_1$ switches to the Finished state by setting the DIR towards its follower. $r_2$ sees the DIR color and sets $r_2.Color$ to CONF. $r_1$ sees the CONF color, it switches to color OFF. $r_2$ becomes the new Leader once it sees the color OFF at $r_1$. In total, this needs at most 4 epochs because of the sequence of colors (DIR, CONF and OFF) and final state change at $r_2$. □

**Theorem 2.** *The algorithm IND runs in $O(m^2)$ epochs.*

*Proof.* For an MIS of size $n$, we can have a chain of length at most $n$. For each increase in a chain of length $i$, it takes at most $7i$ epochs from Lemma 5 and 6. Also, we can have at most $m$ leadership transfers. From Lemma 7, each transfer of leadership takes at most 4 epochs. So in total we need, $\sum_{i=1}^{m} 7i + 4m = O(m^2)$ epochs. Therefore, after at most $O(m^2)$ epochs, an MIS of vertices of the graph becomes filled. □

The corollary below follows from Corollary 1 in [9].

**Corollary 2.** *(i) Assume that there are no inactive intervals between the LCM cycles and that every LCM cycle of every robot takes at most $t_{max}$ time. Then the running time of the IND algorithm is $O(m^2 t_{max})$. (ii) The IND algorithm needs $O(m^2)$ LCM cycles under an FSYNC scheduler.*

## 4   Algorithm for MIS Filling with Multiple Doors

The MULTIND is largely similar to algorithm IND with a few modifications. It works under a *Semi Synchronous* (SSYNC) scheduler, where a subset of robots is activated in each round, and each activated robot finishes its LCM cycle in the same round. We define an epoch similarly as the minimum number of rounds where all robots are activated at least once. Note that an epoch can have a variable number of rounds. The graph has a maximum of $k$ number of Doors. The robots do not have any knowledge about the number of Doors. The visibility range of the robots is five hops. Each robot has $O(\log(\Delta + k))$ bits of persistent memory and $\Delta + k + 7$ colors where $\Delta$ number of DIR colors, CONF, CONFC, CONF2, CONF3, MOV, ON, OFF and $k$ number of WAIT colors denoted by WAIT-1, WAIT-2, ..., WAIT-$k$ representing that a robot is waiting as well as their rank. All the robots entering from a particular Door can only display the WAIT color corresponding to that Door. The WAIT colors can be compared against each other to establish dominance between the robots. Initially, a robot has color ON when placed at the Door. The MULTIND runs in $O(m^2)$ epochs.

## 4.1 The MULTIND Algorithm

The Leader robots need to avoid collision with other Leader robots and the follower robots in chains. The robots use the hierarchy among the $k$ WAIT colors to avoid collision with another Leader robot. The Leader robots also avoid cutting through a chain to avoid collision with follower robots in another chain. In the multiple Door situation, the Leaders display the WAIT-$i$ color instead of the WAIT color used in a single Door case. In the *Look* phase, if a Leader robot $r_i$ with color WAIT-$i$ sees any other Leader $r_j$ with color WAIT-$j$ such that $j < i$, then we say that the Leader robot $r_j$ *dominates* Leader robot $r_i$. The Leader robot $r_j$ is said to be the *dominating* and the Leader robot $r_i$ is said to be *dominated*. Note that another Leader robot can dominate a dominating Leader robot at the same time. If no other robot dominates a Leader robot, it can choose a target.

The rest of the model is the same as in the Single Door case. The robots can be in any one these states during execution: *None, Leader, Follower, Finished*. When the robots appear at the Door, they are initialized with the None state and color ON. We define $\mathcal{P}_k(v_i, v_j)$ as the set of vertices that are part of all the paths from $v_i$ to $v_j$ of length $k$. Note that there can be multiple paths of length $k$, and we include vertices of all those paths. The additional rules apply to the Leader as follows.

**Movement of a Leader Robot:** If a chain is in Packed state, the corresponding leader $r_L$ chooses a free vertex in $V_{r_L}^2$ as a target in one of the following ways.

- If $V_{r_L}^5$ has a dominating leader $r_L'$ of some other chain, then a free vertex $v \in V_{r_L}^2$ is chosen such that $v \notin \mathcal{P}_5(r_L, r_L')$. If no such vertex $v$ is found, then $r_L$ transfers the leadership to its successor by pointing the direction towards the successor and goes into the Finished state by changing its color to OFF.
- If $r_L$ is the leader dominating some other leader $r_L'$ in $V_{r_L}^5$, then a free vertex $v \in V_{r_L}^2$ is chosen such that every vertex $v' \in N_v^2 \cap V_{r_L}^2$ is either unoccupied or occupied with robot in Finished state. If no vertex $v$ is present, $r_L$ transfers its leadership to its successor.
- If $V_{r_L}^5$ does not have any other leader robot, then a free vertex is chosen as a target such that every vertex $v' \in N_v^2 \cap V_{r_L}^2$ is either unoccupied or occupied with a robot in Finished state. If no free vertex exists, $r_L$ transfers the leadership to its successor.

After the target is fixed for a leader $r_L$, it communicates the target to its successor using DIR colors. After getting confirmation from its successor, it moves to the target with the color MOV and waits for the chain to be in the Packed state with the respective WAIT color. Additionally, irrespective of whether a Leader robot is dominated, the target chosen is such that the Leader robot does not cut through a chain while moving to that target vertex. We ensure that the vertex between the target vertex and the current vertex does not have more

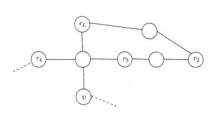

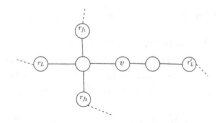

**Fig. 2.** A chain $r_L - r_2 - r_3 - r_4 \cdots$ does not cross itself

**Fig. 3.** Two chain do not cross each other

than one of its neighboring vertices occupied, at least one of which is an active robot (a robot not displaying OFF color) other than the current Leader robot. As shown in Fig. 3, $r_L$ and $r'_L$ are the dominating and dominated Leader at an instance, and $v$ is a free vertex two hops away from $r_L$. $r_{f_1}$ and $r_{f_2}$ are some followers of some other chain. $r_L$ does not reach to $v$ even if $r_L$ is the dominating Leader, as $r_{f_1}$ and $r_{f_2}$ is not in Finished state. As shown in Fig. 2, a Leader $r_L$ sees a free vertex $v$ two hops away. But $r_L$ cannot reach to $v$ (self cross) as the two hops neighboring vertices are occupied by $r_3$ and $r_4$.

## 4.2   Analysis of MULTIND Algorithm

We establish correctness of the MULTIND algorithm by the following lemmas and theorems.

**Lemma 8.** *A Leader robot always moves to and occupies free vertices.*

*Proof.* Since the robots have a visibility range of 5 hops, a Leader robot can choose a target vertex such that none of its neighboring vertices are occupied by a robot that is not a Leader. Also two leaders do not occupy adjacent vertices. For a leader $r_L$, if there is another Leader $r'_L$ within 5 hops visibility of $r_L$, one of them dominates the other. Let $r_L$ be the dominating robot. MULTIND algorithm ensures that $r'_L$ does not choose a vertex as its target in the path where it is being dominated. So, either $r'_L$ finds a target on some other path where is not getting dominated or transfers the leadership (in case of no such target is there). So, the dominating Leader $r_L$ can choose a free vertex as its target from the path $\mathcal{P}_5(r_L, r'_L)$. Hence, the robots cannot choose target vertices such that they are adjacent. □

**Lemma 9.** *The Robots do not collide.*

Consequently, we can state the following corollary.

**Corollary 3.** *A chain does not cross itself. Moreover, two chains do not cross each other.*

**Lemma 10.** *No two robots in the Finished state occupy adjacent vertices.*

**Theorem 3.** *Let $G$ be a connected graph $G$ with $k$ Doors. Algorithm MULTIND fills an MIS of vertices in $O(m^2)$ epochs of $G$ under an SSYNC scheduler, without collisions, by mobile luminous robots having the following capabilities: visibility range of 5 hops, persistent storage of $O(\log(\Delta + k))$ bits, and $\Delta + k + 7$ colors.*

*Proof.* By Lemma 9 and Lemma 10, the filled vertices in $G$ form a MIS and the filling is done without collisions. The robots require $O(\log(\Delta + k))$ bits of memory to store the following: *State* (4 states: 2 bits), *Target* (directions to the target vertex: $2\lceil \log \Delta \rceil + \lceil \log k \rceil$ bits), *NextTarget* (directions to the vertex to which the robot has to move after the *Target* vertex is reached: $2\lceil \log \Delta \rceil$ bits).

The colors used by the robots are $\Delta$ colors to show the directions to the target of the robot, that also acts as a special color to switch to the Finished state. Initially, when a robot is place for the first time at Door, it is colored with ON. There are $k$ numbers of WAIT colors - one for each Door. There are three additional colors (CONF, CONFC, and CONF2) for confirming that the robot saw the signaled direction and confirmations of the Predecessor or the Successor, one color MOV used during the movement, one color CONF3 to indicate that the chain is in Packed state and the OFF color.

Consider a graph where all the $k$ vertices that are connected to Doors node form a clique. In this case, only the Leader corresponding to the highest color Door would occupy one of the nodes of the $k$-clique. In this particular case, only one of the Doors remain active, and robots at all other Doors would go into the Finished state. So the multi-Door case would behave as a single Door case and thus replicate the time-bound for a single Door case. Thus, from Theorem 1 it follows that, MULTIND solves the *MIS Filling problem* in a graph with multiple Doors in $O(m^2)$ epochs.                                                       □

## 5   Discussion

**On the Requirement of the Colors:** We need at least $\Delta$ colors. If there are less than $\Delta$ colors, there exists two ports that are marked by the same color, resulting in a collision on movement to an already visited port, or one port may never get visited. In the absence of $k$ colors corresponding to the $k$ Doors, it may lead to a deadlock or collision between different chains. We use only a constant number of colors for communicating the directions for the robot movements.

While we need $\Delta$ colors to communicate the port numbers, we can minimize the number of colors at the cost of increasing round complexity. We can use Hoffman encoding to reduce the number of colors used for the port numbers from $\Delta$ to a constant number of colors. Now, a sequence of colors would represent a port number instead of a particular color and increases time complexity by a factor of the size of the largest encoding.

**One Hop vs Two Hops Movement:** Our model considers that a robot moves two hops in one LCM cycle. While moving, the color of the robot is set to MOV. However, we can easily avoid this 2 hops movement of a robot by replacing the color MOV with two colors, MOV1 and MOV2. After a robot chooses its target, it sets its color to MOV1 for the first hop and then changes its color to MOV2 before reaching the target.

**Minimality of Visibility Range:** For the single Door case, a robot having a visibility range of two fails to avoid placing robots in adjacent nodes. Consider a graph as shown in Fig. 4(i). Initially, a robot appears at the Door vertex $A$; then it moves to $C$ a vertex two hops away. If the robots only have a visibility range of two, then $r_1$ can go to $E$ without realizing that $E$ and $C$ are connected, resulting in a configuration that is not an independent set. With a visibility range of two, a robot cannot determine whether a robot is present at the neighbor of the target vertex. Hence we need a visibility range of three for a single Door case.

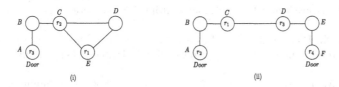

**Fig. 4.** Robots occupy adjacent nodes for (i) a single Door and visibility range two; (ii) multiple Doors and visibility range four.

Similar to the single Door case, consider the graph in Fig. 4(ii) for the multi-Door case. If the robots have a visibility range of four, then the robots at $A$ and $F$ may simultaneously move to occupy $C$ and $D$ and result in a configuration with robots occupying adjacent vertices.

## 6   Conclusion

In this paper, we presented and analyzed two algorithms for solving two flavors of the problem of filling a maximal independent set of vertices in an arbitrary connected graph using luminous mobile robots. The first algorithm IND for graphs with a single Door works under an asynchronous scheduler. It uses robots having three hops of visibility range, $\Delta + 8$ number of colors, and $O(\log \Delta)$ bits of persistent storage and solves the problem in $O(m^2)$ epochs. The second algorithm, MULTIND, works in graphs with $k$ $(> 1)$ Doors. It forms an MIS under a semi-synchronous scheduler using robots with five hops of visibility range, $\Delta + k + 7$ number of colors, and having $O(\log(\Delta + k))$ bits of persistent storage, completing in $O(m^2)$ epochs. It is open to extend this algorithm to the generalized asynchronous scheduler. The model of attaching robot splitting Doors to a graph is a new direction in multi-robot coordination problems, and many other graph problems can be explored under the same model.

# References

1. Albers, S., Henzinger, M.R.: Exploring unknown environments. SIAM J. Comput. **29**(4), 1164–1188 (2000)
2. Augustine, J., Moses Jr, W.K.: Dispersion of mobile robots: a study of memory-time trade-offs. In: Proceedings of the 19th International Conference on Distributed Computing and Networking, pp. 1–10 (2018)
3. Barrameda, E.M., Das, S., Santoro, N.: Deployment of asynchronous robotic sensors in unknown orthogonal environments. In: Fekete, S.P. (ed.) Algorithmic Aspects of Wireless Sensor Networks, pp. 125–140. Springer, Heidelberg (2008). https://doi.org/10.1007/978-3-540-92862-1_11
4. Barrameda, E.M., Das, S., Santoro, N.: Uniform dispersal of asynchronous finite-state mobile robots in presence of holes. In: Flocchini, P., Gao, J., Kranakis, E., Meyer auf der Heide, F. (eds.) ALGOSENSORS 2013. LNCS, vol. 8243, pp. 228–243. Springer, Heidelberg (2014). https://doi.org/10.1007/978-3-642-45346-5_17
5. Bose, K., Kundu, M.K., Adhikary, R., Sau, B.: Arbitrary pattern formation by asynchronous opaque robots with lights. Theor. Comput. Sci. **849**, 138–158 (2020)
6. d'Angelo, G., Di Stefano, G., Klasing, R., Navarra, A.: Gathering of robots on anonymous grids and trees without multiplicity detection. Theor. Comput. Sci. **610**, 158–168 (2016)
7. Défago, X., Potop-Butucaru, M., Raipin-Parvédy, P.: Self-stabilizing gathering of mobile robots under crash or byzantine faults. Distrib. Comput. **33**(5), 393–421 (2020)
8. Hideg, A., Lukovszki, T.: Uniform dispersal of robots with minimum visibility range. In: Fernández Anta, A., Jurdzinski, T., Mosteiro, M.A., Zhang, Y. (eds.) ALGOSENSORS 2017. LNCS, vol. 10718, pp. 155–167. Springer, Cham (2017). https://doi.org/10.1007/978-3-319-72751-6_12
9. Hideg, A., Lukovszki, T.: asynchronous filling by myopic luminous robots. In: Pinotti, C.M., Navarra, A., Bagchi, A. (eds.) ALGOSENSORS 2020. LNCS, vol. 12503, pp. 108–123. Springer, Cham (2020). https://doi.org/10.1007/978-3-030-62401-9_8
10. Hideg, A., Lukovszki, T., Forstner, B.: Filling arbitrary connected areas by silent robots with minimum visibility range. In: Gilbert, S., Hughes, D., Krishnamachari, B. (eds.) ALGOSENSORS 2018. LNCS, vol. 11410, pp. 193–205. Springer, Cham (2019). https://doi.org/10.1007/978-3-030-14094-6_13
11. Hsiang, T.-R., Arkin, E.M., Bender, M.A., Fekete, S.P., Mitchell, J.S.B.: Algorithms for rapidly dispersing robot swarms in unknown environments. In: Boissonnat, J.-D., Burdick, J., Goldberg, K., Hutchinson, S. (eds.) Algorithmic Foundations of Robotics V. STAR, vol. 7, pp. 77–93. Springer, Heidelberg (2004). https://doi.org/10.1007/978-3-540-45058-0_6
12. Kamei, S., Lamani, A., Ooshita, F., Tixeuil, S., Wada, K.: Gathering on rings for myopic asynchronous robots with lights. arXiv preprint arXiv:1911.04757 (2019)
13. Kamei, S., Tixeuil, S.: An asynchronous maximum independent set algorithm by myopic luminous robots on grids. CoRR abs/2012.03399 (2020)
14. Kshemkalyani, A.D., Ali, F.: Efficient dispersion of mobile robots on graphs. In: Proceedings of the 20th International Conference on Distributed Computing and Networking, pp. 218–227 (2019)
15. Poudel, P., Sharma, G.: Fast uniform scattering on a grid for asynchronous oblivious robots. In: Devismes, S., Mittal, N. (eds.) SSS 2020. LNCS, vol. 12514, pp. 211–228. Springer, Cham (2020). https://doi.org/10.1007/978-3-030-64348-5_17

16. Pramanick, S., Samala, S.V., Pattanayak, D., Mandal, P.S.: Filling MIS vertices by myopic luminous robots. CoRR abs/2107.04885 (2021). arxiv.org/abs/2107.04885
17. Suzuki, I., Yamashita, M.: Distributed anonymous mobile robots: formation of geometric patterns. SIAM J. Comput. **28**(4), 1347–1363 (1999). https://doi.org/10.1137/S009753979628292X

# WANMS: A Makespan, Energy, and Reliability Aware Scheduling Algorithm for Workflow Scheduling in Multi-processor Systems

Atharva Tekawade and Suman Banerjee[✉]

Department of Computer Science and Engineering, Indian Institute of Technology Jammu, Jammu 181221, India
{2018uee0137,suman.banerjee}@iitjammu.ac.in

**Abstract.** A scientific workflow is modeled as a *Directed Acyclic Graph* where the nodes represent individual tasks and the directed edges represent the dependency relationship between two tasks. Scheduling a workflow to achieve certain goal(s) (e.g.; minimize makespan, cost, penalty, and energy; maximize reliability, processor utilization, etc.) remains an active area of research. In this paper, we propose an efficient scheduling algorithm for workflows in heterogeneous multi-processor systems that takes into account makespan, energy consumption, and reliability. We name our methodology as **W**ait **A**ware **N**ormalized **M**etric **S**cheduling (henceforth mentioned as **WANMS**). The proposed approach is a list scheduling algorithm consisting of two phases namely, Task Ordering and Allocation. In the task ordering phase, it tries to find out an optimal ordering of the tasks based on the maximum execution cost starting from the current node. In the allocation phase, a task is assigned to a processor based on a normalized linear combination of finish time of the tasks and reliability or based on energy depending on the wait time of the task. Additionally, WANMS is designed in such a way so that it can satisfy any given reliability constraint while minimizing makespan and energy. The proposed algorithm has been analyzed to understand its time and space requirements. Experimental evaluations on the real-world and randomly generated workflows show that WANMS dominates state-of-the-art algorithms in terms of both makespan and energy in most cases and at least one objective in the rest of the cases. In particular, we observe that for high reliability constraints the schedule produced by WANMS Algorithm leads to up to 18% improvement in makespan and 13% improvement in energy on an average compared to the existing algorithms.

**Keywords:** Multi-processor system · Scheduling algorithm · Task graph · DAG · Workflow scheduling

A. R. Molla et al. (Eds.): ICDCIT 2023, LNCS 13776, pp. 20–35, 2023.
https://doi.org/10.1007/978-3-031-24848-1_2

# 1 Introduction

Many real-world complex functionalities of real-time applications in different domains including *industrial automation* [15], *robotics, cyber physical system* [1] etc. requires multiprocessor systems to execute. In all these cases, the application that typically runs consists of many tasks and there exists dependency between two of them. Such workflows often modeled as directed acyclic graph (henceforth mentioned as DAG) where each node represents a task and there is a directed edge from one task to the other task if the second one has a dependency on the first one [6]. These scientific workflows are common in different disciplines including Epigenomics in Biotechnology [17], Cybershake in Earthquake Engineering [4] and many more. As there are inherent dependency among the tasks, so failure of one task may leads to the failure of the entire workflow. Hence, reliability is a very important criteria that one scheduling algorithm needs to fulfill.

Another important criteria of a scheduling algorithm is to minimize the energy consumption. More importantly, this criteria become prevalent for battery operated devices where the availability of energy is limited such as mobile robots, drones etc. Hence, developing energy efficient and reliable scheduling algorithms remains an important research directions. There exists a large body of work on this problem [2,3,7,9,14]. For any scheduler with high reliability, the purpose is to minimize the faults as much as possible during execution. These faults may be for different reasons including network failure, hardware failure, temperature variation and so on. In the literature, the reliability has been modeled in the context of a multi-processor system by exponential distribution [8]. The transit fault during the time interval $t$ can be given as $e^{-\lambda \cdot t}$ where $\lambda$ is the parameter of the exponential distribution and this signifies that the reliability decreases with the increase of the execution time.

Other important aspects of a scheduling algorithm include makespan and energy consumption. The makespan is defined as the difference between the finish time of the last job and the start time of the first job in the schedule. Several scheduling algorithms have been proposed whose goal is to minimize the makespan of the schedule. For most cases, the HEFT algorithm [10] leads to a schedule that results into minimum makespan value. Also, there are other scheduling algorithms that considers reliability and energy consumption such as Least Energy Cost (LEC) [5,12] and Maximum Reliability (MR) [5,13]. Also there are several studies that considers all reliability, energy consumption, and makespan [9,12,14,16]. In the literature, the state-of-the-art DAG scheduling algorithms are HEFT, LEC, MR, LRDSA [9], and ESRG [12]. All these algorithms serve some purpose while working on the same input. The most closest to our study is ODS [5] which considers all three objectives: makespan, reliability and energy consumption.

In this paper, we have proposed a novel scheduling algorithm for scheduling workflows in a multi-processor system. In particular, the contributions of this paper is as follows:

- We study the problem of scheduling a workflow in a multi-processor system considering the energy consumption and makespan under a given reliability constraint.
- We propose a scheduling algorithm that allocates task nodes based on a task's wait time according to a linear combination of its finish time and reliability or on its energy consumption. Formally, we call our algorithm as **W**ait **A**ware **N**ormalized **M**etric **S**cheduling (WANMS). Additionally WANMS is designed so as to satisfy a given reliability constraint.
- We perform an extensive set of experiments to evaluate the performance of the proposed scheduling algorithm and compare with many existing solutions.

The rest of the paper is organized as follows. Section 2 describes the systems model and defines the problem formally. Section 3 contains the description of the proposed scheduling algorithm. Section 4 describes the experimental evaluation of the proposed scheduling algorithm. Finally, Sect. 5 concludes our study and talks about future research directions.

## 2    Systems Model and Problem Formulation

In this section, we describe the system's model and describe our problem formally. For any positive integer $n$, $[n]$ denotes the set $\{1, 2, \ldots, n\}$. Initially, we start by describing a multiprocessor system.

***Multiprocessor System.*** We consider a set of $m$ heterogeneous processors, denoted by $\mathcal{P} = \{u_1, u_2, \ldots, u_m\}$ and each connected with all the other processors resulting a complete graph. Each processor runs for a given range of frequencies. Let $f_{u_k,\min}$ and $f_{u_k,\max}$ denote the minimum and maximum frequencies that the $k^{th}$ processor runs for. For convenience, we normalize the maximum frequency to one i.e., $f_{u_k,\max} = 1$ for all $k \in [m]$. The rest of parameters for the processors will be discussed subsequently.

***Task Graph.*** The application under consideration can be modeled as a DAG denoted by $G(V, E)$, where $V$ denotes the set of vertices: $\{v_1, v_2, v_3, \ldots, v_n\}$ and each task node $v_i$, $i \in [n]$, represents a task of the application. $E$ denotes the set of edges of our task graph. A directed edge between task node $v_i$ to $v_j$ indicates a precedence relationship between $v_i$ and $v_j$ i.e., $v_j$ cannot start unless it has received the necessary output data from $v_i$. The weight of the edge $(v_i, v_j)$ denoted by $w(v_i, v_j)$ gives an idea of the communication time between the two tasks.

Now, we present a few definitions and terminology that will be subsequently required for the rest of the paper.

**Definition 1 (Predecessor of a Task).** *For any task $v_i$, its predecessor task(s) is denoted by $pred(v_i)$ and defined as the set of incoming neighbors of the node $v_i$ in the task graph $G(V, E)$. Mathematically, $pred(v_i) = \{v_j : (v_j, v_i) \in E(G)\}$.*

**Definition 2 (Successor of a Task).** *For any task $v_i$, its successor task(s) is denoted by $succ(v_i)$ and defined as the set of outgoing neighbor(s) nodes of the node $v_i$ in the task graph $G(V, E)$. Mathematically, $succ(v_i) = \{v_j : (v_i, v_j) \in E(G)\}$.*

**Definition 3 (Entry Task).** *The entry task in the task graph is denoted by $v_{entry}$. It is a redundant node added to have a proper notion of first task. For all $v_i$ such that $pred(v_i) = \emptyset$, we add a directed edge from $v_{entry}$ to $v_i$ with zero weight.*

**Definition 4 (Exit Task).** *The exit task is denoted by $v_{exit}$. It is a redundant node added to have a proper notion of last task. For all $v_i$ such that $succ(v_i) = \emptyset$, we add a directed edge from $v_i$ to $v_{exit}$ with zero weight.*

***Timing Metrics.*** Here, we discuss the start, execution, finish times associated with executing a task on a processor. Let $T_s[v_i, u_k]$ and $T_f[v_i, u_k]$ denote the start and finish times of executing the task $v_i$ on the processor $u_k$. Let $T_{exec}[v_i, u_k]$ denote the execution time of executing $v_i$ on $u_k$ at the maximum frequency, which can be determined through WCET analysis method (Worst Case Execution Time) during the analysis phase [11]. Running the processor at a lower frequency $f$ decreases the time proportionately. Once a task starts on a processor, it will run to completion i.e., we do not assume pre-emption. This leads us to the Eq. 1:

$$T_f[v_i, u_k] = T_s[v_i, u_k] + \frac{T_{exec}[v_i, u_k]}{f} \tag{1}$$

Due to the task dependencies, a task needs to transfer data to its successor nodes. Let the time required to communicate the data between tasks $v_i$ to $v_j$ be denoted by $T_{comm}[v_i, v_j]$. As discussed earlier, the weight of the edge $(v_i, v_j)$ denotes the communication time from the task $v_i$ to $v_j$. Additionally, we assume that the communication time is negligible if both tasks are scheduled on the same processor. This leads us to Eq. 2.

$$T_{comm}[v_i, v_j] = \begin{cases} w(v_i, v_j), & \text{if } v_i, v_j \text{ are allocated on different processors} \\ 0, & \text{otherwise} \end{cases} \tag{2}$$

A task can start on a processor if and only if both the below conditions hold:

- It has received output from all of its predecessor task(s).
- The processor on which it is scheduled is not executing another task i.e., it is idle.

This leads us to Eq. 3.

$$T_s[v_i, u_k] = \begin{cases} 0, & \text{if } v_i = v_{entry} \\ \max\{avail[u_k], \max_{v_j \in pred(v_i)}\{T_f[v_j, u_{k'}] + T_{comm}[v_j, v_i]\}\}, & \text{otherwise} \end{cases} \tag{3}$$

where $avail[u_k]$ denotes the earliest time that the processor $u_k$ is free after executing its previous task and $v_j$ is assumed to be scheduled on processor $u_{k'}$. The *makespan* is defined as the total time required to complete the execution of the task graph. Since $v_{exit}$ is the last task to be scheduled and $v_{entry}$ is the first task, the *makespan* can be given by Eq. 4, assuming that $v_{exit}$ is executed on $u_k$ and $v_{entry}$ on $u_{k'}$ and combined with Eq. 3.

$$makespan = T_f[v_{exit}, u_k] - T_s[v_{entry}, u_{k'}] = T_f[v_{exit}, u_k] \tag{4}$$

**Energy.** The power consumption of a processor consists of frequency dependent dynamic consumption, frequency independent dynamic consumption and static consumption components [5]. The frequency dependant dynamic component is the dominant one and can be written as:

$$P = \gamma \cdot c \cdot v^2 \cdot f \tag{5}$$

where, $\gamma$ is the activity factor, $c$ is the loading capacitance, $v$ is the supply voltage, and $f$ is the operating frequency. Since $f \propto v$, we see that $P \propto f^\alpha$. Let $P_{u_k}$ denote the sum of the frequency independent dynamic consumption and static consumption components and $P_{u_k}^f$ denote the overall power consumed by $u_k$ when operated at frequency $f$ which can be written as:

$$P_{u_k}^f = P_{u_k} + c_{u_k} \cdot f^{\alpha_{u_k}} \tag{6}$$

where, $c_{u_k}, \alpha_{u_k}$ denote the processor constants for $u_k$.

Energy consumed will be obtained by taking the product of power and execution time as shown below.

$$E_{v_i, u_k}^f = P_{u_k}^f \cdot \frac{T_{exec}[v_i, u_k]}{f} \tag{7}$$

where, $E_{v_i, u_k}^f$ denotes the energy when the task $v_i$ is executed on $u_k$ with frequency $f$. Now, the total energy consumption is defined as the sum of the energies consumed by each processor which is given in Eq. 8.

$$E(\mathbf{k}, \mathbf{f}) = \sum_{i=1}^{n} E_{v_i, \mathbf{k}[i]}^{\mathbf{f}[i]} \tag{8}$$

where, $\mathbf{k}$ and $\mathbf{f}$ are both vectors of $n$-*dimension*. $\mathbf{k}$ is a vector whose $i^{th}$ element tells us which processor $v_i$ is scheduled on and $\mathbf{f}$ is a vector whose $i^{th}$ element tells us which frequency the processor is operating on.

**Reliability.** As in many other works [5,12] we study dominant transient faults related to processor frequency which can be modeled by exponential distribution as shown in Eq. 9.

$$\lambda_{u_k}(f) = \lambda_{u_k} \cdot 10^{\frac{d_{u_k}(1-f)}{1-f_{u_k,min}}} \tag{9}$$

where, $\lambda_{u_k}$ denotes the average number of faults per second at the maximum frequency and $d_{u_k}$ is a constant specific to the processor. The reliability is modeled using a Poisson distribution, with parameter $\lambda_{u_k}(f)$. Let, $\mathcal{R}^f_{v_i,u_k}$ denote the reliability when the task $v_i$ is executed on the processor $u_k$ operated with the frequency $f$ and this is given by Eq. 10.

$$\mathcal{R}^f_{v_i,u_k} = e^{-\lambda_{u_k}(f) \cdot \frac{T_{exec}[v_i,u_k]}{f}} \tag{10}$$

The task graph executes successfully when all the tasks execute successfully. Hence, the reliability of the task graph is the product of the reliability of all the tasks, assuming failures are independent as given by Eq. 11.

$$\mathcal{R}(\mathbf{k},\mathbf{f}) = \prod_{i=1}^{n} \mathcal{R}^{\mathbf{f}[i]}_{v_i,\mathbf{k}[i]} \tag{11}$$

From Eq. 9 and 10, we can easily observe that both $\lambda_{u_k}(f)$ and $\mathcal{R}^f_{v_i,u_k}$ are increasing functions of $f$. Hence for maximum reliability, each processor must run at its maximum frequency. Furthermore, for each task $v_i$, there is a processor with minimum value $-\lambda_{u_k} \cdot T_{exec}[v_i,u_k]$ that gives maximum reliability value $\mathcal{R}_{v_i,\max}$. Hence, the maximum reliability of the application is denoted by $\mathcal{R}_{\max}$ and expressed in Eq. 12.

$$\mathcal{R}_{\max} = \prod_{i=1}^{n} \mathcal{R}_{v_i,\max} \tag{12}$$

**Problem Formulation.** Given a task graph $G(V,E)$ and a heterogenous processor platform $\mathcal{P}$, we wish to allocate tasks to processors that minimizes makespan and energy under a given reliability constraint ($\mathcal{R}_{req} \leq \mathcal{R}_{\max}$). The problem can now be mathematically formulated as:

$$\mathbf{Minimize} \begin{cases} makespan \\ E(\mathbf{k},\mathbf{f}) \end{cases}$$

$$\mathbf{Subject\ To:}\ \mathcal{R}(\mathbf{k},\mathbf{f}) \geq \mathcal{R}_{req}$$

From the formulation of the problem it is clear that this boils down to the problem of finding the appropriate values of the vectors: $\mathbf{k},\mathbf{f}$. In our solution approach, instead of dealing with both vectors simultaneously, we fix them one by one.

## 3    Proposed Scheduling Algorithm

In this section, we describe the proposed scheduling algorithm. As mentioned previously, we first begin by determining the task-processor allocation. Our proposed approach is a list scheduling heuristic divided into two phases, namely Task Ordering and Processor Allocation. After the processor allocation, we fix the operating frequencies of the processors. First we describe the task ordering phase.

## 3.1   Task Ordering

First we state the notion of up-rank value in Definition 5 of a task and this has been used subsequently.

**Definition 5 (Up-rank Value).** *Given a workflow $G(V, E)$, the up-rank values of the tasks can be computed by traversing the graph in a bottom up manner. For any task $v_i$, let $urv(v_i)$ denotes its up-rank value and this can be computed using Eq. 13.*

$$urv(v_i) = \begin{cases} \frac{1}{m} \cdot \sum_{k=1}^{m} T_{exec}[v_i, u_k], & if \ v_i = v_{exit} \\ \frac{1}{m} \cdot \sum_{k=1}^{m} T_{exec}[v_i, u_k] + \\ \max_{v_j \in succ(v_i)}\{w(v_i, v_j) + urv(v_j)\}, & otherwise \end{cases} \tag{13}$$

Basically, the up-rank value of the task $v_i$ gives an estimate of the maximum duration execution path from $v_i$ to $v_{exit}$. Now, the task order can be found by sorting the tasks in decreasing order based on the computed up-rank value. We denote this ordering of the tasks by $\rho$. Here, we want to highlight that the obtained ordering is in favour of lower makespan but may not be optimal.

## 3.2   Processor Allocation

In this sub-section, we assume that the processors are running at their maximum frequency and task ordering is known from the first phase. Now we begin allocating tasks one-by-one in order. After allocation of a processor to a task till its finish, its time requirement is divided into two phases:

– Idle Phase: In this case, the task is waiting to obtain the output from its predecessor task(s).
– Execution Phase: In this case, the task is getting executed on the processor.

Combining the time associated with the above two phases, we get the wait time of a task which is stated in Definition 6.

**Definition 6 (Task Wait Time).** *Given a workflow $G(V, E)$, the wait time of a task is defined as the amount of time for which a task is allocated to a processor. For any task $v_i$, let $W(v_i)$ denotes its wait time and this can be computed using Eq. 14.*

$$W(v_i) = \max_{v_j \in pred(v_i)}\{w(v_j, v_i)\} + \frac{1}{m} \cdot \sum_{k=1}^{m} T_{exec}[v_i, u_k] \tag{14}$$

The first part in Eq. 14 gives an idea of how much time a task has to remain idle on a processor till all the output(s) from all its predecessor task(s) are received and the second part is the average execution time over all the processors. Naturally, tasks having longer waiting time are prone to be on their allocated processor for longer durations. Thus, to make the makespan shorter, better idea is to allocate these tasks to the processors with lower finish time. We sort the tasks based on decreasing order of their waiting time in an array $W[]$. Then, we allocate the first $\ell(0 \le \ell \le n)$ tasks based on a weighted normalized linear combination of task's finish time with weight $\alpha$ and reliability with weight $(1 - \alpha)$, where the normalization is done using min-max normalization. The remaining tasks are allocated on the basis of their energy consumption. Formally, we name our algorithm as the *Wait Aware Normalized Metric Scheduling (WANMS)*. Algorithm 1 presents this idea in the form of pseudocode. Here $T_{v_i,\min}, T_{v_i,\max}, \mathcal{R}_{v_i,\min}, \mathcal{R}_{v_i,\max}$ denote the minimum and maximum values of finish time and reliability, respectively, of executing task $v_i$ over all processors.

---

**Algorithm 1: WANMS scheduling algorithm**

**Input:** Task graph, processor parameters, reliability constraint.
**Output:** Task-processor Mapping vector($\mathbf{k}$).

1   Compute the up-rank values $urv(v_i)$ using Equation No. 13. ;
2   Compute the ordering $\rho$ by sorting the up-rank values in descending order. ;
3   Calculate $W(v_i)$ according to Equation No. 14;
4   Sort the tasks in decreasing order of waiting time value in array W[]
     and break ties based on the ordering in $\rho$;
5   *makespan* $\longleftarrow$ 0; *reliability* $\longleftarrow$ 1; *energy* $\longleftarrow$ 0;
6   **for** $v_i \in \rho$ **do**
7      **if** $v_i \in [W[1], W[\ell]]$ **then**
8         Assign $v_i$ to processor satisfying

$$u_k^* \longleftarrow \min_{k \in [m]} \{\alpha \cdot \frac{T_f[v_i, u_k] - T_{v_i,\min}}{T_{v_i,\max} - T_{v_i,\min}} + (1 - \alpha) \cdot (\frac{\mathcal{R}_{v_i,\max} - \mathcal{R}_{v_i,u_k}^{f_{u_k},\max}}{\mathcal{R}_{v_i,\max} - \mathcal{R}_{v_i,\min}})\} ;$$

9      **else**
10         Assign $v_i$ to processor satisfying $u_k^* \longleftarrow \min_{k \in [m]} E_{v_i, u_k}^{f_{u_k}, \max} ;$
11      $\mathbf{k}[i] \longleftarrow u_k^*;$
12      *energy* $\longleftarrow$ *energy* $+ E_{v_i, u_k^*}^{f_{u_k^*}, \max};$
13      *reliability* $\longleftarrow$ *reliability* $\cdot \mathcal{R}_{v_i, u_k^*}^{f_{u_k^*}, \max};$
14      **if** $v_i = v_{exit}$ **then**
15         *makespan* $\longleftarrow T_f[v_i, u_k^*];$
16   **if** *reliability* $\ge \mathcal{R}_{req}$ **then**
17      **return** $\mathbf{k}, makespan, reliability, energy$ ;

---

### 3.3   Analysis Based on the Parameters of Algorithm 1

By varying the parameters of our algorithm $\ell$ from 0 to $n$ and the weighting factor $\alpha$ from 0 to 1, we can get a family of solutions. For example, if $\ell = n$ and $\alpha = 0$, we get the maximum reliability allocation. Hence, for any reliability constraint, there always exists an allocation by the WANMS Algorithm that satisfies the given constraint. The value of $\alpha$ can be set empirically depending on the reliability constraint. If the reliability constraint is very tight, we choose small values of $\alpha$ giving more weight to reliability metric than the finish time. In our experiments described in Sect. 4 we vary $\alpha$ from 0.0 to 1.0 in steps of 0.1. On the contrary, the ODS Algorithm cannot achieve high reliability constraints. This is because ODS Algorithm allocates tasks to the processors depending on the out-degree based on either of the below metrics:

$$
u_k^* = \begin{cases} \min_{k \in [m]} T_f[v_i, u_k] + \alpha \cdot (1 - \mathcal{R}_{v_i, u_k}^{f_{u_k}, \max}) \cdot T_{exec}[v_i, u_k] & \text{if } v_i \in [OD[1], OD[\ell]] \\ \min_{k \in [m]} E_{v_i, u_k}^{f_{u_k}, \max} & \text{otherwise} \end{cases}
$$

where, $OD[]$ is an array in which tasks are sorted in decreasing order of out-degrees. Since the reliability metric $(\mathcal{R}_{v_i, u_k}^{f_{u_k}, \max})$ always occurs in combination with $T_f[v_i, u_k]$ and $T_{exec}[v_i, u_k]$, we can see that $\mathcal{R}_{v_i, \max}$ and hence $\mathcal{R}_{\max}$ can never be achieved. Like WANMS, ODS also produces a family of solutions by varying $\ell, \alpha$. Lastly, if $\ell = n$ and $\alpha = 1$, WANMS boils down to HEFT and if $l = 0$, we allocate all tasks only based on energy consumption and this is precisely what LEC does.

Having determined the processor allocation ($\mathbf{k}$), we proceed to find the operational frequencies ($\mathbf{f}$) for the various processors. For this, we use the SOEA method that gives the optimum energy consumption under a given reliability constraint and processor allocation [5]. Each solution from WANMS satisfying the reliability constraint is fed into SOEA which reduces the operating frequencies from maximum value consequently bringing the reliability down to exactly $\mathcal{R}_{req}$.

### 3.4  Complexity Analysis

Calculating the up-rank values and wait times for a task $v_i$ takes $\mathcal{O}(m + |succ(v_i)|)$ and $\mathcal{O}(m + |pred(v_i)|)$ as seen from Eq. 13 and 14. Since $|succ(v_i)|, |pred(v_i)| \leq n$, the time required for one task is almost $\mathcal{O}(m + n)$. Hence, the total time required for all the tasks will be $\mathcal{O}(n \cdot (m + n))$. Then, sorting takes additional $\mathcal{O}(n \cdot \log n)$. For each task-processor pair, calculating the finish time takes $\mathcal{O}(|pred(v_i)|)$ or almost $\mathcal{O}(n)$ time as seen from Eq. 1 and 3. Hence, over all task-processor pairs, we get a complexity of $\mathcal{O}(n^2 \cdot m)$. The complexity of SOEA is $\mathcal{O}(\log(L_o/\epsilon) \cdot m \cdot \log(L_f/\epsilon) \cdot n)$, where $L_o, L_f$ relate to the frequency range under consideration and $\epsilon$ denotes the accuracy of the reliability constraint [5]. Finally, by varying the constants $\ell, \alpha$ the overall time complexity of WANMS+SOEA is $\mathcal{O}(L_\alpha \cdot n^2 \cdot m(n + \log(L_o/\epsilon) \cdot \log(L_f/\epsilon)))$, where $L_\alpha$ denotes the number of iterations over $\alpha$. As for the space complexity, we need extra $\mathcal{O}(n)$ space to store the up-rank and wait time values.

## 4  Experimental Evaluation

In this section, we describe the experimental evaluation of the proposed solution approach. First, we start by describing the example task graphs.

## 4.1 Task Graphs

We perform the experimental evaluation of our proposed scheduling algorithm on the below mentioned workflows which are widey used in literature for comparison [5,11,12].

- **Fast Fourier Transform (FFT):** FFT applications exhibits a high degree of parallelism and has been used in many previous studies. For any given positive integer $\rho$, the number of nodes is $n = (2 + \rho) \times 2^\rho - 1$ [10]. In our experiments, we consider task graphs with $\rho = 4$ or $n = 95$.
- **Gaussian Elimination (GE):** Compared to FFT, GE applications exhibit low degree of parallelism. For a given positive integer $\rho$, the number of tasks can be given by the following equation: $N = (\rho^2 + \rho - 2)/2$. In our experiments, we consider task graphs with $\rho = 16$ or $n = 135$.
- **Random Workflow:** As the name suggests, random graphs are generated in a random fashion. The degree of parallelism is random. In our experiments, we consider $n = 100$.

## 4.2 Experimental Setup

We implement the proposed solution approach with on a workbench system with i5 $10^{th}$ generation processor and 32 GB memory in Python 3.8.10. Processor parameters are set randomly in the ranges: $P_{u_k} \in [0.4, 0.8]$, $c_{u_k} \in [0.8, 1.3]$, $f \in [0.3, 1.0]$, $\alpha_{u_k} \in [2.7, 3.0]$, $\lambda_{u_k} \in [10^{-6}, 10^{-5}]$, $d_{u_k} \in [1, 3]$ which reflect the real-world characteristics, such as for Intel Mobile Pentium III and ARM Cortex-A9 as in [5]. The number of processors is set uniformly at random from the interval $[20, 40]$. Additionally, $\epsilon$ in SOEA is set to $10^{-5}$ as in [5]. We set the task graph parameters as in [5,11] where $T_{exec}[v_i, u_k] \in [10, 100]$ and $w(v_i, v_j) \in [10, 100]$.

## 4.3 Algorithms Compared

We compare the performance of the proposed scheduling algorithm with the following existing solutions from the literature:

- **Out Degree Scheduling (ODS)** [5]: This is one of the state-of-the-art algorithm for workflow scheduling in multiprocessor system that takes makespan, energy and reliability into consideration by allocating task nodes based on their out-degree. ODS has a similar time and space complexity as WANMS.
- **Energy Efficient Scheduling with Reliability Goal (ESRG)** [12]: This is another state-of-the-art algorithm that minimizes for energy under reliability constraint. Many existing studies have also compared their results with this method.
- **Heterogeneous Earliest Finish Time (HEFT)** [10]: This is one of the earliest study on the workflow scheduling in multiprocessor system and has been used by many authors for performance comparison of their proposed algorithm. This algorithm minimizes makespan by allocating a task to the processor with least finish time at each step.

- **Maximum Reliability (MR)** [5,13]: As the name suggests, this algorithm assigns each task to the processor that leads to the maximum reliability ($\mathcal{R}_{v_i,\max}$) without considering makespan or energy consumption.
- **Least Energy Cost (LEC)** [5,12]: As the name suggests, this algorithm assigns each task to the processor that leads to the least energy consumption without considering makespan or reliability.

We would like to highlight that we do not consider LRDSA for comparison because it assumes that all processors run at the same frequency. The energy can be decreased by changing this frequency, but it is not able to precisely control the value of reliability, hence making it impossible to take reliability as a constraint [5].

### 4.4  Experimental Description

The reliability constraints for each experiment are set as $\eta \cdot \mathcal{R}_{\max}$, where $\eta$ is set from 0.9 to 0.99 in steps of 0.01 (low reliability constraint) and from 0.991 to 0.999 in steps of 0.001 (high reliability constraint). For each constraint, the experiment is run for 10 instances. In general HEFT, LEC and ODS Algorithms cannot achieve high reliability constraints. In such cases, we take the maximum value of $\eta$ as long as the constraint is satisfied from all the instances run and replace the algorithm with MR for the remaining values of $\eta$. In both the cases, we report the average value over all the instances that satisfies the constraint. From the set of solutions generated by ODS and WANMS, we take the solutions which satisfy the reliability constraint and further pass it through SOEA for energy optimization. Out of these solutions, we choose the ones with the least energy. This set of experiments is to demonstrate that WANMS always achieves the least energy and gives significant improvement for high reliability cases.

Additionally, we choose the solution with best makespan from ODS+SOEA and denote its corresponding energy by $E$. Then, we choose the solution with best makespan value from the WANMS+SOEA with energy $\leq E$. This set of experiments is to demonstrate the fact that WANMS gives the least makespan value and additionally dominates ODS in terms of energy. We call this the best makespan solution. Since HEFT, LEC and MR Algorithm do not consider frequencies, we also pass them through SOEA to determine the operational frequencies. ESRG Algorithm determines both frequencies and processor-allocation together.

### 4.5  Experimental Result with Discussions

Now, we describe the experimental results along with the observations. Figure 1 shows the reliability ratio vs. makespan and reliability ratio vs. energy consumption plots for the FFT workflow with $\rho = 4$. In particular, in Fig. 1(a) and (b), we show the plots for the best makespan solutions and in Fig. 1(c) and (d) we show the plots for the best energy consumption solution. From Fig. 1(a) we can conclude that if we increase the reliability ratio up to $\eta = 0.995$ for WANMS the makespan decreases and beyond this limit makespan increases sharply. We also observe that among all the algorithms WANMS leads to the least makespan value. As an example, for the low reliability solutions, when the reliability ratio value is 0.9 the makespan and energy values for the ODS Algorithm is 778 and 3345, respectively. The same for the WANMS Algorithm is 728 and 2604, respectively. So, the improvement in terms of makespan and energy is approximately 6.42% and 22.15%, respectively. However, when the value of the reliability ratio has been increased to 0.997, we observe that the ESRG is the best among the existing solution methodologies. For this, the makespan and energy values are 958 and 3063, respectively. The same for WANMS Algorithm is 750 and 2700, respectively. The improvement in terms of makespan and energy are 21.7% and 11.8%, respectively. When we are considering the best energy solutions, for the reliability ratio 0.9 the makespan and the energy values obtained by ODS, LEC and the proposed algorithm are exactly coinciding with each other. Hence, the proposed methodology is as good as the state-of-the-art methods. When we increase the value of the reliability ratio to 0.997, the makespan and energy consumption by the ESRG Algorithm is 958 and 3063, respectively. The same for WANMS Algorithm is 780 and 2246. So the improvement in the front of makespan and energy is 18.58% and 26.6%, respectively.

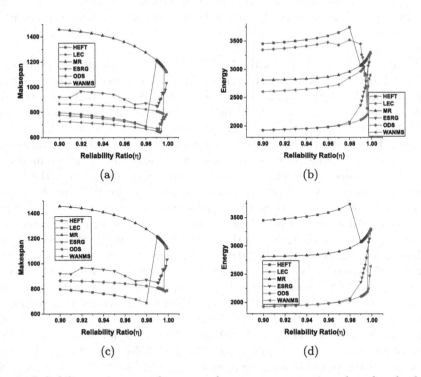

**Fig. 1.** Reliability ratio vs. makespan and energy consumption plots for the best makespan and energy solutions for the FFT Workflow ($\rho = 4$)

Figure 2 shows the reliability ratio vs. makespan and reliability ratio vs. energy consumption plots for the GE workflow with $\rho = 16$. From Fig. 2(a), we can observe that, in case of the best makespan solutions the schedule generated by WANMS Algorithm leads to the least makespan value for any reliability ratio. When we are considering the best makespan solution and the value of $\eta$ is set to 0.9, the makespan and the energy consumption by the ODS Algorithm is 1619 and 4676, respectively. The same for the WANMS Algorithm is 1533 and 4262, respectively. The improvement in the front of makespan and energy is 5.3% and 8.8%, respectively. When the value of the reliability ratio is increased to 0.991, the ESRG Algorithm is found to be the best among the existing solution approaches. For this algorithm the makespan and energy values are 2280 and 4995, respectively. The same for the WANMS Algorithm is 1984 and 3098, respectively. In this case the improvement in the front of makespan and energy is 36.3% and 18%, respectively. When we are considering the best energy solutions, we find that WANMS coincides with LEC achieving the lowest energy value slightly better than ESRG followed by ODS. For the reliability ratio value is 0.9, the makespan and the energy values for WANMS are 2068 and 2673, respectively. The same obtained by ESRG is 2059 and 2690, respectively. However, when the reliability ratio has been increased to 0.991 the makespan and energy consumption values for ESRG are 2280 and 4995, respectively. The same for the proposed approach is 1948 and 3098, respectively. So, the improvement in the makespan and energy consumption front is approximately 14.5% and 18%, respectively.

Figure 3 shows the plots for the RANDOM workflow for the best makespan and reliability solutions. In this workflow also our observations are consistent with the previous two workflows. For the best makespan solutions when the value of $\eta$ is set to 0.9, the value of the makespan and the energy for ODS are 1549 and 3391, respectively. However, the same for the WANMS Algorithm is 1513 and 3179, respectively. So, the improvement in case of makespan and energy is 2.3% and 6.2%, respectively. When the value of the reliability ratio is increased to 0.996, the value of the makespan and the energy consumption of the ESRG Algorithm will be 1705 and 2925, respectively. The same for WANMS Algorithm is 1468 and 2700, respectively. So the improvement in makespan and energy front is 13.9% and 7.6%, respectively. Equivalently, when we are considering the best energy solution and the value of the reliability ratio is 0.9 the value of the makespan and the energy for the ODS Algorithm is 1651 and 2233, respectively. However, the same for WANMS and LEC are 1650 and 2143, respectively. So, in this case the WANMS again achieves the least energy and additionally dominates ODS in terms of makespan. For the high reliability solutions when the reliability ratio has been increased to 0.996, we observe that the performance of the ESRG Algorithm is the best among the existing methods. So, for this algorithm the value of the makespan and energy are 1705 and 3056, respectively. However, the same for the proposed solution approach is 1561 and 2494, respectively. So the improvement in the makespan and energy consumption front is 8.45% and 18.4%, respectively.

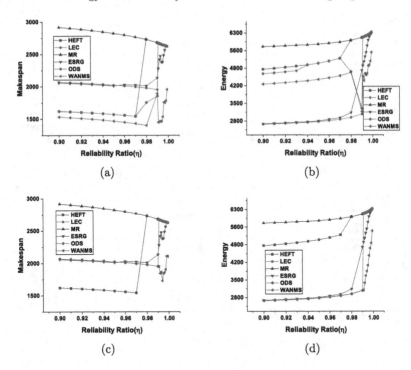

**Fig. 2.** Reliability ratio vs. makespan and energy consumption plots for the best makespan and energy solutions for the GE Workflow ($\rho = 16$)

In conclusion, we see from the best makespan solutions that WANMS gives the least makespan value among all the methods, and always outperforms the ODS Algorithm in terms of energy consumption. From the best energy solutions, we see that WANMS performs at least as good as the other methods. For high reliability constraints, both best makespan and best energy solutions significantly dominate the next best ESRG method in terms of both makespan and energy. In general, the makespan decreases with increasing reliability constraint, but as the constraint becomes tighter, lesser solutions generated by WANMS will satisfy the constraint and hence the best makespan solution out of these may be inferior when compared to a looser constraint. The energy consumption increases as the reliability constraint becomes higher.

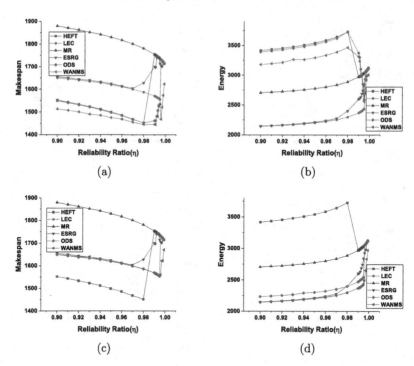

**Fig. 3.** Reliability ratio vs. makespan and energy consumption plots for the best makespan and energy solutions for the Random Workflow ($n = 100$)

## 5   Concluding Remarks

In this paper, we have proposed a scheduling algorithm for workflows in a multi-processor system whose goal is to minimize makespan and energy consumption for a given reliability constraint. We have analyzed the proposed approach to understand its time and space requirements. We perform a number of experiments with real-world workflows to show the effectiveness of the proposed solution approach. We have also observed that our proposed solution approach outperforms many of the existing solution approaches. In the future, we will try to include fault-tolerance in our model for reliability constraints greater than $\mathcal{R}_{max}$ and also study the problem of assigning frequencies to processors in this case.

## References

1. Capota, E.A., Stangaciu, C.S., Micea, M.V., Curiac, D.I.: Towards mixed criticality task scheduling in cyber physical systems: challenges and perspectives. J. Syst. Softw. **156**, 204–216 (2019)
2. Chang, W., Pröbstl, A., Goswami, D., Zamani, M., Chakraborty, S.: Battery-and aging-aware embedded control systems for electric vehicles. In: 2014 IEEE Real-Time Systems Symposium, pp. 238–248. IEEE (2014)

3. Dogan, A., Ozguner, F.: Matching and scheduling algorithms for minimizing execution time and failure probability of applications in heterogeneous computing. IEEE Trans. Parallel Distrib. Syst. **13**(3), 308–323 (2002)
4. Farid, M., Latip, R., Hussin, M., Hamid, N.A.W.A.: Scheduling scientific workflow using multi-objective algorithm with fuzzy resource utilization in multi-cloud environment. IEEE Access **8**, 24309–24322 (2020)
5. Huang, J., Li, R., Jiao, X., Jiang, Y., Chang, W.: Dynamic DAG scheduling on multiprocessor systems: reliability, energy, and makespan. IEEE Trans. Comput. Aided Des. Integr. Circ. Syst. **39**(11), 3336–3347 (2020)
6. Kwok, Y.K., Ahmad, I.: Benchmarking and comparison of the task graph scheduling algorithms. J. Parallel Distrib. Comput. **59**(3), 381–422 (1999)
7. Ma, Y., Zhou, J., Chantem, T., Dick, R.P., Wang, S., Hu, X.S.: Online resource management for improving reliability of real-time systems on "big-little" type MPSoCs. IEEE Trans. Comput.-Aided Des. Integr. Circ. Syst. **39**(1), 88–100 (2018)
8. Shatz, S.M., Wang, J.P.: Models and algorithms for reliability-oriented task-allocation in redundant distributed-computer systems. IEEE Trans. Reliab. **38**(1), 16–27 (1989)
9. Tang, X., Li, K., Qiu, M., Sha, E.H.M.: A hierarchical reliability-driven scheduling algorithm in grid systems. J. Parallel Distrib. Comput. **72**(4), 525–535 (2012)
10. Topcuoglu, H., Hariri, S., Wu, M.Y.: Performance-effective and low-complexity task scheduling for heterogeneous computing. IEEE Trans. Parallel Distrib. Syst. **13**(3), 260–274 (2002)
11. Xie, G., Chen, Y., Liu, Y., Wei, Y., Li, R., Li, K.: Resource consumption cost minimization of reliable parallel applications on heterogeneous embedded systems. IEEE Trans. Ind. Inf. **13**(4), 1629–1640 (2016)
12. Xie, G., Chen, Y., Xiao, X., Xu, C., Li, R., Li, K.: Energy-efficient fault-tolerant scheduling of reliable parallel applications on heterogeneous distributed embedded systems. IEEE Trans. Sustain. Comput. **3**(3), 167–181 (2017)
13. Xie, G., et al.: Minimizing redundancy to satisfy reliability requirement for a parallel application on heterogeneous service-oriented systems. IEEE Trans. Serv. Comput. **13**(5), 871–886 (2017)
14. Zhao, B., Aydin, H., Zhu, D.: On maximizing reliability of real-time embedded applications under hard energy constraint. IEEE Trans. Ind. Inf. **6**(3), 316–328 (2010)
15. Zhou, J., et al.: Security-critical energy-aware task scheduling for heterogeneous real-time MPSoCs in IoT. IEEE Trans. Serv. Comput. **13**(4), 745–758 (2019)
16. Zhou, J., et al.: Resource management for improving soft-error and lifetime reliability of real-time MPSoCs. IEEE Trans. Comput. Aided Des. Integr. Circ. Syst. **38**(12), 2215–2228 (2018)
17. Zhou, X., Zhang, G., Sun, J., Zhou, J., Wei, T., Hu, S.: Minimizing cost and makespan for workflow scheduling in cloud using fuzzy dominance sort based HEFT. Future Gener. Comput. Syst. **93**, 278–289 (2019)

# Multiple Criteria Decision Making-Based Task Offloading and Scheduling in Fog Environment

Nidhi Kumari[⊠][iD] and Prasanta K. Jana[iD]

Department of Computer Science and Engineering, Indian Institute of Technology (ISM) Dhanbad, Dhanbad 826004, Jharkhand, India
nidhi3rdjan1995@gmail.com, prasantajana@iitism.ac.in

**Abstract.** Fog computing is a three-tier architecture that provides an emerging technology aiming to reduce the delay and energy consumption between IoT (end) devices and the cloud. The fog layer is close to IoT devices; hence the tasks of time-sensitive applications are offloaded from the end devices to the fog nodes. Efficient offloading and scheduling of the tasks (i.e., the order in which tasks are executed at a fog node) jointly minimize waiting and response time. Given a set of fog nodes and a set of tasks, how to select a fog node and how to effectively schedule the tasks to minimize delay is a challenging problem due to heterogeneous nature of the fog environment. To deal with this challenge, we need to jointly offload and schedule the tasks by ranking the fog nodes and the tasks respectively. Although some papers have addressed task offloading and scheduling jointly, none of them have used performance-based ranking. In this paper, we propose a scheme that uses the multilevel Multiple Criteria Decision Making (MCDM) technique for fog node selection during offloading and determining order of task execution in scheduling. The proposed scheme is based on Entropy-based Technique for Order of Preference by Similarity to Ideal Solution (E-TOPSIS), which incorporates delay, energy, and reliability to rank the fog nodes as well as tasks. Through extensive simulations, we show that the proposed scheme outperforms some existing (baseline) algorithms.

**Keywords:** Fog computing · Task offloading · Task scheduling · Delay · Energy · Reliability · MCDM techniques

## 1 Introduction

Internet of Things (IoT) is one of the most extensively used technology today, making our lives simpler and more manageable by introducing connected intelligence into our homes, offices, automobiles, etc. By 2025, the number of connected IoT devices is expected to reach 75.44 billion [11]. These devices have limited storage, computing power, and battery life, making them difficult to process the data. Thus, they require cloud computing which has limitless storage and computing power. However, the distance between the IoT devices (end devices) and

A. R. Molla et al. (Eds.): ICDCIT 2023, LNCS 13776, pp. 36–50, 2023.
https://doi.org/10.1007/978-3-031-24848-1_3

the cloud results in high latency, security issues, and network congestion which affects delay-sensitive applications such as smart transportation, smart health care, and fire detection & fighting [15]. Therefore, these applications do not function well on Cloud-IoT architecture [2]. Thus fog computing was introduced in which fog nodes (FNs) are deployed at the edge of the network of the IoT devices. Fog forms a mini cloud close to the IoT devices, with lower processing power, energy, and storage capacity compared to cloud servers [13]. When the tasks are generated at the IoT devices, they are usually offloaded to the FNs for their execution rather than sending them to the far cloud and therefore, the delay and energy consumption are reduced significantly [20,23].

Task offloading is a mechanism for transferring a full or fraction of a locally originated task to an FN or a remote cloud for processing. Given a set of $m$ FNs and a set of $n$ ($n >> m$) tasks, selecting the best FN for a task to be offloaded is challenging. Moreover, multiple tasks may be offloaded to a single FN at a time. So determining the order of their execution on the offloaded FN (referred to as task scheduling) is equally important as it impacts various quality of service (QoS) parameters, including delay, energy consumption and reliability. It is noteworthy that in a fog environment, task offloading and task scheduling are the two main operations in order to provide IoT services efficiently. Depending on the application scenario, selecting the best FN and order of execution of tasks on an FN can be pretty different, which creates a decision-making problem [12]. To deal with this, various performance parameters must be included for choosing a correct alternative, and in this situation, the MCDM (Multiple Criteria Decision Making) approaches can be very effective. This is an efficient method that has been successfully applied to solve various decision-making problems in different domains such as grid computing, cloud computing, and many real-time applications, e.g., employee recruitment and determining the path for humanoid robots [4,24].

In this paper, we consider a joint problem of task offloading and scheduling and propose a method that uses the MCDM technique. The objective is to minimize delay and energy consumption and also to maximize the reliability of FNs' operation. We use Entropy-based Technique for Order of Preference by Similarity to Ideal Solution (E-TOPSIS) to find the best FN during offloading and ranking of the tasks during scheduling. The proposed algorithm is extensively simulated and the results are compared with two baseline algorithms to show the effectiveness with respect to various performance metrics such as delay and energy consumption.

Many research works have been carried out on the individual aspects of task offloading and task scheduling in fog environment, a review of which can be found in [1,3,6,10,13,21]. Quite a few papers [5,7,9,14,17] have also dealt with these issues jointly on fog platform. For examples, Guo et al. [5] proposed a recursive computation offloading algorithm in a fog radio access network that jointly optimized the offloading and scheduling by reducing the average execution delay of tasks. Ju et al. [9] offloaded the task based on their deadline and used Particle Swarm Optimization (PSO) to schedule them. As a result, the proposed solution reduced the number of missed-deadline requests and the completion time. The authors of [7] proposed an Energy-Effective Task Offloading (EETO) method that prioritizes tasks based on multiple QoS criteria and used the Lyapunov optimization technique to make efficient scheduling decisions for those prioritized tasks. Sellami et al. [17] proposed a deep reinforcement learning approach for task offloading and scheduling in an SDN-enabled IoT network, reducing latency and assuring energy efficiency. In the vehicular network, the mobility-aware task offloading takes optimal time to decide where and when to offload the tasks to minimize system costs [14]. They also proposed a Cooperative Task Offloading Scheduling strategy which prioritizes the tasks and schedules them according to their requirement. However, none of them have considered a performance-based selection through a decision-making system such as MCDM or MADM (Multiple Attribute Decision Making) to the best of our knowledge. Moreover, the existing works have not considered a delay, energy, and reliability as the decision-making factors together.

The rest of the paper is organized as follows: Sect. 2 presents the system models. The proposed method is presented in Sect. 3. Section 4 discusses the change in the path of results due to the migration of IoT devices. The extensive simulation of the proposed algorithm is presented in Sect. 5, and finally, we conclude our paper in Sect. 6.

## 2   System Model

This section introduces the software-defined network (SDN) based IoT-enabled fog architecture with core components: the Cloud, Controller, FNs, and IoT devices, Fig. 1. The SDN is a combination of the data plane and the control plane, which stores the information of the FNs. As a result, the controller is responsible for offloading decisions. The tasks which are not time-sensitive and require huge storage and resources are offloaded to the cloud directly. The execution of time-sensitive tasks is carried out locally and remotely on FNs. Here, we only consider the offloading of time-sensitive tasks. We assume '$n$' number of IoT devices and '$m$' number of FNs, where $n >> m$. We also consider that the FNs are heterogeneous and create clusters. Each cluster has an access point (AP) that builds a connection between the IoT devices and the fog devices. The IoT devices generate different tasks, and a fraction of them is offloaded to their nearby cluster. In this work, we consider that the network functions in the time-frame manner, each of frame length '$t$' [23], where task offloading and task scheduling operates in alternate time frames. The tasks are generated continuously, independent of the time frame. Initially, the controller decides the optimal FN, and the tasks are offloaded. The tasks are merely queued and do not execute on the FNs during task offloading. In the subsequent time frame, the FNs independently execute task scheduling and determine the order of execution of their tasks. At this time frame offloading decision-making process is halted. The subsequent sections describe the task offloading, delay, energy, and reliability models.

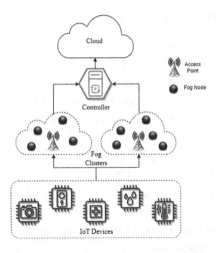

**Fig. 1.** SDN-based IoT-enabled fog architecture.

## 2.1 Task Offloading and Scheduling Model

Consider a set of $n$ IoT devices and a set of $m$ FNs. Let us assume that there exist multiple clusters of FNs. The tasks generated by the IoT devices are offloaded to the FN of the nearby cluster. Let each $i^{th}$ device generate a task of size $l_i$, $i \in \{1, 2, \ldots, n\}$. Any task to be offloaded is expected to be indivisible, which implies that it is offloaded partially or fully to a single FN only. Let $\alpha$ be the fraction of a task to be offloaded to an FN where $\alpha = 0$ value indicates no offloading (i.e., the task is completely executed on the local device itself), 1 represents the full offloading, and the value between 0 to 1 is partial offloading. Let $\eta_j$ be the CPU cycles of $j^{th}$ FN for processing a bit of data, $\omega_j$ be the CPU frequencies of FN $j$, and $\theta_j$ be the energy consumption per CPU cycle of $j^{th}$ FN. Let all such information be kept with the controller. As $n >> m$ multiple tasks are offloaded to a single FN of a cluster at an instance of time. The tasks must be scheduled based on their ranking. We have not considered downloading time and energy consumption during this period. The common terminologies used in this paper are summarized in Table 1.

## 2.2 Delay Model

Delay is the most critical offloading criterion as it is the main reason to introduce the middleware technology, fog computing. As we have already mentioned that a task may be offloaded fully or partially, the delay is calculated as the sum of the local processing time, transmission time and remote processing time of the task. The transmission time for downloading results is assumed to be negligible. Now, the time required to process the data locally at the $i^{th}$ IoT device is given as follows:

**Table 1.** Parameters used for simulation.

| Notation | Descriptions |
|---|---|
| $n$ | Total number of IoT devices |
| $m$ | Total number of FNs |
| $l_i$ | size of the task $i$ generated by $i^{th}$ IoT device |
| $\eta_i$ | CPU cycles for processing 1 bit of data at IoT device $i$ |
| $\eta_j$ | CPU cycles for processing 1 bit of data at FN $j$ |
| $\omega_i$ | CPU frequency of IoT device $i$ |
| $\omega_j$ | CPU frequency of FN $j$ |
| $\theta_i$ | Energy consumption per CPU cycle of IoT device $i$ |
| $\theta_j$ | Energy consumption per CPU cycle of IoT device $j$ |
| $\beta$ | Shape parameter of Weibull distribution |
| $\gamma$ | Scale parameter |
| $\varpi_0$ | White noise power |
| $g_n^j$ | Channel gain |
| $p_t$ | Transmission power |
| $\sigma$ | Path loss factor |
| $r_{ij}$ | Data transfer rate to transfer $i^{th}$ task to $j^{th}$ FN |

$$d_i = \frac{(1-\alpha)l_i\eta_i}{\omega_i} \tag{1}$$

where, $\alpha$ is the fraction of the task, $l_i$ is the size of the $i^{th}$ task generated by $i^{th}$ IoT device, $\eta_i$ and $\omega_i$ is the CPU cycle and CPU frequency of $i^{th}$ IoT device respectively.

The transmission time is calculated by Eq. 2, where $r_{ij}$ is data rate to transfer the $i^{th}$ task to $j^{th}$ FN and it is calculated by Shannon formula given in Eq. 3 [22]

$$d_t = \frac{\alpha l_i}{r_{ij}} \tag{2}$$

$$r_{ij} = B_j \log_2(1 + \frac{p_t * g_n^j}{\varpi_0}) \tag{3}$$

where, $B_j$ is the bandwidth occupied by FN, $p_t$ is the transmission power, $g_n^j$ is the channel gain between IoT device and FN, and $\varpi_0$ is the white noise power. The delay of the task offloaded to the FN is calculated as follows:

$$d_j = \frac{\alpha l_i \eta_j}{\omega_j} \tag{4}$$

Following Eq. 1, 2 and 4 the total delay $(D)$ of a task is calculated as follows:

$$D = d_i + d_t + d_j \tag{5}$$

## 2.3   Energy Model

Energy is the most important goal for all types of devices, especially those that are battery-powered. Different types of services use different amounts of energy. The increase in energy consumption during the partial offloading of the task is compensated by the sub-task executed at the local device. Let $E_i$ be the energy consumption of the $i^{th}$ IoT device due to local execution is given as:

$$E_i = (1 - \alpha)l_i\eta_i\theta_i \tag{6}$$

Further, the partial task is offloaded to the FN. Hence, the energy consumption to transfer a task is $E_t$ and energy consumption of the $j^{th}$ FN to execute the partially offloaded task is represented by $E_j$.

$$E_t = \frac{\alpha l_j}{r_{ij}} \tag{7}$$

$$E_j = \alpha l_j \eta_j \theta_j \tag{8}$$

Therefore, the total energy consumption $(E)$ to execute a task is given as the sum of $E_i$, $E_t$ and $E_j$ through Eq. 6, 7 and 8 respectively.

$$E = E_i + E_t + E_j \tag{9}$$

## 2.4   Reliability Model

In reliability analysis, we use the Weibull distribution [16], which is one of the most often used probability distributions for measuring the chance of failure of an electronic device such as FN. Failure density function $f(\tau)$ is the probability of failure occurring over time which is represented in Eq. 10. Here, the $\tau$ is equivalent to the total delay $D$.

$$f(\tau) = \frac{\beta}{\gamma}\left(\frac{\tau}{\gamma}\right)^{\beta-1}e^{-(\tau/\gamma)^\beta} \tag{10}$$

Let reliability function $R(\tau)$ be the probability of an item operating for a certain amount of time without failure and the failure rate or hazard function $h(\tau)$ be the frequency of a device failure in a unit amount of time. Then $R(\tau)$ and $h(\tau)$ are calculated using the following equations

$$R(\tau) = e^{-(\tau/\gamma)^\beta} \tag{11}$$

$$h(\tau) = \frac{f(\tau)}{R(\tau)} \tag{12}$$

$$= \frac{\beta}{\gamma}\left(\frac{\tau}{\gamma}\right)^{\beta-1} \tag{13}$$

where, $\beta$ is the shape parameter, $\gamma$ is the scale parameter or characteristic life (or mean time to failure (MTTF))

$$\beta = \begin{cases} \text{device failure rate decreases with time,} & \beta < 1 \\ \text{device failure rate is constant,} & \beta = 1 \\ \text{device failure rate increases with time} & \beta > 1 \end{cases}$$

## 3  Proposed Method

As mentioned in the system model, the network functions in the time-frame manner in which task offloading and task scheduling operates in alternate time frames. The offloading and scheduling decisions are taken in every odd and even time frame, respectively. The proposed method is applied in each time frame as follows. We use delay, energy consumption and reliability as the criteria (parameters). First, we obtain the weights for all these criteria by applying Entropy Weight Method (EWM). The weights are then fed to E-TOPSIS to find the ranks of the FNs in the odd time frame and the task is offloaded to the highest ranked FN in the cluster. The same weights are also fed to E-TOPSIS in the even time frame to obtain the ranks of the tasks. The tasks are then executed in ascending order of the ranks. In the subsequent sections, we detail the proposed method with illustrative examples as needed for task offloading only. An illustration of the proposed method for task scheduling can be similarly made.

### 3.1  Evaluate Relative Weights Using EWM

The EWM is based on Shannon entropy [18], which derives relative weights of criteria. Shannon entropy is a probabilistic measure of information uncertainty. According to the information theory, weight and information entropy are inversely related i.e., lower information entropy indicates a higher weight to the criteria. The weight calculated by the entropy-based method is input to the TOPSIS, and it is detailed as follows [19].

**Step 1:** Build a decision matrix $A_{mk}$ for $m$ number of FNs and $k$ number of decision criteria as:

$$A_{mn} = \begin{array}{c} \\ FN_1 \\ FN_2 \\ \\ FN_m \end{array} \begin{array}{c} C_1 \quad C_2 \quad \ldots \quad C_k \\ \begin{bmatrix} a_{11} & a_{12} & \cdots & a_{1k} \\ a_{21} & a_{22} & \cdots & a_{2k} \\ \vdots & \vdots & \ddots & \vdots \\ a_{m1} & a_{m2} & \cdots & a_{mk} \end{bmatrix} \end{array}$$

**Illustration 1:** Let us consider a task offloading example with five FNs (FN1, FN2,..., FN5) and three criteria delay, energy consumption and reliability. We also assume a random decision-making matrix as shown in Table 2.

**Table 2.** Decision matrix.

| Fog nodes | Delay | Energy | Reliability |
|-----------|-------|--------|-------------|
| FN1 | 3.912 | 11.383 | 0.858 |
| FN2 | 1.100 | 2.379 | 0.987 |
| FN3 | 6.636 | 14.350 | 0.643 |
| FN4 | 1.209 | 3.584 | 0.985 |
| FN5 | 1.705 | 7.579 | 0.971 |

**Step 2:** The range of values for each criterion is different. The decision matrix is normalized in order to make a consistent comparison using below formula.

$$n_{jl} = \frac{a_{jl}}{\sqrt{\sum_{j=1}^{n} a_{jl}^2}} \tag{14}$$

**Step 3:** Probability of each criterion for the specific alternatives is calculated using characteristic proportion as follows:

$$c_{jl} = \frac{n_{jl}}{\sqrt{\sum_{j=1}^{n} n_{jl}}} \tag{15}$$

**Step 4:** With the value of characteristic proportion, the entropy value is calculated as follows:

$$e_l = \frac{-1}{\ln(k)} \sum_{j=1}^{n} c_{jl} \cdot \ln(c_{jl}) \tag{16}$$

**Step 5:** Degree of divergence for each criterion is calculated as follows:

$$dd_l = 1 - e_l \tag{17}$$

**Step 6:** To determine the weight of each criterion, entropy weight is determined using the degree of divergence as given below.

$$w_e(l) = \frac{dd_l}{\sum_{l=1}^{m}(dd_l)} \text{ where, } \sum_{l=1}^{m} w_e(l) = 1 \tag{18}$$

**Illustration 2:** The entropy weight of each criterion obtained by Eq. 18 is presented in Table 3.

**Table 3.** Obtained entropy weights.

| Fog nodes | Delay | Energy | Reliability |
|---|---|---|---|
| entropy weight $(w_e(l))$ | 0.565 | 0.407 | 0.026 |

### 3.2 TOPSIS for Ranking the FNs

TOPSIS is a systematic method that is reasonably simple and fast. The rank of the FNs is determined using the TOPSIS.

**Step1:** Create the weighted normalized decision matrix $(X_{jl})$ using the normalized matrix and weighted entropy as follows (Table 4):

$$x_{jl} = n_{jl} * w_e(l) \tag{19}$$

**Table 4.** Weighted decision matrix.

| Fog nodes | Delay | Energy | Reliability |
|-----------|-------|--------|-------------|
| FN1 | 0.274 | 0.228 | 0.011 |
| FN2 | 0.077 | 0.047 | 0.013 |
| FN3 | 0.465 | 0.288 | 0.008 |
| FN4 | 0.084 | 0.072 | 0.013 |
| FN5 | 0.119 | 0.152 | 0.013 |

**Step2:** Determine the positive ideal solution (PI) for indicating the best FN and the negative ideal solution (NI) for indicating worst FN as given below.

$$PI = \max\{x_{1l}, \ldots, x_{kl}\}\text{for beneficial criteria} \qquad (20)$$
$$NI = \min\{x_{1l}, \ldots, x_{kl}\} \qquad (21)$$

These values of $PI$ and $NI$ are just opposite for the cost criteria.

**Step3:** Determine the separation measure from a positive and negative ideal solution, $SP_j$ and $SN_j$, respectively using following equations.

$$SP_j = \sqrt{\sum_{j=1}^{m}(x_{jl} - PI)^2} \text{ and } SN_j = \sqrt{\sum_{j=1}^{m}(x_{jl} - NI)^2} \qquad (22)$$

**Step4:** Rank the nodes based on the value of relative closeness (the node with the highest relative closeness value is selected because it is nearest to the best ideal solution and farthest from the worst ideal solution) as follows:

$$C_j = \frac{SN_j}{SP_j + SN_j} \qquad (23)$$

**Illustration 3:** The above steps are used by the weighted normalized matrix and Table 5 presents the rank of FNs using E-TOPSIS. The requested task is now offloaded to the highest ranked FN i.e., $FN5$.

**Table 5.** Final result of E-TOPSIS.

| Fog nodes | $SP_j$ | $SN_j$ | $C_j$ | Ranks |
|-----------|--------|--------|-------|-------|
| FN1 | 0.190325 | 0.431071 | 0.693714 | 3.0 |
| FN2 | 0.064822 | 0.543896 | 0.893510 | 2.0 |
| FN3 | 0.480181 | 0.360597 | 0.428885 | 5.0 |
| FN4 | 0.369519 | 0.399471 | 0.519475 | 4.0 |
| FN5 | 0.018439 | 0.589615 | 0.969676 | 1.0 |

## 4  Migration

After offloading the task to the best FN (using the aforementioned method), the IoT device may move to another location. In this situation, the task is computed at the offloaded node (prior cluster), and the result is migrated from the previous AP to the current AP. Compared to the result's migration, migrating the task to another cluster will cause higher delay and energy usage. We consider the following two cases for the change of location of IoT devices which are shown in Fig 2. Steps 1, 2, and 3 depict case 1, in which an IoT device moves around AP-A, while steps a, b, c, d, and e depict case 2, in which an IoT device moves to AP-B at any moment after offloading a task to an FN in fog cluster 1. The details of the cases are as follows.

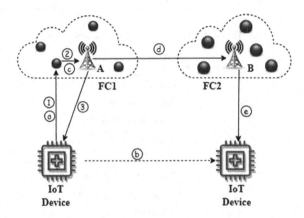

**Fig. 2.** Migration of IoT device and their result.

**Case 1:** IoT device moves around the AP-A

1. Offload task to the FN selected by the controller in fog cluster 1.
2. Send the result to the AP-A.
3. AP-A sends the results back to the IoT device.

**Case 2:** IoT device moves anytime after offloading task to FN of fog cluster 1, towards the AP-B

1. Offload task to the FN
2. IoT device moves from fog cluster 1 (prior cluster) to fog cluster 2 (current cluster)
3. Result send to AP-A
4. Migrate result to AP-B
5. AP-B sends the result back to the IoT device.

## 5   Results and Discussions

In this section, we investigate the performance of our proposed method through extensive simulations and compare the results with the state-of-the-art Fair Task Offloading (FTO) algorithm [25] and Priority-aware Genetic Algorithm (PGA) algorithm [8] for offloading and scheduling respectively. We assume that the transmission power to offload the task is 100 mW and channel noise power is $-100$ dBm. The channel gain parameter is defined as $(d_n^j)^{-\sigma}$, where $d_n^j$ is the distance between the IoT device and the FNs and $\sigma = 4$ is the path loss factor. The rest of the simulation parameters are given in Table 6. We have simulated the results 20 times and the average value is taken into account. The ranks evaluated using E-TOPSIS and Preference Ranking Organization Method for Enrichment Evaluations (PROMETHEE II) are indicated as TopRanks and ProRanks, respectively in Table 7 and Table 8.

**Table 6.** Simulation parameters.

| Parameters | Values |
|---|---|
| $l_i$ | 0.5–16 MB |
| $\eta_i$ | 100 cycle/bit |
| $\eta_j$ | 200–2000 cycle/bit |
| $\omega_i$ | 1 GHz |
| $\omega_j$ | 1–15 GHz |
| $\theta_i$ | $1.5 * 10^{-9}$ J/cycle |
| $\theta_j$ | $[1–10]*10^{-9}$ J/cycle |
| $\varpi_0$ | $-100$ dBm |
| $B_j$ | 10 MHz |
| $p_t$ | 100 mW |
| $\sigma$ | 4 |

### 5.1   Task Offloading

For simplicity, we consider a cluster of 5 FNs and a single IoT device that has to offload its task to any one of the efficient FNs in the cluster. According to E-TOPSIS the highest ranked FN is $FN5$ and thus, the task is offloaded to it. The $FN3$ is the highest ranked using PROMETHEE II. The delay and energy consumption are shown in Fig. 3 with varying sizes of the tasks. It is clearly observed that our proposed method outperforms both the FTO and PROMETHEE II.

Table 7. Ranks of FNs in a cluster.

| Fog nodes | Delay | Energy | Reliability | TopRanks | ProRanks |
|-----------|-------|--------|-------------|----------|----------|
| FN1 | 0.19691 | 1.319 | 0.999612 | 2 | 4 |
| FN2 | 0.254802 | 2.364 | 0.999351 | 4 | 3 |
| FN3 | 0.247804 | 2.989 | 0.999386 | 5 | 1 |
| FN4 | 0.3866 | 1.196 | 0.998507 | 3 | 2 |
| FN5 | 0.090124 | 0.483 | 0.999919 | 1 | 5 |

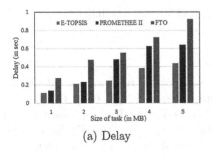

(a) Delay

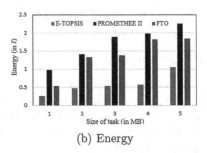

(b) Energy

Fig. 3. Delay and Energy consumption for task offloading.

## 5.2  Task Scheduling

Here we consider five tasks in the queue for a single FN, and the E-TOPSIS determines the order of execution of the tasks. The order of tasks execution using E-TOPSIS is $T1 \rightarrow T3 \rightarrow T4 \rightarrow T2 \rightarrow T5$ and PROMETHEE II is $T5 \rightarrow T4 \rightarrow T2 \rightarrow T1 \rightarrow T3$ which are presented in Table 8. The performance evaluation based on delay and energy consumption is compared with PGA and is presented in Fig. 4 with varying numbers of tasks. This is clear from the figure that E-TOPSIS outperforms.

Table 8. Ranks of tasks at an FN.

| Tasks | Delay | Energy | Reliability | TopRanks | ProRanks |
|-------|-------|--------|-------------|----------|----------|
| T1 | 0.068409 | 3.212 | 0.999953 | 1 | 4 |
| T2 | 0.132022 | 4.917 | 0.999826 | 4 | 3 |
| T3 | 0.103413 | 1.47 | 0.999893 | 2 | 5 |
| T4 | 0.111324 | 5.725 | 0.999876 | 3 | 2 |
| T5 | 0.320172 | 4.021 | 0.998975 | 5 | 1 |

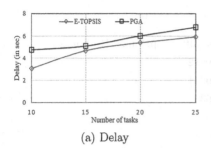

(a) Delay

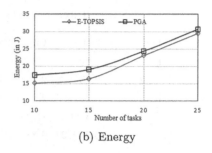

(b) Energy

**Fig. 4.** Delay and Energy consumption for task scheduling.

### 5.3   Complexity

Let $m$ be the number of FNs and $k$ be the number of criterion. For calculating the delay, energy consumption and reliability of each FN, it takes $O(m)$ time. Building the decision matrix takes $O(mk)$ time and normalization takes $O(m^2k)$ time. Further, the EWM is done in $O(k)$ time. After calculating EWM, the calculation of weighted normalized matrix takes $O(mk)$ time. Then for calculating Eqs. 20–23, it takes $O(mk)$ time. Finally, sorting of $m$ FNs takes $O(mlogm)$ time. Therefore, the offloading takes $O(m^2k)$. Similarly, scheduling of $n$ tasks from $n$ IoT devices takes $O(n^2k)$ time. Thus, the proposed scheme requires $O(m^2k)$ time assuming $n = m$.

## 6   Conclusion

We have presented a common method for task offloading and scheduling in an IoT-enabled fog network. The method is based on MCDM, which uses EWM for calculating criteria weights and TOPSIS for ranking the FNs and tasks as well. Through simulation results, we have shown that the proposed method outperforms two baseline algorithms with respect to delay and energy consumption. We have also explained how the results of the executed tasks to be migrated due to change of location of any IoT devices with two case studies.

## References

1.  Aazam, M., Zeadally, S., Harras, K.A.: Offloading in fog computing for IoT: review, enabling technologies, and research opportunities. Future Gener. Comput. Syst. **87**, 278–289 (2018)
2.  Adhikari, M., Mukherjee, M., Srirama, S.N.: DPTO: a deadline and priority-aware task offloading in fog computing framework leveraging multilevel feedback queueing. IEEE Internet Things J. **7**(7), 5773–5782 (2019)
3.  Alizadeh, M.R., Khajehvand, V., Rahmani, A.M., Akbari, E.: Task scheduling approaches in fog computing: a systematic review. Int. J. Commun Syst **33**(16), e4583 (2020)

4. Chiu, W.Y., Yen, G.G., Juan, T.K.: Minimum manhattan distance approach to multiple criteria decision making in multiobjective optimization problems. IEEE Trans. Evol. Comput. **20**(6), 972–985 (2016)
5. Guo, K., Sheng, M., Quek, T.Q., Qiu, Z.: Task offloading and scheduling in fog ran: a parallel communication and computation perspective. IEEE Wirel. Commun. Lett. **9**(2), 215–218 (2019)
6. Hamdi, A.M.A., Hussain, F.K., Hussain, O.K.: Task offloading in vehicular fog computing: state-of-the-art and open issues. Future Gener. Comput. Syst. **133**, 201–212 (2022)
7. Hazra, A., Adhikari, M., Amgoth, T., Srirama, S.N.: Joint computation offloading and scheduling optimization of IoT applications in fog networks. IEEE Trans. Netw. Sci. Eng. **7**(4), 3266–3278 (2020)
8. Hoseiny, F., Azizi, S., Shojafar, M., Ahmadiazar, F., Tafazolli, R.: PGA: a priority-aware genetic algorithm for task scheduling in heterogeneous fog-cloud computing. In: IEEE INFOCOM 2021-IEEE Conference on Computer Communications Workshops (INFOCOM WKSHPS), pp. 1–6. IEEE (2021)
9. Ju, C., Ma, Y., Yin, Z., Zhang, F.: An request offloading and scheduling approach base on particle swarm optimization algorithm in IoT-fog networks. In: 2021 13th International Conference on Communication Software and Networks (ICCSN), pp. 185–188. IEEE (2021)
10. Kaur, N., Kumar, A., Kumar, R.: A systematic review on task scheduling in fog computing: taxonomy, tools, challenges, and future directions. Concurrency Comput. Pract. Experience **33**(21), e6432 (2021)
11. Kishor, A., Chakarbarty, C.: Task offloading in fog computing for using smart ant colony optimization. Wireless Pers. Commun. **127**, 1683–1704 (2021)
12. Kishor, A., Chakraborty, C., Jeberson, W.: Reinforcement learning for medical information processing over heterogeneous networks. Multimedia Tools Appl. **80**(16), 23983–24004 (2021). https://doi.org/10.1007/s11042-021-10840-0
13. Kumari, N., Yadav, A., Jana, P.K.: Task offloading in fog computing: a survey of algorithms and optimization techniques. Comput. Netw. **214**, 109137 (2022)
14. Lakhan, A., Memon, M.S., Elhoseny, M., Mohammed, M.A., Qabulio, M., Abdel-Basset, M., et al.: Cost-efficient mobility offloading and task scheduling for microservices IoVT applications in container-based fog cloud network. Clust. Comput. **25**(3), 2061–2083 (2022)
15. Mouradian, C., Naboulsi, D., Yangui, S., Glitho, R.H., Morrow, M.J., Polakos, P.A.: A comprehensive survey on fog computing: state-of-the-art and research challenges. IEEE Commun. Surv. Tutorials **20**(1), 416–464 (2017)
16. Rausand, M., Hoyland, A.: System Reliability Theory: Models, Statistical Methods, and Applications, vol. 396. Wiley, Hoboken (2003)
17. Sellami, B., Hakiri, A., Yahia, S.B., Berthou, P.: Energy-aware task scheduling and offloading using deep reinforcement learning in SDN-enabled IoT network. Comput. Netw. **210**, 108957 (2022)
18. Shannon, C.E.: A mathematical theory of communication. Bell Syst. Tech. J. **27**(3), 379–423 (1948)
19. Tomar, A., Jana, P.K.: Mobile charging of wireless sensor networks for internet of things: a multi-attribute decision making approach. In: Fahrnberger, G., Gopinathan, S., Parida, L. (eds.) ICDCIT 2019. LNCS, vol. 11319, pp. 309–324. Springer, Cham (2019). https://doi.org/10.1007/978-3-030-05366-6_26
20. Wu, H.Y., Lee, C.R.: Energy efficient scheduling for heterogeneous fog computing architectures. In: 2018 IEEE 42nd Annual Computer Software and Applications Conference (COMPSAC), vol. 1, pp. 555–560. IEEE (2018)

21. Yang, X., Rahmani, N.: Task scheduling mechanisms in fog computing: review, trends, and perspectives. Kybernetes (2020)
22. Yang, Y., Liu, Z., Yang, X., Wang, K., Hong, X., Ge, X.: POMT: paired offloading of multiple tasks in heterogeneous fog networks. IEEE Internet Things J. **6**(5), 8658–8669 (2019)
23. Yang, Y., Zhao, S., Zhang, W., Chen, Y., Luo, X., Wang, J.: Debts: delay energy balanced task scheduling in homogeneous fog networks. IEEE Internet Things J. **5**(3), 2094–2106 (2018)
24. Youssef, A.E.: An integrated MCDM approach for cloud service selection based on TOPSIS and BWM. IEEE Access **8**, 71851–71865 (2020)
25. Zhang, G., Shen, F., Yang, Y., Qian, H., Yao, W.: Fair task offloading among fog nodes in fog computing networks. In: 2018 IEEE International Conference on Communications (ICC), pp. 1–6. IEEE (2018)

# Static Data Race Detection in Multi-task Programs for Industrial Robots

Ameena K. Ashraf[✉] and Meenakshi D'Souza[✉]

International Institute of Information Technology, Bangalore, India
{ameena.ashraf,meenakshi}@iiitb.ac.in

**Abstract.** An industrial robot is an automatic multi-purpose manipulator, programmable in three or more axes. A program written in a high-level programming language controls these robots, many of these programs involve multiple tasks controlling different robots. Data races are a common problem in concurrent and multi-threaded programming and they are of big concern for the multi-task industrial robotics programmers too. We present a static analysis method for detecting data races in multi-task programs for industrial robots. We propose a technique based on a relation that models when two or more statements from a task occur in between two or more statements in another task. Our static analysis is preceded by a manual, dynamic analysis step for ensuring consistency among tasks for one of the constructs which involves a task waiting for a particular duration. We define a set of not-occurs in-between rules to detect whether two statements in different tasks may race with each other. We have developed a prototype implementation of our tool for the Rapid programming language that is used to program industrial robots of ABB. Rapid has all the features of a typical programming language for industrial robots and hence our race detection framework will generalize to any programming language for industrial robots.

**Keywords:** Data race · Multi-task programs · Industrial robots

## 1 Introduction

Industrial robots are controlled by programs that handle inputs and signals from the controller and provide control commands for robot arm movements. Apart from high level programming languages, a popular class of programming languages and frameworks have emerged to program these robots [4,5,13]. Many robotics applications need parallelism where two or more robots execute their functionality simultaneously. ROS (Robotics Operating System) [9] also has constructs like *nodes* along with a publish/subscribe mechanism which is used by multi-threaded programs for inter-thread communication. Such executions are meant to manipulate objects and tools in the real world, making them safety critical in nature. Safety standards for industrial robots [1] also mandate use of

© The Author(s), under exclusive license to Springer Nature Switzerland AG 2023
A. R. Molla et al. (Eds.): ICDCIT 2023, LNCS 13776, pp. 51–66, 2023.
https://doi.org/10.1007/978-3-031-24848-1_4

rigorous testing and verification techniques to ensure that the robot operations are reliable.

Two or more tasks in a multi-task program for industrial robots execute in a pseudo parallel fashion, by sharing and exchanging information on work objects, digital signals, variables, and actions to be done in parallel or in sequence etc. Robot programming languages like Rapid [13], KRL [4], VAL3 [5] etc. support multi-tasking. Each task has a set of modules and the programs associated with the tasks read from/write to a set of persistent variables which are global variables shared by the tasks. The languages provide various constructs for synchronization and interleaved execution of the tasks.

Multi-task programs written in these languages also have concurrency problems like data races [2] while reading from and writing to shared data resources. A data race occurs when two or more tasks in a multi-task program access the same memory location concurrently, and at least one of the accesses is for writing, and the tasks are not using any exclusive locks or other mechanisms to control their accesses to that memory. Since these robots operate in safety critical environments, data races can lead to unexpected accidents.

We propose a static data race detection technique for multi-task programs that manipulate industrial robots using a notion of "not-occurs-in-between" relation on statements or block of statements among tasks in a multi-task program. A statement or a block of statements $s_2$ in task $t_2$ does not occur in between a statement or a block of statements $s_1$ in task $t_1$ if it is not possible for $t_2$ running $s_2$ to preempt $t_1$ running $s_1$. If $s_1$ and $s_2$ cannot occur in between each other, data race will not happen among $s_1$ and $s_2$. We have defined conditions that ensure that a statement cannot occur in between another based on the constructs available in Rapid programming language. These "rules" take care of many of the typical synchronization constructs that programs for industrial robots use.

Along with our static data race detection framework, we introduce a dynamic analysis step in order to ensure wait time consistency among tasks. Some programming languages for industrial robots have a *wait time* construct that ensures that a particular task waits for the given time unit. The duration of the wait is decided by the programmer and is not amenable to static analysis as it is specific to a particular application and its requirements. We ensure that if such a construct is present in a program, a pre-processing step of checking that it is consistent and if not, fine tuning the wait time to ensure consistency, is done before our static analysis framework is invoked.

We have implemented a race detector framework for the Rapid programming language of ABB [13]. Typical programming languages for industrial robots like [4,5] have special data-types to identify points in the 3D plane, Move instructions, programming constructs, tasks and multi-tasking features like synchronization. Hence our definitions and algorithms will generalize to other programming languages for industrial robots. Our framework parses a multi-task Rapid program to extract all the relevant information and then implements a set of rules on this information to identify or rule out the presence of data races. Our technique is sound in the sense that it identifies all possible potential data races and when it concludes that the given program is non-racy, it is indeed provably free of any data races.

The paper is organized as follows. Related work is discussed in Sect. 2. Multi-task Rapid programs, their structure and data races are explained through an example in Sect. 3. Semantics of multi-task programs are explained in Sect. 4. Our algorithm and the prototype implementation is explained in Sect. 5 and experiments are discussed in Sect. 6. Section 7 finally concludes the paper.

## 2 Related Work

Data races are a common issue in all concurrent and multi-threaded programs. The techniques for detecting data races can be broadly classified as being *dynamic* or *static* based on whether they execute the program or not respectively. While dynamic race detection techniques find real data races, they might not succeed in finding all of them. Static race detection algorithms have been proposed as *sound* techniques that are capable of detecting all potential data races [16].

Early work in this area is by Leslie Lamport where an algorithm for detecting data races was proposed based on a "happens-before" relation [8]. A partially ordered happens-before relation is checked to be consistent with a total ordering of events and the paper proposes notions like clock synchronization that have been found to be useful in checking for inconsistencies related to races. [3] defines the notion of high level data races. Their dynamic race detection approach is mainly based on a property called "view consistency" in multi-threaded programs, which permits detecting high level data races that can lead to an inconsistent program state.

The Eraser algorithm [7,10] which has been implemented in the Visual Threads tool [6] to analyze C and C++ programs, is another example of a *dynamic* algorithm that examines a program execution trace for locking patterns and variable accesses in order to predict potential data races. The algorithm maintains a lock set for each variable, which is the set of locks that have been owned by all threads accessing the variable in the past. New accesses lead to refinement of the lock sets and they use the changes in the lock sets to detect data races. Each new access causes a refinement of the lock set with the set of locks currently owned by the accessing thread. The set is initialized to the set of locks owned by the first accessing thread. If the set ever becomes empty, a data race is possible.

[11] provided practical methods to analyze data races and transactionality in Priority Ceiling Protocol(PCP) programs. Their analysis of linear equalities can be considered as one instance of an analysis framework which generalizes the functional approach from programs with procedures to programs with procedures, interrupts, priorities and resources following the PCP protocol.

Recently, [16] proposed a way to *statically* detect data races in RTOS (Real Time Operating System) applications that use a variety of non-standard synchronization constructs. They present a relation that formalizes when two individual statements may interleave with other scheduler commands and be racy with each other in a multi-threaded RTOS program. Their analysis consists of a pre-analysis followed by the main analysis for checking a set of rules or conditions, where the occurrence of the pattern of at least one rule is enough to assure that the given multi-threaded RTOS program is free from potential data races.

Programs controlling industrial robots like those written in Rapid [13] and VAL3 [5] are multi-task programs that run in the RTOS of the underlying controller platform. Our proposed algorithm extends the technique in [16] to the setting of programs for industrial robots with complex data types, motion commands and multiple tasks, resulting in a sound, static race detection algorithm for detecting data races. In addition, to cater to real-time wait constructs, we run a pre-processing dynamic analysis to ensure that wait time between tasks is set correctly.

## 3    Multi-task Industrial Robot Programs: An Overview

As mentioned earlier, many languages for manipulating industrial robots support multiple tasks that can execute in a (semi-)parallel fashion. We present a brief comparison of three languages that are popular for programming industrial robots with multiple tasks in Table 1. We found that Rapid programming language of ABB is the most widely used and has several different constructs to write complex functionality for programs with multiple tasks. Our technique has been prototyped to work on Rapid and can extend to all the other languages too. We now describe the Rapid programming language in detail.

A program in Rapid can support multiple tasks to enable one or more robots to execute different functionalities [14]. Each task $t$ has a name, an optional task to run in the foreground of $t$, an optional type and a program to execute when it runs. The task in the foreground of $t$ has higher priority than $t$ and if no such foreground task is set, $t$ has highest priority. A program associated with a task is grouped into modules and executes the functionality of an industrial robot. The program is written in Rapid and involves variables including persistent variables which are global variables shared by tasks, target points for robot movements, movement instructions and other program constructs [13]. A programming language like Rapid supports up to ten tasks, with at most one of them being involved in motion/movement functionality at a particular time during execution. If no foreground tasks are configured, then all the tasks have the same priority. In such a case, the statements of each task are executed in a round robin way, interleaved such that one instruction from each task is executed at a time.

Rapid also supports multi-move programs where more than one task can be a motion task [15]. Task priorities cannot be configured in multi-move programs, they are otherwise similar to multi-task programs. Multi-move programs are broadly classified into independent, semi-coordinated and coordinated synchronized movements based on whether several robots are working independently or on the same work object or working with different work objects respectively.

**Table 1.** Comparison among three multi-task programming languages

| Feature | Rapid | KRL | VAL3 |
|---|---|---|---|
| Program | Program modules and system modules in a task | A SRC file and a DAT file of the same name | A sequence of instructions, local variables and parameters |
| Data-types | num, bool, string, robtarget | integer, real, Boolean, character, AXIS, E6AXIS, FRAME, POS and E6POS | bool, num, string, dio, aio, sio |
| Variable | num, bool, string and PERS variables | Starts with '$' | Identified by name, type size and scope |
| Signal | Digital inputs and digital outputs | Starts with '$' | Present |
| Constant | Initialized with CONST | Initialized in a data list, preceded by the keyword CONST | No previous declaration is needed |
| Motion Programming | linear, circular, joint motions | point-to-point, linear and circular motion | Present |
| Inputs, Outputs | SETAO, SETDO, SETGO | ANIN, ANOUT, PULSE, SIGNAL | Present |
| Multi-tasking | Multiple tasks and multi-move tasks | Present | Synchronous and asynchronous tasks |
| Global variables, Sub-programs, Functions | Present | Present | Present |

Variables declared as persistent variables in a Rapid program are considered as global variables and they are available for use by all tasks, i.e., each task can read the values of the variables, make local copies of them, initialize them locally and write back values to the persistent variables. Like all other multi-tasking programs, multi-task programs in Rapid for industrial robots also support standard features for inter-task communication. More details are given in Sect. 4.

```
1  BEGINTASK <Task1, Null>
2     MODULE Module1
3        CONST robtarget Target_10:=[[326.494, 0, 634.740],
          [0.104, 0, 0.983, 0],[-1, 0, -1, 0],[9E+09, 9E+09, 9E
          +09, 9E+09,9E+09, 9E+09]];
```

```
4     CONST robtarget Target_20:=[[474.642, 0, 298.999],
      [0.087, 0, 0.996, 0], ...];
5     CONST robtarget Target_30:=[[17.446, 0, 764.740],
      [0.809, 0, 0.587, 0],[0, -1, 0, 4], ...;
6     PERS bool obstacle:=FALSE;
7     PROC MAIN()
8     IF do1=1 THEN
9         obstacle:=TRUE;
10        Path_backward;
11    ELSE
12        Path_forward;
13    ENDIF
14    ENDPROC
15    PROC Path_forward()
16     MoveL Target_10,v1000,z100,MyTool\WObj:=wobj0;
17     MoveL Target_20,v1000,z100,MyTool\WObj:=wobj0;
18    ENDPROC
19    PROC Path_backward()
20     MoveL Target_10,v1000,z100,MyTool\WObj:=wobj0;
21     MoveL Target_30,v1000,z100,MyTool\WObj:=wobj0;
22    ENDPROC
23   ENDMODULE
24  ENDTASK
25  BEGINTASK<Task2, Null>
26   MODULE Module1
27     CONST robtarget Target_10:=[[195.6535, -502.4544,
      902.740],[0.617, 0, 0.786,0],[-1, 0, -1, 4],[9E+09, 9E
      +09, 9E+09, 9E+ 09,9E+09, 9E+09]];
28     CONST robtarget Target_20:=[[576.642, -502.4544,
      408.999],[0.087, 0, 0.996, 0],[-1, 0, 0, 0], ...;
29     CONST robtarget Target_30:=[[],[],[],[]];
30     PERS bool obstacle;
31     PROC main()
32     IF obstacle=FALSE THEN
33         Path_forward;
34     ELSE
35         Path_backward;
36     ENDIF
37     ENDPROC
38     ENDMODULE
39  ENDTASK
```

**Listing 1.1.** A Rapid program with two tasks

Consider a multi-task Rapid program given in Listing 1.1, with two tasks Task1
and Task2 of equal priorities executed by two different robots. The two robots
check for the presence of an obstacle in line 8 (represented by the digital signal
do1) and if the obstacle is not found, a particular robot can move forward (line
12). If an obstacle is found (do1=1) then the robot checking for the obstacle
moves backward (line 10). Forward movement in the presence of an obstacle

will cause an accident. The robots communicate the detection of an obstacle by updating a common, persistent variable that they share called `obstacle`. The two tasks essentially execute the same check for the presence of an obstacle, update the value to indicate the presence/absence and decide to move forward or backward based on the value.

## 3.1   Data Races in Multi-task Programs

When two or more tasks communicate with each other by reading from and writing to values of global, persistent variables, they need to ensure that read and write accesses to the variables are *protected*. Unprotected read/write accesses can lead to *data races* resulting in undesired errors. Along with the programming languages listed in Table 1, ROS (as an operating system) also provides features for multi-task programming and synchronization/communication among tasks. Programmers often misunderstand, misuse or omit these constructs resulting in unprotected accesses to global variables.

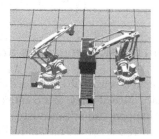

**Fig. 1.** Simulation: Robots executing the program in Listing 1.1

For example, in the program in Listing 1.1, since priorities are not set, the two tasks execute their respective programs in an interleaved fashion. `Task1` begins execution—after initializing `obstacle` to false (line 6), followed by `Task2` declare the same (line 30), `Task1` checks for the presence of an obstacle (line 8) and before the next statement updating the obstacle can be executed, `Task2` resumes its execution (line 32). It could have been the case that `do1=1` and `Task1` should have executed line 9 to update the value of the variable `obstacle` and move backward, all as a continuous sequence of statements to be executed by `Task1`. Instead, the interleaved executions continue and `Task2` ends up moving forward while the variable `obstacle` is true. This leads to an erroneous move depicted as follows: The `if do1=1` statement of `Task1` returns true but before the value of `obstacle` is set to true, `Task2` takes over and its check for `obstacle` being false also passes. Now the control goes back to `Task1`, which sets `obstacle` to true and the next statement of `Task2` moves forward. This results in the second robot moving forward in the presence of an obstacle, causing undesired behaviour as illustrated in Fig. 1. We term this as a data race on the variable `obstacle` and develop a sound static program analysis techniques to detect such data races.

**Table 2.** Wait and synchronization constructs in rapid

| Wait constructs | Synchronization constructs |
| --- | --- |
| `WaitSyncTask`: To synchronize several program tasks at a special point in each program | `SetDO-ISignalDO`: To set a digital output signal in one task, order and enable interrupts from the same digital output signal in another task |
| `WaitUntil_TestAndSet`: Used together with a normal data object of the type Boolean, as a binary semaphore, to retrieve exclusive access to specific code or system resources | `SetDO-WaitDO`: Used to set the value of a digital output signal in one task and wait until the same digital output is set in another task |
| `WaitUntil`: Used to wait until a logical condition is met. | `Dispatcher`: A persistent string variable containing the name of the routine to execute in another task |
| `WaitTime`: Used to wait for a given unit of time | `IEnable-IDisable`: Used to enable and disable interrupts during program execution |

## 4    Semantics of Multi-task Programs and Data Races

Data races are common in concurrent multi-threaded programming. Two statements are involved in a data race if they are conflicting accesses to a shared memory location and can happen "simultaneously" or one after another. By statically analyzing the data races, we mean analyzing the code during compile time without any run time overheads. The following subsections shows how we statically analyze and detect the data races specifically for multi-task Rapid programs for industrial robots.

We work with a multi-task Rapid program [13] and its semantics given by its *state* as it runs in the controller. The tasks can use a variety of instructions and functions to communicate with other tasks (as given in Table 2). A set of persistent variables, interrupts, and system generated digital inputs and outputs are used for synchronisation among tasks.

Given a multi-task Rapid program $\mathcal{R}$, we consider all the parameters that constitute the in-memory state of $\mathcal{R}$. This includes information about the various tasks (their priority and task in the foreground (if any)), currently executing task and the tasks that are waiting, the values of all the variables including persistent variables, the resources (work objects, flex pendants, digital signals etc.) and the tasks that are holding a resource at a particular point in time, and finally, the location counter in the running task. These parameters will be extracted during parsing and will be used to detect the presence of potential data races by our proposed algorithm.

## 4.1 Data Races in Multi-task Programs

Let $\mathcal{R}$ be a multi-task Rapid program. Consider a statement or a block of statements $s_1$ in task $t_1$ and another statement or block of statements $s_2$ in task $t_2$, consisting of shared variables and the read/write accesses to the variables in $\mathcal{R}$. We say that $s_1$ and $s_2$ are involved in a *data race* in $\mathcal{R}$ if they access a persistent variable and at least one of them is a write access. Also, $s_2$ *occurs-in-between* $s_1$ if there is an execution of $s_1$ in which statement/s of $s_2$ occurs sometime between or before the statement/s of $s_1$.

Typically, wait and synchronization constructs are used by programmers in multi-task programs to ensure that statements or blocks of statements do not occur in between each other. Races can be ruled out by ensuring that statements accessing persistent variables do not occur-in-between as per the above definition by using appropriate wait or synchronization constructs. The example in Sect. 3 was devoid of any such constructs resulting in a data race on the persistent variable `obstacle`.

# 5 Static Race Detection Algorithm

We describe our static analysis based algorithm for detecting data races in multi-task Rapid programs in this section. The crux of our algorithm is a set of rules that can be checked to ensure that (blocks of) statements in two different tasks that access persistent variables do not occur in between each other. Our rules are *sound* in the sense that when they conclude that two (blocks of) statements do not occur-in-between, they will hold in all inter-leavings, ensuring absence of data races between the concerned statements.

## 5.1 Rules for Checking Occurs-in-Between Property

We propose the following rules under which a statement/block of statements $s_1$ in a Rapid program task $t_1$ **cannot** occur in between another statement/block of statements $s_2$ in another program task $t_2$. For each wait and synchronization construct given in Table 2, there is a corresponding rule that prescribes when $s_1$ and $s_2$ **do not** occur-in-between as per the semantics of the construct.

- **C1 (Synchronizing between tasks):** Identical synchronization points are set in the code by using the `WaitSyncTask` instruction after executing $s_1$ in $t_1$ and before executing $s_2$ in $t_2$.
- **C2 (Accessing shared resources by using a flag):** $s_1$ in $t_1$ and $s_2$ in $t_2$ are enclosed in a block beginning with setting the persistent variable flag to true with the instruction `WaitUntilTestAndSet` and ending with resetting the flag to false and there will not be any instruction in between $s_1$ or before $s_1$ in $t_1$ which will cause the program control to transfer from $t_1$ to $t_2$.
- **C3 (Polling among tasks):** Each of the following conditions must hold. 1. A Boolean persistent variable is set to true before the execution of $s_1$ in task $t_1$. 2. In task $t_2$, before executing $s_2$, a `WaitUntil` instruction is used along with the same Boolean persistent variable as used in $t_1$.

- **C4 (Waiting using interrupts):** Each of the following conditions must hold.
  1. A persistent digital signal is set in the program of first task using `SetDO` instruction after $s_1$. 2. In $t_2$, before executing $s_2$, an `ISignalDO` instruction is declared with the same persistent digital signal as declared in $t_1$ and it is used as an interrupt to start the second task $t_2$.
- **C5 (SetDO-WaitDO block):** Each of the following conditions must hold.
  1. There is a `SetDO` instruction after executing $s_1$ in $t_1$ with a persistent digital signal. 2. There is a `WaitDO` instruction in task $t_2$ before executing $s_2$ with the same persistent digital signal with the same value as in task $t_1$.
- **C6 (Using a dispatcher):** A digital signal is used by $t_1$ after the execution of $s_1$ to call another task $t_2$, indicating specifically that a particular routine which may contain $s_2$ could be executed in the called task $(t_2)$.
- **C7 (IEnable-IDisable block):** $s_2$ will be inside an `IEnable-IDisable` block for a common interrupt.

Our algorithm parses a multi-task Rapid program detecting persistent variables, wait and synchronization constructs and applies the above rules to check for data races. The soundness of our rules, proved in the next section, guarantee that the algorithm will never miss a potential data race.

## 5.2   Proof of Soundness of the Rules

Let $\mathcal{R}$ be a multi-task Rapid program. Let $s_1$ and $s_2$ be two instructions (can be a block of statements also) in tasks $t_1$ and $t_2$ respectively in $\mathcal{R}$, that satisfy one of the seven conditions above. Then we argue that $s_2$ will not occur-in-between $s_1$. Proof (by contradiction) of soundness for conditions **C1**, **C2** and **C5** are given below. Proof of soundness for the rest of the rules are similar.

Soundness of rule **C1**: Suppose $s_1$ and $s_2$ satisfy the condition **C1**, and suppose there is an execution of $\mathcal{R}$ in which $s_2$ occurs in between $s_1$. Let us say $s_1$ is executed by $t_1$ and $s_2$ is executed by $t_2$. The only way in which $s_2$ can occur in between $s_1$ is if the synchronization points are set in the code by using the `WaitSyncTask` instruction before executing $s_1$ in $t_1$ and before executing $s_2$ in $t_2$ with the same synchronisation identity points. But this is not possible since the condition says that synchronization points are set in the code by using the `WaitSyncTask` instruction after executing $s_1$ in $t_1$ and before executing $s_2$ in $t_2$ with the same synchronisation identity points.

Soundness of rule **C2**: Suppose $s_1$ in task $t_1$ and $s_2$ in task $t_2$ satisfy the condition **C1**, and suppose there is an execution of $\mathcal{R}$ in which $s_2$ occurs in between $s_1$. The only way in which $s_2$ can occur in between $s_1$ is as follows:

- $s_1$ in $t_1$ and $s_2$ in $t_2$ are enclosed in a block beginning with setting of a persistent variable `flag` to `true` with the instruction `WaitUntilTestAndSet` and ending with resetting the `flag` to `false`. Also, there will be a `Wait` instruction in between $s_1$ or before $s_1$ in $t_1$ which will cause the program control goes from $t_1$ to $t_2$. This is not possible since the condition says that $s_1$

in $t_1$ and $s_2$ in $t_2$ are enclosed in a block beginning with setting the persistent variable `flag` to `true` with the instruction `WaitUntilTestAndSet` and ending with resetting the `flag` to `false` and there will not be any `Wait` instruction that will meet the above property.

Soundness of rule **C5**: Suppose $s_1$ and $s_2$ satisfy the condition **C5**, and suppose there is an execution of $\mathcal{R}$ in which $s_2$ occurs in between $s_1$. Let us say $s_1$ is executed by $t_1$ and $s_2$ is executed by $t_2$. The only way in which $s_2$ can occur in between $s_1$ is if there is a `SetDO` instruction after executing $s_1$ with a persistent digital signal in task $t_1$ and there is a `WaitDO` instruction in task $t_2$ before executing $s_1$ with the same persistent `digital signal` with a different value as set by $t_1$. This is not possible since the condition says that there should be a `SetDO` instruction after executing $s_1$ in $t_1$ with a persistent `digital signal` and there is a `WaitDO` instruction in $t_2$ before executing $s_2$ with the same persistent digital signal with the same value as set by $t_1$.

Our static race detector algorithm checks the occurrence of each rule in a multi-task Rapid program considering all pairs of tasks one by one. Since **C1** and **C2** consists of checking the occurrence of the same `Wait` instructions in all the tasks, it can be applied on any multi-task or multi-move Rapid program, with or without priorities among tasks. Rest of the rules are applicable only in programs with two tasks as $s_2$ in $t_2$ completely depends on $s_1$ in $t_1$.

## 5.3   Implementation

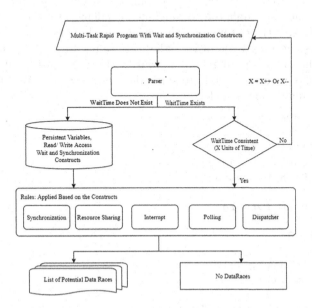

**Fig. 2.** Static data race detector framework

Figure 2 outlines the steps of our static data race detector framework. Input to our framework is a syntactically correct multi-task Rapid program, which is parsed to extract all the persistent variables, statements that access these variables, the type of access (read/write), synchronization constructs, wait constructs etc. as given in the program. The output from the parser is directly given to a static data race detector where it checks for the occurrence of all the seven rules one by one. Occurrence of the pattern of at least one rule in the input multi-task program is sufficient to say that the program is data race free. Otherwise, the static race detector outputs the list of possible potential data races among tasks along with the list of statements.

One of the constructs, `WaitTime`, takes as argument a time value (in seconds) specifying the duration for which a task needs to wait before resuming its execution. We understand from the online developer's forum of programmers of industrial robots that this construct is typically not used by programmers but is present in the syntax of the programming language. Since this construct takes a real-time value as argument, we cannot propose sound rules for statically detecting data races in programs with this construct. We run a *pre-processing* step that involves a dynamic wait time consistency check before giving the program to parser.

Dynamic `WaitTime` consistency checking involves the following steps. Assume we have a multi-task Rapid program with wait construct `WaitTime` of $x$ units in one of the tasks. We need to ensure that $x$ units of waiting by a particular task is sufficient for other tasks to finish executing the statements that are programmed to execute, and there are no unwanted inter-leavings that occur. In order to ensure this, we need to check the program execution logs (through a *watch window*) of an actual simulation (through flex pendent outputs) for the values of shared variables. If the outputs match, we can conclude that the `WaitTime` of $x$ units is consistent among the tasks. Otherwise we need to change the value of $x$ and repeat the process until $x$ become consistent among tasks. After this step, we can give the program directly to static data race detector. We illustrate this step using a small example below.

```
 1  BEGINTASK <Task1 , Null >
 2      PROC MAIN ()
 3      IF do1 =1 THEN
 4          obstacle := TRUE ;
 5          Path_backward ;
 6      ELSE
 7          Path_forward ;
 8      ENDIF
 9      ENDPROC
10  ENDTASK
11  BEGINTASK <Task2 , Null >
12      PROC main ()
13      WaitTime 2;
14      IF obstacle = FALSE THEN
15          Path_forward ;
```

```
16      ELSE
17          Path_backward;
18      ENDIF
19      ENDPROC
20  ENDTASK
```

**Listing 1.2.** A Rapid program with two tasks with WaitTime

In the example given in Listing 1.2, there is a **WaitTime** of 2 units in Task 2 before checking the obstacle value in line no 13. This **WaitTime** is inconsistent since it is not enough for Task 1 to set the obstacle value to True within that time. As a result, robot 2 collides with the obstacle and we conclude that there is WaitTime inconsistency among the tasks. So we manually update the WaitTime by one unit each in an iterative manner. In this example, a WaitTime of 5 units happens to be consistent among the tasks. This WaitTime consistent program is given to the parser.

The output from the parser is directly given to a static data race detector where it checks for the occurrence of all the the seven rules one by one. Occurrence of at least one rule in the input multi-task program is sufficient to say that the program is data race race free. Otherwise static race detector outputs the list of possible data races among tasks.

The check is done per pair of tasks that share a wait or synchronization construct, as defined in the input program. When there are more than two tasks that share a variable, Rapid provides synchronization points using the WaitSyncTask command which is checked by rule **C1**.

**Table 3.** Experimental results

| Example program | Synchronization/Wait construct used | Rule applied |
|---|---|---|
| Obstacle program | WaitUntil_TestAndSet | C2 |
| Quadrant program | SetDO-WaitDO | C5 |
| Flex loader | WaitSyncTask | C1 |
| Exhaust pipe | WaitSyncTask | C1 |

## 6   Experimental Results

We have implemented the static race detector algorithm depicted in Fig. 2 in Java by first writing a parser for Rapid programs that have multiple tasks, followed by checking of the proposed rules. The complete implementation along with examples illustrating each of the seven rules are available at https://github.com/GithubAmeena/Robotic-static-race-detector.git. While the prototype implementation has been done for Rapid, it can smoothly be extended to other programming languages that support multi-task programs. The prototype implementation was tried out on several different examples of multi-task Rapid programs including all the examples available with the licensed version of Robot-Studio and some in-house examples. Table 3 shows our experimental results in some of the programs in Rapid. Our implementation was able to successfully analyze each of the programs for data races. We now illustrate the working of our prototype implementation on a few of them in detail.

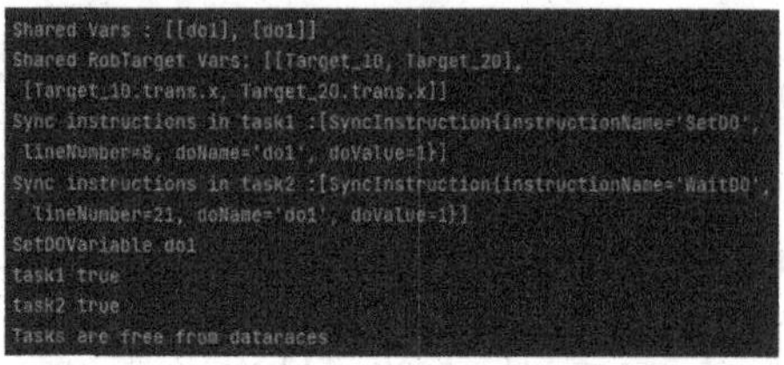

**Fig. 3.** Race detector output of rule C5 applied in quadrant program

*Quadrant program* given in Listing 1.3 is a multi-task Rapid program consisting of two tasks that we worked on internally. Task $t_1$ is a motion task and $t_2$ is a non-motion task, both $t_1$ and $t_2$ runs in parallel since priorities are not set among them. In $t_1$ there is an if statement in lines 6 and 7 (consider these statements as $s_1$ which checks for the values of the x-coordinates of two points in the 3D plane (both are declared as persistent variable among the tasks). At the same time, $t_2$ resets the values of the same coordinates to zero in lines 22 and 23 (consider this statements as $s_2$) and as a result, data race will happen since $s_2$ in $t_2$ occurs-in-between $s_1$ in $t_1$ and a wrong path will be drawn during simulation. In order to avoid this racy situation, we applied rule **C5** where a SetDO instruction was included after executing $s_1$ (line 8) with a persistent digital signal do1 in $t_1$ and a corresponding WaitDO instruction was included in $t_2$ (line 21) before executing $s_2$ with the same persistent digital signal do1 with the same value do1=1 as set by $t_1$. Our tool subsequently reported absence of any potential data race as seen in Fig. 3.

*Obstacle program* is the one discussed in Sect. 3. In this program, if we consider the if statement inside main of $t_1$ as $s_1$ and the if statement inside main of $t_2$ as $s_2$, then the problem here is the occurrence of $s_2$ in between $s_1$. In order to avoid this unexpected behaviour, we can use any of the seven conditions as we discussed in Sect. 5. For example, if we apply **C2** in the program by putting $s_1$ in $t_1$ and $s_2$ in $t_2$ inside a block beginning with setting the persistent variable flag to true with the instruction WaitUntilTestAndSet and ending with resetting the flag to false, and there will not be any instruction in between $s_1$ or before $s_1$ in $t_1$ which will cause the program control to change from $t_1$ to $t_2$. As a result $s_2$ will not occur-in-between $s_1$ and the program will be data race free.

*Flex loader* is a real-time demo example available in ABB RobotStudio [12] consisting of four tasks (robots) working in a synchronized way using the concept of TaskList. Tasklist is a list of available tasks to do a particular job at a given point in time. Here all the four tasks are grouped into four different task lists and along with the wait construct `WaitSyncTask` in all appropriate places in the program. Our tool did not report any potential data race in this example. *Exhaust Pipe* is also a real-time demo example available in ABB RobotStudio [12] consisting of three tasks in a synchronized way using the concept of `TaskList`. Here we applied rule **C1** to again conclude that there are no potential data races.

```
1  BEGINTASK<Task1, Null>
2    MODULE Module1
3    CONST robtarget Target_10 :=[[326.494, 425, 634.740],
        [0.104, 0, 0.983, 0], [-1, 0, -1, 0], [9E+09, 9E+09, 9E
        +09, 9E+09, 9E+09, 9E+09]];
4    CONST robtarget Target_20 :=[[-474.642, 265, 298.999],
        [0.087, 0, 0.996, 0], [-1, 0, 0, 0], [9E+09, 9E+09, 9E
        +09, 9E+09, 9E+09, 9E+09]];
5  % PROC MAIN()
6      IF Target_10.trans.x>0 and Target_20.trans.x<0
7          Path_Q1_Q2;
8      SetDO, do1,1;
9    ENDPROC
10   PROC Path_Q1_Q2()
11     MoveJ Target_10,v1000,z100,MyTool\WObj:=wobj0;
12     MoveJ Target_20,v1000,z100,MyTool\WObj:=wobj0;
13     ENDPROC
14   ENDMODULE
15 ENDTASK
16 BEGINTASK<Task2, Null>
17   MODULE Module1
18   PERS robtarget Target_10.trans.x ;
19   PERS robtarget Target_20.trans.x ;
20   PROC MAIN()
21     WaitDO, do1,1;
22         Target_10.trans.x:=0;
23         Target_20.trans.x:=0;
24   ENDPROC
25   ENDMODULE
26 ENDTASK
```

**Listing 1.3.** Quadrant Program Highlighting Rule C5 among Tasks

## 7  Conclusion and Future Work

We have proposed a static data race detection technique for multi-task programs controlling industrial robots using an occurs-in-between relation for (block of) statements in a task and another statement/block of statements in a second task. The static race detector framework is prototyped with ABB's Rapid programming language and can be used to detect potential data races in a sound way. While our framework is prototyped on Rapid, the technique can be extended to other robot programming languages that support multi-tasking too.

# References

1. ANSI/RIA R15.06-1999 American National Standard for Industrial Robots and Robot Systems - Safety Requirements (revision of ANSI/ R15.06-1992 (2012)
2. Adve, S.V.: Data races are evil with no exceptions: technical perspective. Commun. ACM **53**(11), 84 (2010)
3. Artho, C., Havelund, K., Biere, A.: High-level data races. Softw. Test. Verification Reliab. **13**, 207–227 (2003). https://doi.org/10.1002/stvr.281
4. Braumann, J., Brell-Cokcan, S.: Parametric robot control. Integrated CAD/CAM for architectural design (2011)
5. Faverges, S.: VAL3 REFERENCE MANUAL Version 5.3 (2006)
6. Harrow, J.J.: Runtime checking of multithreaded applications with visual threads. In: Havelund, K., Penix, J., Visser, W. (eds.) SPIN 2000. LNCS, vol. 1885, pp. 331–342. Springer, Heidelberg (2000). https://doi.org/10.1007/10722468_20
7. Havelund, K., Rosu, G.: Monitoring Java programs with Java PathExplorer. Electron. Notes Theor. Comput. Sci. **55**, 200–217 (2001)
8. Lamport, L.: Time, clocks, and the ordering of events in a distributed system. Commun. ACM **21**(7), 558–565 (1978)
9. O'Kane, J.M.: A gentle introduction to ROS (2014)
10. Savage, S., Burrows, M., Nelson, G., Sobalvarro, P., Anderson, T.: Eraser: a dynamic data race detector for multi-threaded programs. ACM Trans. Comput. Syst. **15**(4), 391–411 (1997)
11. Schwarz, M.D., Seidl, H., Vojdani, V., Lammich, P., Müller-Olm, M.: Static analysis of interrupt-driven programs synchronized via the priority ceiling protocol. In: Proceedings of the 38th Annual ACM SIGPLAN-SIGACT Symposium on Principles of Programming Languages, pp. 93–104. POPL 2011, Association for Computing Machinery, New York, NY, USA (2011). https://doi.org/10.1145/1926385.1926398
12. Simulation, Software, O.P.: Robotstudio 2021.2 (2021). https://new.abb.com/products/robotics/robotstudio/downloads
13. Sweden, A.: Technical reference manual: RAPID instructions, functions and data types. RobotWare 5.13 (2014). https://library.e.abb.com/public/688894b98123f87bc1257cc50044e809/Technical%20reference%20manual_RAPID_3HAC16581-1_revJ_en.pdf. Accessed 12 Sept 2021
14. Sweden, A.: Application Manual Controller Software IRC5, robotware 6.03, abb, 2016. Robotware 6.03 (2016)
15. Sweden, A.: Application manual MultiMove RobotWare 6.08, ABB, 2018. RobotWare 6.08 (2018)
16. Tulsyan, R., Pai, R., D'Souza, D.: Static race detection for RTOS applications (2020)

# Ordered Scheduling in Control-Flow Distributed Transactional Memory

Pavan Poudel[1], Shishir Rai[2], Swapnil Guragain[2], and Gokarna Sharma[2]($\boxtimes$)

[1] ATGWORK, Norcross, GA, USA
poudelpavan@gmail.com
[2] Kent State University, Kent, OH, USA
{srai,sguragai,gsharma2}@kent.edu

**Abstract.** Consider the *control-flow* model of transaction execution in a distributed system modeled as a communication graph where shared objects positioned at nodes of the graph are immobile but the transactions accessing the objects send requests to the nodes where objects are located to read/write those objects. The control-flow model offers benefits to applications in which the movement of shared objects is costly due to their sizes and security purposes. In this paper, we study the *ordered scheduling* problem of committing *dependent* transactions according to their predefined priorities in this model. The considered problem naturally arises in areas, such as loop parallelization and state-machine-based computing, where producing executions equivalent to a priority order is needed to satisfy certain properties. Specifically, we study ordered scheduling considering two performance metrics fundamental to any distributed system: (i) *execution time* - total time to commit all the transactions and (ii) *communication cost* - the total distance traversed in accessing required shared objects. We design scheduling algorithms that are individually or simultaneously efficient for both the metrics and rigorously evaluate them through several benchmarks on random and grid graphs, validating their efficiency. To our best knowledge, this is the first study of ordered scheduling in the control-flow model of transaction execution.

## 1 Introduction

Concurrent processes (threads) need to synchronize to avoid introducing inconsistencies while accessing shared data objects. Traditional mechanisms of locks and barriers have well-known downsides, including deadlock, priority inversion, reliance on programmer conventions, and vulnerability to failure or delay. *Transactional memory* (TM) [16,37] has emerged as an attractive alternative. Using TM, program code is split into *transactions*, blocks of code that appear to execute atomically. Transactions are executed *speculatively*: synchronization conflicts (or failures) may cause an executing transaction to *abort*: its effects are rolled back and the transaction is restarted. In the absence of conflicts (or failures), a transaction typically *commits*, causing its effects to become visible to all threads. Several commercial processors support TM, e.g., Intel's Haswell [22] and IBM's Blue Gene/Q [14], zEnterprise EC12 [27], and Power8 [8].

TM has been studied extensively for *multiprocessors*, where processors operate on a single shared memory and the latency to access (read/write) shared memory is the same (and negligible) for each processor. However, recently, the computing trend is shifting

A. R. Molla et al. (Eds.): ICDCIT 2023, LNCS 13776, pp. 67–83, 2023.
https://doi.org/10.1007/978-3-031-24848-1_5

toward *distributed multiprocessors*, where the memory access latency varies depending on the processor in which the thread executes and the physical segment of memory that stores the requested memory location. Therefore, the recent research focus is on how to support TM in distributed multiprocessors. Some proposals in this direction include TM$^2$C [13], NEMO [26], cluster-TM [3,24], GPU-TM [10], and HYFLOW [38].

TM is beneficial in distributed systems where data is spread across multiple nodes. For example, distributed data centers can use TM to simplify the burden of distributed synchronization and provide more reliable and efficient program execution while accessing data from remote nodes. *Distributed TM* (DTM) designed for such systems need to execute transactions effectively by taking into consideration the system's infrastructure. The network structure can play a crucial role in the DTM performance, since the data transactions access has to be reached across the network in a timely manner.

In this paper, we study ordered scheduling (ORDS) problem in distributed multiprocessors. We model distributed multiprocessors as an $n$-node connected, undirected, and weighted graph $G$, where each node denotes a processor and each edge denotes a communication link between processors. A set of $w$ shared objects $S := \{S_1, S_2, \ldots, S_w\}$ reside on the (possibly different) nodes of $G$. We consider the *control-flow* model [33], where objects are immobile but transactions send access requests to the nodes the required objects are located. Consider a set $T := \{T(v_1, age_1), T(v_2, age_2), \ldots\}$ of transactions mapped (arbitrarily) to the nodes of $G$ with each $T(v_i, age_i)$ accessing an arbitrary subset of the shared objects $S(T(v_i, age_i)) \subseteq S$, where $age$ is an externally provided parameter that is unique for each transaction providing a priority order. We say transaction $T(v_i, age_i)$ is *dependent* on $T(v_j, age_j), age_j < age_i$, if at least an object read/write by $T(v_i, age_i)$ is being written by $T(v_j, age_j)$. The ORDS problem is to commit the dependent transactions in the $age$ order. For example, transaction $T(v_i, age_i)$ that depends on $T(v_j, age_j), age_j < age_i$, commits only after $T(v_j, age_j)$ has been committed. Non-dependent transactions can execute and commit in parallel.

ORDS naturally arises in applications where producing (dependent) executions equivalent to a priority order is needed to satisfy/guarantee certain properties. Example applications include speculative loop parallelization and distributed computation using state machine approach [31]. In loop parallelization [32], loops designed to run sequentially are parallelized by executing their operations concurrently using TM. Providing an order matching the sequential one is fundamental to enforce equivalent semantics for both the parallel and sequential code. Regarding state machine approach [19], many distributed systems order tasks before executing them to guarantee that a single state machine abstraction always evolves consistently on distinct nodes, e.g., Paxos [23].

ORDS has been studied heavily in multiprocessors [11,31] where execution time is the only metric of interest. However, those studies focused on empirical studies and they do not extend to distributed multiprocessors as they do not consider latency. Recently, Poudel *et al.* [29] studied for the first time the ORDS problem in a distributed multiprocessor. However, they considered the *data-flow* model where transactions are immobile but the objects are mobile. Since the data-flow model is direct opposite of the control-flow model, the contributions in [29] do not apply to the control-flow model.

**Contributions.** In this paper, we design ORDS scheduling algorithms in the control-flow model and establish complementary results compared to [29]. We consider the *synchronous* communication model [6,7] where time is divided into discrete steps. We optimize two performance metrics: (i) *execution time* – the total time to execute and commit all the transactions, and (ii) *communication cost* – the total distance messages travel to access shared objects. A transaction's execution finishes as soon as it commits. The presented algorithms determine the time step when each transaction executes and commits. We measure the efficiency using a widely-studied notion of *competitiveness* – the ratio of total time (communication cost) for a designed algorithm to the minimum time (communication cost) achievable by an optimal scheduling algorithm.

Specifically, we have the following five contributions:

1. We provide an impossibility result showing that the optimal execution time and optimal communication cost can not be achieved simultaneously. (Sect. 3)
2. For the offline version, we provide two algorithms, one with optimal execution time and another with 2-competitive on communication cost. (Sect. 4)
3. For the partial dynamic version with the knowledge of transactions and their priorities but not the shared objects, we provide an $O(\log^2 n)$–competitive algorithm for both execution time and communication cost. (Sect. 5)
4. For the fully dynamic version with transactions arriving over time, we provide an $O(D)$–competitive algorithm for both execution time and communication cost, where $D$ is the diameter of the graph $G$. (Sect. 6)
5. We implement and rigorously evaluate the designed algorithms through micro-benchmarks and complex STAMP benchmarks on random and grid graphs, which validate the efficiency of the designed algorithms. (Sect. 7)

**Techniques.** For the offline version, the optimal time algorithm sends access requests in parallel following the shortest paths in $G$. The 2-competitive communication cost algorithm sends (combined) access requests through a minimum Steiner tree that connects the graph nodes containing the required objects.

In the partial dynamic version (with the knowledge of transactions and their priorities but not the shared objects), the proposed algorithm exploits the concept of *distributed directory protocols* [17,35]. Particularly, the directory protocol technique based on the hierarchical partitioning of the graph into clusters is used. This technique guarantees that the object access cost for a transaction is within an $O(\log^2 n)$ factor from the cost of minimum Steiner tree for that transaction. The directory protocol technique is then extended to the dynamic version guaranteeing $O(D)$-competitiveness without knowing transactions and their priorities a priori. This bound is interesting since the hierarchical partitioning technique used in the partial dynamic version is shown to only provide $O(D \log^2 n)$-competitive bound for the fully dynamic version. Therefore, the dynamic algorithm uses the directory protocol running on a spanning tree.

**Related Work.** Gonzalez-Mesa *et al.* [11] introduced the ORDS problem for multi-processors and Saad *et al.* [31] presented three improved algorithms and evaluated them through empirical studies. Transaction scheduling with no predefined ordering is widely-studied in multiprocessors providing provable upper and lower bounds, and impossibility results [1,34], besides several other scheduling algorithms that were only evaluated experimentally [39]. The multiprocessor ideas are not suitable for distributed multiprocessors as they do not deal with a crucial metric, communication cost.

Many previous studies on transaction scheduling in distributed multiprocessors, e.g., [2,4–7,35,36], considered the data-flow model. The papers [17,35,40] focused on minimizing communication cost. Execution time minimization is considered by Zhang et al. [40]. Busch et al. [4] considered minimizing both execution time and communication cost. Busch et al. [5] considered special topologies (e.g., grid, line, clique, star, hypercube, butterfly, and cluster) and provided offline algorithms minimizing execution time and communication cost. Recently, Busch et al. [7] provided dynamic (online) algorithms. However, all these works have no predefined ordering requirement.

Some papers considered the hybrid model that combines data-flow with control-flow. Hendler et al. [15] studied a lease based hybrid DTM which dynamically determines whether to migrate transactions to the nodes that own the leases or to demand the acquisition of these leases by the node that originated the transaction. Palmieri et al. [28] presented a comparative study of data-flow versus control-flow models.

## 2   Model and Preliminaries

**Graph.** We consider a distributed multiprocessor $G = (V, E, \mathfrak{w})$ of $n$ nodes (representing processing nodes) $V = \{v_1, v_2, \ldots, v_n\}$, edges (representing communication links between nodes) $E \subseteq V \times V$, and edge weight function $\mathfrak{w} : E \to \mathbb{Z}^+$. A *path* $p$ in $G$ is a sequence of nodes (with respective edges between adjacent nodes) with $\mathsf{length}(p) = \sum_{e \in p} \mathfrak{w}(e)$. We assume that $G$ is connected and $\mathsf{dist}(u, v)$ denotes the shortest path length (distance) between two nodes $u, v \in G$. The *diameter* $D := \max_{u,v \in G} \mathsf{dist}(u, v)$, the maximum shortest path distance between two nodes $u, v \in G$. The communication links are *bidirectional* – messages can be sent in both directions. Both the nodes and links are non-faulty and the links deliver messages in FIFO order. There is no bandwidth restriction on the edges, i.e., the messages can be of any size and any number of messages can traverse an edge at any time. The $k$-*neighborhood* of a node $u \in G$ is the set of nodes which are at distance $\leq k$ from $u$.

**Communication Model.** We consider the synchronous communication model where time is divided into discrete steps such that at each time step a node receives messages, performs a local computation, and then transmits messages to adjacent nodes [5–7]. For an edge $e = (u, v) \in E$, it takes $\mathfrak{w}(e)$ time steps to transfer a message $msg$ from $u$ to $v$ (and vice-versa); the *communication cost* contributed by $msg$ is $\mathfrak{w}(e)$.

**Transactions.** Let $S = \{S_1, S_2, \ldots, S_w\}$ denote the $w$ shared objects residing on nodes of $G$. Each object has some value which can be read/written. The node of $G$ where an object $S_i$ is currently positioned is called the *owner* of $S_i$, denoted as $owner(S_i)$. A transaction $T(v_i, age_i)$ is an atomic block of code mapped at node $v_i$ which requires a set of objects $S(T(v_i, age_i)) \subseteq S$ and has priority $age_i$. To simplify the analysis, we assume that each object has a single copy (for both read/write). We assume that each node runs a single thread and issues transactions sequentially.

**Control-Flow Model.** The model works in two steps:

i.  **Object Access Phase:** Transaction $T(v_i, age_i)$ sends access request to the owner node of each object $S_i \in S(T(v_i, age_i))$ and the owner node of $S_i$ replies back a *success* or *failure* message to $v_i$. A success message for $S_i$ means that $T(v_i, age_i)$ was able to read/write $S_i$, whereas a failure message means denied access.

ii. **Validation Phase:** If transaction $T(v_i, age_i)$ receives *success* message from owner node of each $S_i \in \mathcal{S}(T(v_i, age_i))$, then it commits. If $T(v_i, age_i)$ receives at least a *failure* message, then it either aborts or waits.

**Transaction Execution and Conflicts.** For an access request received for $S_j$ from $T(v_i, age_i)$, $owner(S_j)$ handles that request by allowing $T(v_i, age_i)$ to read or write (update) $S_j$ and replies a *success* message back to $v_i$. If $owner(S_j)$ receives two access requests for object $S_j$ at the same time and at least one of them is a write request, *conflict* is said to be occurred between transactions accessing $S_j$. $owner(S_j)$ handles such type of simultaneous access requests by denying at least one request. In case $owner(S_j)$ denies the access request, it replies a *failure* message back to node $v_i$.

**Performance Metrics.** Let $\mathcal{E}$ be an execution schedule following an algorithm $\mathcal{A}$.

**Definition 1** *(Execution Time). For a set of transactions $T$, the total time for $\mathcal{E}$ is the time elapsed until the last transaction finishes its execution in $\mathcal{E}$. The execution time of algorithm $\mathcal{A}$ is the maximum time over all possible executions for $T$.*

**Definition 2** *(Communication Cost). For a set of transactions $T$, the communication cost of $\mathcal{E}$ is the sum of the distances messages travel during $\mathcal{E}$. The communication cost of $\mathcal{A}$ is the maximum cost over all possible executions for $T$.*

**The ORDS Problem.** Each transaction $T(v_i, age_i)$ is assigned age, $age_i$, before it is activated, and the age signifies the transaction commit order under dependencies. Following [11,29,31], parameter *age* is (i) *unique* – no two transactions can have the same age, (ii) *non-modifiable* – it never changes once assigned, and (iii) *externally determined* – it does not depend on transaction execution.

For transaction $T(v_i, age_i)$, $\mathcal{S}(T(v_i, age_i)) := write(\mathcal{S}(T(v_i, age_i))) \cup read(\mathcal{S}(T(v_i, age_i)))$. We say $T(v_i, age_i)$ is *dependent* on $T(v_j, age_j), age_j < age_i$, if $(write(\mathcal{S}(T(v_i, age_i))) \cap \mathcal{S}(T(v_j, age_j)) \neq \emptyset) \vee (read(\mathcal{S}(T(v_i, age_i))) \cap write(\mathcal{S}(T(v_j, age_j))) \neq \emptyset)$. I.e., at least an object read/write by $T(v_i, age_i)$ is being written by $T(v_j, age_j)$. $T(v_i, age_i)$, if dependent on $T(v_j, age_j)$, can commit only after $T(v_j, age_j)$ commits. Formally,

**Definition 3** *(The ORDS problem). Given a set of transactions $T := \{T(v_1, age_1), T (v_2, age_2), \ldots\}$ mapped (arbitrarily) to the nodes of $G$, commit dependent transactions in $T$ in the increasing order of age in the control-flow model.*

## 3    Impossibility Result

Consider a star graph $G$ as shown in Fig. 1 with eight rays going out from the center node. Let there be three nodes on each ray (except the center node). Additionally, let the end nodes of consecutive rays are connected. Suppose there are six objects $a, b, c, d, e$, and $f$ positioned on six consecutive end nodes, and a transaction $T$ is mapped at the center node and it requests all six objects. All edges have unit weight.

**Theorem 1.** *There are transaction scheduling instances for which execution time and communication cost cannot be minimized simultaneously in the control-flow model.*

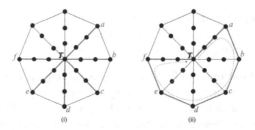

**Fig. 1.** (i) Transaction $T$ accessing objects in parallel through blue colored paths, (ii) $T$ accessing objects sequentially again through blue colored paths. (Color figure online)

*Proof.* When transaction $T$ sends object access requests in parallel, they can be reached in 3 steps. In next 3 steps, $T$ gets reply messages from all the object nodes, and in one additional step, it can execute and commit. This gives optimal execution time of 7 steps. However, total communication cost becomes 36 (3 steps to reach requests to objects and 3 steps to receive replies back). Alternatively, let $T$ sends the object access request (combined) first to $a$ and then to $b, c, d, e, f$ in order. Moreover, the reply from $a$ is also sent together with the (combined) request towards $b$ and so on with others. Thus, when the request reaches $f$, the replies from $a, b, c, d, e$ also reach there. Now, from $f$, all the replies traverse the ray connecting $f$ and $T$. The communication cost becomes 11 steps, which is optimal. Total execution time becomes $11 + 1 = 12$, which is sub-optimal. $\square$

## 4   Offline Algorithms

In this section, we study the offline version of the ORDS problem. We present two algorithms, one called OFFEXEC that achieves optimal execution time and another called OFFCOMM that is 2-competitive on communication cost.

**Execution Time Algorithm** OFFEXEC. OFFEXEC accesses required objects for each transaction in parallel. All transactions in $\mathcal{T}$ are initiated at time step $t = 0$. Therefore, at $t = 0$, all the transactions in $\mathcal{T}$ send requests to access the required objects to the respective owner nodes following the shortest paths. Each owner node then replies *success* message for every request (after performing the read/write operation) respecting the age order and dependency of the transactions at corresponding owner node.

For transaction $T(v_i, age_i)$ at node $v_i$, let $\mathcal{S}(T(v_i, age_i)) \subseteq \mathcal{S}$ be the set of objects it needs. $T(v_i, age_i)$ sends corresponding access requests to $owner(S_j)$ of each object $S_j \in \mathcal{S}(T(v_i, age_i))$ following the shortest path from $v_i$ to $owner(S_j)$. After the access request reaches $owner(S_j)$, $owner(S_j)$ sends *success* message back to $v_i$ as soon as $T(v_i, age_i)$ is able to read/write that object respecting the age order. Specifically, there can be two cases: (i) There is no $T(v_k, age_k), age_k < age_i$, in $\mathcal{T}$ which also wants to access $S_j$, then $owner(S_j)$ immediately sends *success* message back to $v_i$ (ii) There is another transaction $T(v_k, age_k), age_k < age_i$, in $\mathcal{T}$ that conflicts with $T(v_i, age_i)$ while accessing $S_j$, then $owner(S_j)$ sends *success* message to $v_k$ first and to $v_i$ in the next time step. When $v_i$ receives *success* messages from all $owner(S_j)$, $T(v_i, age_i)$ finishes its execution and commits.

Let $t_i^{S_j}$ be the time step at which $owner(S_j)$ of object $S_j \in S(T(v_i, age_i))$ replies *success* message back to node $v_i$ corresponding to the request sent by $T(v_i, age_i)$. Then, $t_i^{S_j} = \max\{t_{prev(T(v_i, age_i))}^{S_j} + 1, \text{dist}(v_i, owner(S_j))\}$, where $t_{prev(T(v_i, age_i))}^{S_j}$ is the time step at which $owner(S_j)$ replies to the dependent transaction of $T(v_i, age_i)$ that is immediately previous to $T(v_i, age_i)$ in the age order. For the lowest aged transaction $T(v_1, age_1)$, $t_1^{S_j} = \text{dist}(v_1, owner(S_j))$.

Let $CT_i$ be the time step at which transaction $T(v_i, age_i) \in T$ commits. Then,

$$CT_i = \begin{cases} CT_{prev(T(v_i, age_i))} + 1, & \text{if } t_i' < CT_{prev(T(v_i, age_i))} \\ t_i' + 1, & \text{otherwise.} \end{cases}$$

where $CT_{prev(T(v_i, age_i))}$ is the time at which the transaction dependent to $T(v_i, age_i)$ that is immediately previous to $T(v_i, age_i)$ in the age order commits and

$$t_i' = \max_{S_j \in S(T(v_i, age_i))} (t_i^{S_j} + \text{dist}(v_i, owner(S_j))).$$

For the lowest aged transaction $T(v_1, age_1)$, $CT_1 = \max_{S_j \in S(T(v_1, age_1))} 2 \cdot \text{dist}(v_1, owner(S_j)) + 1$.

**Theorem 2.** OFFEXEC *achieves optimal execution time.*

*Proof.* The execution time depends on two factors. First, how long does a transaction take to access required objects and second, when does each transaction commit? In OFFEXEC, each transaction accesses required object using the shortest path in $G$ which is thus optimal. Now, we need to show that each transaction commits at the earliest possible time. First, let there is no conflict between any transactions in $T$. Then all the transactions can access required objects in parallel and as soon as each transaction receives success messages from the owner nodes of each required object, it can commit. The total execution time becomes

$$\max_{T(v_i, age_i) \in T} \left\{ \max_{S_j \in S(T(v_1, age_1))} 2 \cdot \text{dist}(v_i, owner(S_j)) + 1 \right\}$$

which is optimal.

Now, let there are conflicts between transactions in $T$ when accessing objects. Let $T = \{T(v_1, age_1), T(v_2, age_2), \ldots, T(v_n, age_n)\}$ be the set of transactions. Let a dependency graph $H = (V_H, E_H)$ holds the dependency between the conflicting transactions where the nodes $V_H$ represent transactions in $T$ and the directed edges $E_H$ represent dependencies between the transactions. The edge $(T(v_i, age_i), T(v_j, age_j)) \in E_H$, where $age_i < age_j$, represents a dependency between $T(v_i, age_i)$ and $T(v_j, age_j)$ such that $T(v_j, age_j)$ can commit only after $T(v_i, age_i)$ commits. The ORDS problem requires the dependent transactions to commit in their age order. The diameter $D_H$ of $H$ provides the longest chain of dependent transactions and the total execution time of any optimal algorithm will be the time required by all the transactions that belong to $D_H$ to commit. During the execution of OFFEXEC, for each transaction $T(v_i, age_i)$, if there is no any dependent transaction in $H$ or all the dependent transactions in $H$ have already been committed, then $T(v_i, age_i)$ can commit as soon as it receives success messages from the owner nodes of all required objects. Note that both, object access requests and success messages, are sent through the shortest paths in $G$. When the highest age transaction that belongs to $D_H$ of $H$ commits, OFFEXEC finishes. Hence, the total execution time is optimal.                                                                                              □

**Theorem 3.** OFFEXEC *is* $k$-*competitive in communication cost, where* $k$ *is the maximum number of shared objects accessed by a transaction in* $T$.

**Communication Cost Algorithm** OFFCOMM. In OFFCOMM, we convert the execution of each transaction to a Minimum Steiner Tree (MST) [20,21]. Steiner trees have been extensively studied in the context of weighted graphs [12]. Given a graph $G = (V, E)$ and a subset $P \subseteq V$, a *Steiner tree* spans through $P$. The Steiner tree problem in our case is to find a Steiner tree that connects all the vertices of $P$ with the minimum possible total weight. Computing MST is known to be NP-Hard. We follow the algorithm of Takahashi and Matsuyama [18] which provides $2(1 - 1/|P|)$–approximation for MST. The algorithm of [18] constructs a Steiner tree as follows:

- Start from a participant node in $P$.
- Find the next participant that is closest to the current tree.
- Join the closest participant to the closest node of the tree.
- Repeat until all nodes in $P$ are connected.

Now, we discuss how MST is constructed for each transaction in $T$. Let $S(T(v_i, age_i)) \subseteq S$ be the set of objects required by a transaction $T(v_i, age_i) \in T$. Let $P_i \subseteq V$ contains node $v_i$ and the owner node of each object $S_j \in S(T(v_i, age_i))$ (i.e., $P_i := (\forall_{S_j \in S(T(v_i, age_i))} owner(S_j)) \cup v_i$). Now, the problem is to find a MST that connects the nodes in $P_i$ which is constructed by following the algorithm of [18] and is denoted as $MST_i$. Then, $T(v_i, age_i)$ sends object access requests in $MST_i$. The total message cost incurred by transaction $T(v_i, age_i)$ is $2.|MST_i|$. That means, messages visit each edge of $MST_i$ exactly twice, one for sending access request and the other for receiving reply (*success* or *failure*) message from each owner node.

Instead of sending requests individually to access the objects in $S(T(v_i, age_i))$, $T(v_i, age_i)$ sends them collectively in $MST_i$. Each neighboring node recursively sends the request to the next neighbor in $MST_i$ until the request reaches all the owner nodes of the required objects. To be specific, if $v_p, v_q \in MST_i$ be any two owner nodes of objects which share a common path from $v_i$ up to some intermediate node $v_s$, then the requests to $v_p$ and $v_q$ from $v_i$ are sent collectively up to $v_s$ as a single message. The request is then divided into two at $v_s$ and they are forwarded separately towards $v_p$ and $v_q$. When all the access requests reach respective owner nodes, the reply messages are collected in the opposite direction. Here, each intermediate node which had initially sent access requests to the neighboring nodes later collects the reply messages from those neighboring nodes and returns them collectively to the ancestor node. When $v_i$ receives reply messages from all the neighboring nodes in $MST_i$, $T(v_i, age_i)$ commits (provided that all the reply messages are *success* messages).

The OFFCOMM algorithm works as follows. It produces a conflict-free execution schedule. At time step $t = 0$, each transaction $T(v_i, age_i)$ sends access requests to required objects following its corresponding $MST_i$. When the access request reaches

$owner(S_j)$, $owner(S_j)$ sends *success* message back to $v_i$ as soon as $T(v_i, age_i)$ is able to read/write that object respecting the age order of the dependent transactions. Let $\text{dist}_{MST_i}(v_i, v_j)$ represents the distance between nodes $v_i$ and $v_j$ following the shortest path in $MST_i$. Then, for each $T(v_i, age_i) \in \mathcal{T}$, $owner(S_j)$ of each $S_j \in S(T(v_i, age_i))$ replies success message to $v_i$ at time step: $t_i^{S_j} = \max\{t_{prev(T(v_i, age_i))}^{S_j} + 1, \text{dist}_{MST_i}(v_i, owner(S_j))\}$, where $t_{prev(T(v_i, age_i))}^{S_j}$ is the time step at which $owner(S_j)$ replies to the dependent transaction of $T(v_i, age_i)$ that is immediately previous to $T(v_i, age_i)$ in the age order.

The commit time step $CT_i$ for each $T(v_i, age_i)$ is:

$$CT_i = \begin{cases} CT_{prev(T(v_i, age_i))} + 1, & \text{if } t_i' < CT_{prev(T(v_i, age_i))} \\ t_i' + 1, & \text{otherwise.} \end{cases}$$

where $CT_{prev(T(v_i, age_i))}$ is the time at which the transaction dependent to $T(v_i, age_i)$ that is immediately previous to $T(v_i, age_i)$ in the age order commits and $t_i' = \max_{S_j \in S(T(v_i, age_i))}(t_i^{S_j} + \text{dist}_{MST_i}(v_i, owner(S_j)))$.

**Theorem 4.** OFFCOMM *is 2-competitive in communication cost.*

*Proof.* Let $MST_i$ be the minimum cost Steiner tree constructed for transaction $T(v_i, age_i)$ in OFFCOMM. Let $\text{dist}_{MST_i}(v_x, v_y)$ be the shortest path distance between $v_x$ and $v_y$ in $MST_i$. If $\text{dist}(v_x, v_y)$ be the shortest path distance in $G$, then we have: $\text{dist}_{MST_i}(v_x, v_y) \leq 2 \cdot \text{dist}(v_x, v_y)$. Since OFFCOMM follows the shortest paths in respective MSTs for accessing required objects, the communication cost $C_{T(v_i, age_i)}$ of executing each transaction $T(v_i, age_i) \in \mathcal{T}$ is: $C_{T(v_i, age_i)} = 2 \cdot C_{opt}^{T(v_i, age_i)}$, where $C_{opt}^{T(v_i, age_i)}$ is the cost of any optimal communication algorithm for executing $T(v_i, age_i)$ that accesses required objects following the shortest paths in $G$. If $C_{total}$ and $C_{opt}$ be the total communication costs of OFFCOMM and any optimal algorithm, respectively, such that $C_{opt} = \sum_{T \in \mathcal{T}} C_{opt}^T$, then, $C_{total} = \sum_{T \in \mathcal{T}} C_T = \sum_{T \in \mathcal{T}} 2 \cdot C_{opt}^T = 2 \cdot C_{opt}$.  □

**Theorem 5.** OFFCOMM *is $r$-competitive in execution time, where $r$ is the maximum stretch of MST computed for each transaction in $\mathcal{T}$ which is given by:*

$$r = \max_{T(v_i, age_i) \in \mathcal{T}} \left\{ \max_{S_j \in S(T(v_i, age_i))} \frac{\text{dist}_{MST}(v_i, owner(S_j))}{\text{dist}(v_i, owner(S_j))} \right\}.$$

## 5 Partial Dynamic Algorithm

Here we study the partial dynamic version of the ORDS problem, where a priori knowledge on transactions and their priorities is available, but not the shared objects they access and their locations. All transactions arrive at time $t = 0$. Thus, the following two tasks are additional to the offline version:

i. Determine the owner nodes of all the shared objects that a transaction requests.

ii. Determine the node where the next transaction in the commit order is located and the path to reach that node.

We present an efficient algorithm PARTDYN using distributed directory protocol technique [2,9,17,30,35]. We compute two distributed queues, the first helps transactions accessing required objects and the second helps sending commit messages to the next dependent transaction in age order. The first is called *distributed object queue* where *object access tours* are constructed for each transaction. The second is called *distributed transaction queue* that satisfies the commit order of transactions. Each transaction sends commit message to the next transaction in order following the path in its respective transaction tour in the distributed transaction queue. We use the hierarchy-of-clusters-based overlay tree ($\mathcal{OT}$) (discussed next) for the computation of both queues.

**Overlay Tree $\mathcal{OT}$ Construction.** The well-known approaches for $\mathcal{OT}$ construction are based on either a *spanning tree* or a *hierarchy of clusters* on $G$. The spanning tree was used in directory protocols [2,9] and the hierarchy of clusters was used in directory protocols [17,35,36].

Both approaches work, however, hierarchy-of-clusters-based overlay trees are more suitable to control communication costs (and hence the execution time) compared to the spanning-tree-based overlay trees. Therefore, in the following, we discuss the construction of hierarchy-of-clusters-based overlay tree $\mathcal{OT}$. In a high level, divide the graph $G$ into a hierarchy of clusters with $H_1 = \lceil \log D \rceil + 1$ layers such that the clusters sizes grow exponentially (i.e., $2^\ell, 0 \leq \ell \leq H_1$). A *cluster* is a subset of nodes, and its diameter is the maximum distance between any two nodes. The diameter of each cluster at layer $\ell$, where $0 \leq \ell < H_1$, is no more than $f(\ell)$, for some function $f$, and each node participates in no more than $g(\ell)$ clusters at layer $\ell$, for some other function $g$. Moreover, for each node $u$ in $G$, there is a cluster at layer $\ell$ such that the $(2^\ell - 1)$-neighborhood of $u$ is contained in that cluster.

There are known algorithms, such as a *hierarchical sparse cover* of $G$, that give a cluster hierarchy $\mathcal{Z}$ of $H_1$ layers with $f(\ell) = O(\ell \log n)$ and $g(\ell) = O(\log n)$. This construction was used in the directory protocol, SPIRAL, by Sharma et al. [35], where additionally, each layer $\ell$ is decomposed into $H_2 = O(\log n)$ sub-layers of clusters, such that a node participates in all the sub-layers of a layer but in a different cluster within each sub-layer, i.e., at each layer $\ell$ a node $u$ participates in $g(\ell) = O(\log n)$ clusters. Suppose a node in each cluster is designated as the *leader* of the cluster. Connecting the leaders of the clusters in the subsequent levels gives $\mathcal{OT}$.

An *upward path* $p(u)$ for each node $u \in G$ is built by visiting leader nodes in all the clusters that $u$ belongs to starting from layer 0 (the bottom layer in $\mathcal{Z}$) up to layer $H_1$ (the top layer in $\mathcal{Z}$). Within each layer, $H_2$ sub-layers are visited by $p(u)$ according to the order of their sub-layer labels. The upward path $p(u)$ visits two subsequent leaders using shortest paths in $G$ between them. Lets say two paths *intersect* if they have a common node. Using this definition, two upward paths intersect at layer $i$ if they visit the same leader at layer $i$. The lemmas below are satisfied in the construction of [35].

**Lemma 1.** *The upward paths $p(u)$ and $p(v)$ of any two nodes $u, v \in G$ intersect at layer $\min\{H_1, \lceil \log(\text{dist}(u,v)) \rceil + 1\}$.*

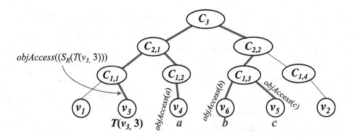

**Fig. 2.** Illustration of computation of distributed object queue for transaction $T(v_3, 3)$ requiring objects $(a, b, c)$. $T(v_3, 3)$ sends $objAccess(S_R(T(v_3, 3)))$ message in its upward path to cluster $C_{1,1}$ which recursively sends it to $C_{2,1}$. $C_{2,1}$ contains the owner node of object $a$ (i.e., $v_4$), thus sends $objAccess(a)$ message to $v_4$. Then, after removing $a$ from $S_R(T(v_3, 3))$, $C_{2,1}$ sends $objAccess(S_R(T(v_3, 3)))$ message to cluster $C_3$. $C_3$ sends the message downward until the requests reach nodes $v_5$ and $v_6$. Later, all three nodes $v_4, v_5$, and $v_6$ reply *success* messages which are combined at clusters $C_{1,3}$ and $C_{2,1}$, and finally reach node $v_3$. Then $T(v_3, 3)$ commits. The edges traversed by the messages are highlighted in red. (Color figure online)

**Lemma 2.** *For any upward path $p(u)$ for any node $u \in G$ from the bottom layer upto layer $\ell$ (and any sub-layer in layer $\ell$), $\mathsf{length}(p(u)) \leq O(2^\ell \log^2 n)$.*

**Computing Distributed Transaction Queue.** We denote the distributed transaction queue by $DTQueue(\mathcal{T})$. To construct $DTQueue(\mathcal{T})$, each transaction $T(v_i, age_i)$ sends a $findT(T(v_i, age_i))$ message in its upward path $p(v_i)$ in $\mathcal{OT}$. The $findT(T(v_i, age_i))$ message contains information about the required objects by $T(v_i, age_i)$ and moves upward until it meets the similar messages sent by it's previous and next conflicting transactions in age order. When two messages $findT(T(v_i, age_i))$ and $findT(T(v_j, age_j))$ meet at some node $v_k$, it can easily be found that whether $T(v_i, age_i)$ and $T(v_j, age_j)$ conflict with each other or not by looking at the information of required objects for each of them. When such meetings happen for all $findT(prev(T(v_i, age_i)))$, $findT(T(v_i, age_i))$, and $findT(next(T(v_i, age_i)))$, $1 \leq i \leq n$, the computation of $DTQueue(\mathcal{T})$ is completed.

The upward paths $p(v_i)$ and $p(v_j)$ for the two consecutive dependent transactions $T(v_i, age_i)$ and $T(v_j, age_j)$ intersect at some node $v_k$ at some layer $l > 0$. Transaction $T(v_i, age_i)$ sends a commit message to $T(v_j, age_j)$ by first sending it upward in $p(v_i)$ up to $v_k$ and then sending the message downward in $p(v_j)$ from $v_k$ up to node $v_j$. The following theorem follows from the hierarchy of clusters based $\mathcal{OT}$.

**Theorem 6.** *If $d$ is the shortest path distance between nodes $v_i, v_j \in G$, then the distance between $v_i, v_j$ following the upward paths $p(v_i)$ and $p(v_j)$ in $\mathcal{OT}$ is $O(d \cdot \log^2 n)$.*

**Computing Distributed Object Queues.** Distributed object queue for each transaction $T(v_i, age_i) \in \mathcal{T}$ is denoted as $DOQueue(T(v_i, age_i))$. $DOQueue(T(v_i, age_i))$ contains object tour(s) to access the object(s) requested by $T(v_i, age_i)$.

$DOQueue(T(v_i, age_i))$ is constructed as follows. Let $\mathcal{S}_R(T(v_i, age_i)) \subseteq \mathcal{S}(T(v_i, age_i))$ be the set of objects required by $T(v_i, age_i)$ that are not present on $v_i$. $T(v_i, age_i)$ sends $objAccess(S_R(T(v_i, age_i)))$ message in its upward path $p(v_i)$.

Let at some level $l > 0$, $objAccess(\mathcal{S}_R(T(v_i, age_i)))$ reaches a cluster with node $v_j$ that contains an object $S_j \in \mathcal{S}_R(T(v_i, age_i))$. Then the leader of the cluster (say $v_l$) forwards $objAccess(S_j)$ to the node $v_j$ downward in the path $p(v_j)$. The leader also removes object $S_j$ from $\mathcal{S}_R(T(v_i, age_i))$ and forwards $objAccess(\mathcal{S}_R(T(v_i, age_i)))$ message upward in the path $p(v_i)$ if $\mathcal{S}_R(T(v_i, age_i))$ is not empty. This process continues until $\mathcal{S}_R(T(v_i, age_i))$ becomes empty and by that time, the computation of $DOQueue(T(v_i, age_i))$ is completed.

Later, during the execution of $T(v_i, age_i)$, when the object access request $objAccess(S_j)$ reaches the owner node of $S_j$, $owner(S_j)$, $T(v_i, age_i)$ performs read or write operation on $S_j$. After the read or write operation is completed, $v_j$ replies a *success* message back following the previous path in the opposite direction (i.e., upward from $v_j$ to the leader node $v_l$ in $p(v_j)$). Each leader node when receives reply messages from the owner nodes of objects, combines them into a single message and sends it back downward in the path $p(v_i)$ to node $v_i$. The leader node waits to combine the reply message until it receives reply messages from all the paths that it has sent previously the access requests. Figure 2 illustrates this idea.

**Algorithm** PARTDYN. PARTDYN starts with computing distributed object queues $DOQueue(T(v_i, age_i))$ for each transaction $T(v_i, age_i) \in \mathcal{T}$ and distributed transaction queue $DTQueue(\mathcal{T})$. $DOQueue(T(v_i, age_i))$ contains object tours to access all the required objects in $\mathcal{S}(T(v_i, age_i))$.

All the transactions that do not depend on any lower aged transactions start execution at time $t = 0$. $T(v_1, age_1)$ starts at $t = 0$ and sends object access requests recursively following object tours in $DOQueue(T(v_1, age_1))$. Then, for each object $S_j \in \mathcal{S}(T(v_1, age_1))$, $objAccess(S_j)$ reaches the owner node $owner(S_j)$. $T(v_1, age_1)$ performs read or write operation on all $S_j$ and a *success* message from each $owner(S_j)$ is replied back following the object tours in the backward direction. $T(v_1, age_1)$ commits after it receives *success* messages from all the owner nodes of required objects (possibly in combined form). Let $T(v_1, age_1)$ commits at time step $t_1 > 0$. $T(v_1, age_1)$ sends commit message $commit(T(v_1, age_1))$ to the next conflicting transaction in age order $next(T(v_1, age_1)) = T(v_k, age_k), age_k > age_i$, by following upward paths in $DTQueue(\mathcal{T})$. When $T(v_k, age_k)$ receives commit messages from all the dependent transactions, $T(v_k, age_k)$ executes and commits at time step $t_k > t_1$ and sends $commit(T(v_k, age_k))$ message to $next(T(v_k, age_k))$. The process continues until the highest aged transaction $T(v_h, age_h)$ commits at some time step $t_h$.

**Theorem 7.** PARTDYN *is* $O(\log^2 n)$-*competitive in both execution time and communication cost.*

## 6   Fully Dynamic Algorithm

Here, we study ORDS with no a priori knowledge on transactions, their priorities, the shared objects they access, and their initial locations. Additionally, transactions arrive at different nodes of $G$ arbitrarily over time. Once a transaction arrives at some node $v_i$, it knows the priority (i.e., age) of that transaction and the objects needed by it. We present an algorithm DYN that achieves $O(D)$ competitive ratio in both execution time and communication cost. Algorithm DYN works on top of a spanning-tree-based overlay tree, denoted as $OT_{ST}$. Let $v_{root}$ be the root node of $\mathcal{OT}_{ST}$. For any node $v$, the upward path $p(v)$ in $\mathcal{OT}_{ST}$ is the path obtained by connecting the parent nodes in $ST$ from node $v$ up to the root $v_{root}$. DYN executes in two phases:

- **Phase 1 – Object Advertisement** in which each node of graph $G$ is advertised with the locations of all the objects.
- **Phase 2 – Transaction Execution** in which transactions are executed and committed according to age order.

**Phase 1 – Object Advertisement.** The object advertisement phase makes each node of $G$ know the locations of all the shared objects. Later, when a transaction at node $v_i$ needs some object $S_j$, $v_i$ can forward object access request to the owner node of that object. The ownership of each object is advertised in the form of a hash map where each key-value pair represents $(objID, nodeID)$, where $objID$ is the ID of an object located at node $v \in V$ and $nodeID$ is the ID of $v$.

Execution starts from leaf nodes of $\mathcal{OT}_{ST}$. Each leaf node $v_l$ sends a hash map $(objID, nodeID)$. If $v_l$ contains no object, $v_l$ sends an empty hash map. Also, if $v_l$ contains more than one object, it sends a hash map with multiple key-value pairs. When a parent node $v_{p1}$ receives hash maps from all its child nodes, $v_{p1}$ merges those into a single hash map and appends new key-val pair(s) if it contains any object(s). The updated hash map is then sent upward to the next parent node $v_{p2}$. $v_{p2}$ again merges all hash maps into a single one after receiving from all the child nodes. This process is repeated until the current node is the root $v_{root}$. When $v_{root}$ receives hash maps from all of its child nodes, it merges them into a single hash map and replies back the updated hash map to all the child nodes recursively. This phase ends when all the leaf nodes receive updated hash map containing all $(objID, nodeID)$ pairs.

**Lemma 3.** *Phase 1 finishes in $O(D)$ time steps with communication cost $O(n)$.*

**Phase 2 – Transaction Execution.** Let $H$ be the height of $\mathcal{OT}_{ST}, H \leq D$. As soon as transaction $T(v_i, age_i)$ is initiated, it sends an arrival message $T_{arrival}(T(v_i, age_i), t_i)$ to $v_{root}$ following the upward path $p(v_i)$, where $t_i$ is the time step at which $T(v_i, age_i)$ arrives at node $v_i$. Let $\mathcal{T}_t(v_{root})$ be a list maintained by $v_{root}$ which contains the information of pending transactions at time step $t$ sorted by arrival time. The arrival message $T_{arrival}(T(v_i, age_i), t_i)$ sent from node $v_i$ reaches $v_{root}$ in $\leq H$ time steps. Thus, when $v_{root}$ receives a transaction arrival message $T_{arrival}(T(v_i, age_i), t_i)$ at some time step $t_r \geq t_i$, it includes $T(v_i, age_i)$ in $\mathcal{T}_t(v_{root})$ at time step $t_i' = t_i + H$.

Let $T(v_x, age_x) \in \mathcal{T}_t(v_{root})$ be the lowest aged transaction in $\mathcal{T}_t(v_{root})$ at time $t$. $v_{root}$ sends $startExec(T(v_x, age_x))$ message to node $v_x$ to execute $T(v_x, age_x)$. $T(v_x, age_x)$ sends object access requests to the owner nodes of $\mathcal{S}(T(v_x, age_x))$. When $T(v_x, age_x)$ successfully accesses all the required objects in $\mathcal{S}(T(v_x, age_x))$, it commits and sends a commit message to $v_{root}$. After that, $v_{root}$ removes $T(v_x, age_x)$ from $\mathcal{T}_t(v_{root})$ and schedules next conflicting transaction in the age order to execute. Note that, $v_{root}$ can schedule multiple transactions together which are not dependent on any lower aged transactions or receive commit messages from all the dependent transactions during the execution. Phase 2 finishes when all transactions in $\mathcal{T}$ commit.

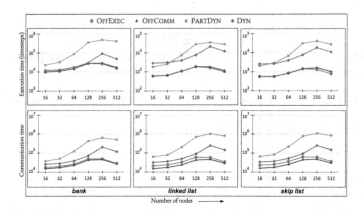

**Fig. 3.** Time and communication (log scale) in micro-benchmarks on random graphs.

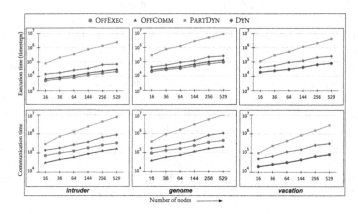

**Fig. 4.** Time and communication (log scale) in STAMP benchmarks on grid graphs.

**Lemma 4.** *In Phase 2, each transaction finishes its execution in $O(D)$ time steps with communication cost $O(D)$-competitive.*

Combining Lemmas 3 and 4, we have,

**Theorem 8.** DYN *is $O(D)$-competitive in both execution time and communication cost.*

*Proof.* DYN executes in two phases, Phase 1 and Phase 2, sequentially. Phase 1 finishes in $O(D)$ time steps. In Phase 2, each transaction in $\mathcal{T}$ spends $O(D)$ time steps to execute and commit. So, for all $n$ transactions in $\mathcal{T}$, it takes $O(n \cdot D)$ time steps to execute and commit. In total, both Phase 1 and Phase 2 of DYN end in $O(D) + O(n \cdot D) = O(n \cdot D)$ time steps. Since, transactions need to follow the age order to commit, any optimal algorithm requires at least $O(n)$ time steps to execute and commit. Hence, DYN is $O(D)$-competitive in execution time. The same analysis works to show $O(D)$-competitive in communication cost. □

# 7   Evaluation

We have implemented OFFEXEC, OFFCOMM, PARTDYN, and DYN and evaluated them using a set of micro- and complex benchmarks. The experiments were performed on an Intel Core i7-7700K processor with 32 GB RAM, simulating two different communication graphs, namely *random* and *grid* whose diameters range from 3 to 6 and 6 to 44, respectively. The results presented are the average of 10 runs.

**Results on micro-benchmarks:** We experimented against three micro-benchmarks *bank, linked list,* and *skip list.* Figure 3 provides the results in random graph.

**Results on STAMP benchmarks:** We experimented against *intruder, genome,* and *vacation* benchmarks from STAMP [25]. Figure 4 provides the results in grid graph.

**Results Discussion.** For both random and grid graphs, OFFEXEC has the minimum execution time (which is optimal) in all the benchmarks. The execution time for OFF-COMM is higher than OFFEXEC but always within factor 2 of optimal. Similarly, in all the benchmarks, OFFCOMM has the minimum communication cost, which is with in factor of 2 from optimal. The experimental results in all the benchmarks show that the execution time of PARTDYN is always within $O(\log^2 n)$ factor compared to OFFEXEC. Moreover, the execution time in DYN is always within $O(D)$ factor. The communication cost results follow the same pattern. In fact, the results are substantially better than the theoretical bounds for both PARTDYN and DYN. In all the results, we can see that DYN has less execution time and less communication cost than PARTDYN. This is because of $D < \log^2 n$ in the experiment.

# 8   Concluding Remarks

In this paper, we have studied the ordered scheduling problem of committing transactions according to their predefined priorities in the control-flow distributed transactional memory, minimizing execution time and communication cost. The control-flow model is important because in many applications, the movement of data is costly due to its size and security purposes. We have provided a range of algorithms considering this problem in the offline and dynamic settings. As a future work, it will be interesting to deploy the algorithms in real distributed system(s) and measure the wall clock results.

**Acknowledgements.** This research was supported by National Science Foundation under Grant No. CAREER CNS-2045597.

# References

1. Attiya, H., Epstein, L., Shachnai, H., Tamir, T.: Transactional contention management as a non-clairvoyant scheduling problem. Algorithmica **57**(1), 44–61 (2010)
2. Attiya, H., Gramoli, V., Milani, A.: Directory protocols for distributed transactional memory. In: Guerraoui, R., Romano, P. (eds.) Transactional Memory. Foundations, Algorithms, Tools, and Applications. LNCS, vol. 8913, pp. 367–391. Springer, Cham (2015). https://doi.org/10.1007/978-3-319-14720-8_17

3. Bocchino, R.L., Adve, V.S., Chamberlain, B.L.: Software transactional memory for large scale clusters. In: PPoPP, pp. 247–258 (2008)

4. Busch, C., Herlihy, M., Popovic, M., Sharma, G.: Impossibility results for distributed transactional memory. In: PODC, pp. 207–215 (2015)

5. Busch, C., Herlihy, M., Popovic, M., Sharma, G.: Fast scheduling in distributed transactional memory. In: SPAA, pp. 173–182. ACM (2017)

6. Busch, C., Herlihy, M., Popovic, M., Sharma, G.: Time-communication impossibility results for distributed transactional memory. Distrib. Comput. **31**(6), 471–487 (2017). https://doi.org/10.1007/s00446-017-0318-y

7. Busch, C., Herlihy, M., Popovic, M., Sharma, G.: Dynamic scheduling in distributed transactional memory. In: IPDPS, pp. 874–883 (2020)

8. Cain, H.W., Michael, M.M., Frey, B., May, C., Williams, D., Le, H.Q.: Robust architectural support for transactional memory in the power architecture. In: ISCA, pp. 225–236 (2013)

9. Demmer, M.J., Herlihy, M.P.: The arrow distributed directory protocol. In: Kutten, S. (ed.) DISC 1998. LNCS, vol. 1499, pp. 119–133. Springer, Heidelberg (1998). https://doi.org/10.1007/BFb0056478

10. Fung, W.W.L., Singh, I., Brownsword, A., Aamodt, T.M.: Hardware transactional memory for GPU architectures. In: MICRO, pp. 296–307 (2011)

11. Gonzalez-Mesa, M.A., Gutiérrez, E., Zapata, E.L., Plata, O.G.: Effective transactional memory execution management for improved concurrency. ACM Trans. Archit. Code Optim. **11**(3), 1–27 (2014)

12. Gouveia, L.E.N., Magnanti, T.L.: Network flow models for designing diameter-constrained minimum-spanning and Steiner trees. Networks **41**(3), 159–173 (2003)

13. Gramoli, V., Guerraoui, R., Trigonakis, V.: Tm$^2$c: a software transactional memory for many-cores. Distrib. Comput. **31**(5), 367–388 (2018). https://doi.org/10.1007/s00446-017-0310-6

14. Haring, R., et al.: The IBM Blue Gene/Q compute chip. IEEE Micro **32**(2), 48–60 (2012)

15. Hendler, D., Naiman, A., Peluso, S., Quaglia, F., Romano, P., Suissa, A.: Exploiting locality in lease-based replicated transactional memory via task migration. In: Afek, Y. (ed.) DISC 2013. LNCS, vol. 8205, pp. 121–133. Springer, Heidelberg (2013). https://doi.org/10.1007/978-3-642-41527-2_9

16. Herlihy, M., Moss, J.E.B.: Transactional memory: architectural support for lock-free data structures. In: ISCA, pp. 289–300 (1993)

17. Herlihy, M., Sun, Y.: Distributed transactional memory for metric-space networks. Distrib. Comput. **20**(3), 195–208 (2007). https://doi.org/10.1007/s00446-007-0037-x

18. Hiromitsu, T., Akira, M.: An approximate solution for the Steiner problem in graphs. Math. Jap. Jpn. DA **24**(6), 573–577 (1980). BIBL. 9 REF. (1980)

19. Hirve, S., Palmieri, R., Ravindran, B.: Archie: a speculative replicated transactional system. In: Middleware, pp. 265–276 (2014)

20. Hwang, F.K.: On Steiner minimal trees with rectilinear distance. SIAM J. Appl. Math. **30**(1), 104–114 (1976). https://www.jstor.org/stable/2100587

21. Hwang, F., Richards, D., Winter, P.: The Steiner Tree Problem. Elsevier Science (1992)

22. Intel (2012). https://www.software.intel.com/en-us/blogs/2012/02/07/transactional-synchronization-in-haswell

23. Lamport, L.: The part-time parliament. ACM Trans. Comput. Syst. **16**(2), 133–169 (1998)

24. Manassiev, K., Mihailescu, M., Amza, C.: Exploiting distributed version concurrency in a transactional memory cluster. In: PPoPP, pp. 198–208 (2006)

25. Minh, C.C., Chung, J., Kozyrakis, C., Olukotun, K.: STAMP: Stanford transactional applications for multi-processing. In: IISWC, pp. 35–46 (2008)

26. Mohamedin, M., Peluso, S., Kishi, M.J., Hassan, A., Palmieri, R.: Nemo: NUMA-aware concurrency control for scalable transactional memory. In: ICPP, pp. 38:1–38:10 (2018)

27. Nakaike, T., Odaira, R., Gaudet, M., Michael, M.M., Tomari, H.: Quantitative comparison of hardware transactional memory for Blue Gene/Q, zEnterprise EC12, Intel Core, and POWER8. In: ISCA, pp. 144–157 (2015)
28. Palmieri, R., Peluso, S., Ravindran, B.: Transaction execution models in partially replicated transactional memory: the case for data-flow and control-flow. In: Guerraoui, R., Romano, P. (eds.) Transactional Memory. Foundations, Algorithms, Tools, and Applications. LNCS, vol. 8913, pp. 341–366. Springer, Cham (2015). https://doi.org/10.1007/978-3-319-14720-8_16
29. Poudel, P., Rai, S., Sharma, G.: Processing distributed transactions in a predefined order. In: ICDCN, pp. 215–224. ACM (2021)
30. Rai, S., Sharma, G., Busch, C., Herlihy, M.: Load balanced distributed directories. Inf. Comput. **285**, 104700 (2021)
31. Saad, M.M., Kishi, M.J., Jing, S., Hans, S., Palmieri, R.: Processing transactions in a predefined order. In: PPOPP, pp. 120–132 (2019)
32. Saad, M.M., Palmieri, R., Ravindran, B.: Lerna: parallelizing dependent loops using speculation. In: SYSTOR, pp. 37–48 (2018)
33. Saad, M.M., Ravindran, B.: Snake: control flow distributed software transactional memory. In: Défago, X., Petit, F., Villain, V. (eds.) SSS 2011. LNCS, vol. 6976, pp. 238–252. Springer, Heidelberg (2011). https://doi.org/10.1007/978-3-642-24550-3_19
34. Sharma, G., Busch, C.: A competitive analysis for balanced transactional memory workloads. Algorithmica **63**(1–2), 296–322 (2012). https://doi.org/10.1007/s00453-011-9532-3
35. Sharma, G., Busch, C.: Distributed transactional memory for general networks. Distrib. Comput. **27**(5), 329–362 (2014). https://doi.org/10.1007/s00446-014-0214-7
36. Sharma, G., Busch, C.: A load balanced directory for distributed shared memory objects. J. Parallel Distrib. Comput. **78**, 6–24 (2015)
37. Shavit, N., Touitou, D.: Software transactional memory. Distrib. Comput. **10**(2), 99–116 (1997)
38. Turcu, A., Ravindran, B., Palmieri, R.: Hyflow2: a high performance distributed transactional memory framework in Scala. In: PPPJ, pp. 79–88 (2013)
39. Yoo, R.M., Lee, H.S.: Adaptive transaction scheduling for transactional memory systems. In: SPAA, pp. 169–178 (2008)
40. Zhang, B., Ravindran, B., Palmieri, R.: Distributed transactional contention management as the traveling salesman problem. In: Halldórsson, M.M. (ed.) SIROCCO 2014. LNCS, vol. 8576, pp. 54–67. Springer, Cham (2014). https://doi.org/10.1007/978-3-319-09620-9_6

# Fault-Tolerant Graph Realizations in the Congested Clique, Revisited

Manish Kumar$^{(\boxtimes)}$ ID

Indian Statistical Institute, Kolkata, India
manishsky27@gmail.com

**Abstract.** We study the graph realization problem in the Congested Clique in a distributed network under the crash-fault model. We focus on the *degree-sequence realization*, each node $v$ is associated with a degree value $d(v)$, and the resulting degree sequence is realizable if it is possible to construct an overlay network with the given degrees. This paper focuses on the message and round complexity of deterministic graph realization in the anonymous network. It has been shown by Kumar et al. [ALGOSENSORS 2022] that the graph realization can be solved using $O(n^2)$ message and $O(f)$ round without the knowledge of $f$, of which $f$ nodes could be faulty ($f < n$). However, their algorithm works for $KT_1$ (Knowledge Till 1 hop) model where nodes know their neighbors' IDs; or in the $KT_0$ (Knowledge Till 0 hop) model, in which each node knows the IDs of all the nodes in the clique, but doesn't know which port is connecting to which node-ID. In this paper, we extend the result to $KT_0$ when the network is anonymous, i.e., the IDs of the neighboring nodes are unknown. We present an algorithm that solves the graph realization problem in the $KT_0$ model with matching performance guarantees as in the $KT_1$ model.

**Keywords:** Graph realizations · Congested-Clique · Distributed algorithm · Fault-tolerant algorithm · Crash fault · Time complexity · Message complexity

## 1 Introduction and Related Work

*Graph Realization* is one of the fundamental problems in distributed computation that has been explored recently. It has been extensively studied in the literature of the sequential setting for over half a century. Generally, graph realization problems involve constructing a graph that attains certain properties. The area is mostly focused on realizing graphs with specified degrees [15,26,27], some other properties like connectivity [18–20], flow [25] and eccentricity [9,36] have also been studied.

*Degree sequence* realization is the most elevated graph realization problem. Mainly, a sequence of degree $D = (d_1, d_2, \ldots, d_n)$ with non-negative numbers is known as realizable if there exists a graph of $n$ nodes whose sequence of degree matches the given degree sequence $D$. In 1960 Erdös and Gallai [15], for the first time established the complete characterization of the problem. They showed that $D$ is realizable if and only if $\sum_{i=1}^{k} d_i \leq k(k-1) + \sum_{i=k+1}^{n} \min(d_i, k)$ for every $k \in [1, n]$. Later in 1962, Havel and Hakimi [26,27] found the definitive solution independently. They gave a

© The Author(s), under exclusive license to Springer Nature Switzerland AG 2023
A. R. Molla et al. (Eds.): ICDCIT 2023, LNCS 13776, pp. 84–97, 2023.
https://doi.org/10.1007/978-3-031-24848-1_6

recursive algorithm that can determine whether a given $D$ is realizable and compute a realizing graph when it is realizable (see Sect. 3 for more detail).

In recent past, Augustine et al. [3] studied the graph realization problem in distributed fault-free networks for the first time. They built the overlay networks by adapting graph realization algorithms. For a given network $V = \{v_1, \ldots, v_n\}$ of $n$ nodes, where each $v_i$ is assigned a degree $d_i$. They created an overlay graph $G(V, E)$ such that $d(v_i) = d_i$ in $G$, and for any edge $e \in E$, at least one of $e$'s end points knows the existence of $e$. This is known as *implicit realization*. Notice that, in implicit realization for any edge, at least one end point must be aware of its existence. Therefore, a node may not fully realize the graph. It is crucial to know the entire overlay graph in most of the P2P overlay applications. This setting is known as explicit realization. In this, both endpoints of any edge in the realized graph are aware of the edge which is a special case of implicit realization.

The problem is easy if there are no node failures. However, the problem becomes more difficult if some nodes in the network are faulty. We extend our work in the faulty setting (crash fault) where nodes may crash at any time during the communication with each other or before/after communication takes place. Fault-tolerant computation has always been a popular area of research in distributed computation. It is becoming more popular with the prevalence of P2P networks that encourage high decentralization. Often such researches are focused on maintaining connectivity [6], recovery, or ensuring the resilience of the network, i.e., the number of faults a network can tolerate [44]. In the area of P2P overlays, a lot of research exists to create overlays that provide structure and stability. These are best captured by overlays such as Chord [43], CAN [42], and Skip Graphs [2]. Overlays are specifically designed to tolerate faults [5,17]. For more details on P2P overlays and their properties, interested readers may follow the surveys [38,39].

Our work is focused on Congested Clique in the distributed network which was introduced by Lokter et al., [37]. It is well-studied, in both the faulty and non-faulty settings [12,40]. The Congested Clique model is explored widely for many fundamental distributed problems [7,8,11,14,22,23,28,29,32,33]. Distributed networks are faulty by nature. There are various faulty models such as clean crash, crash fault, omission failure, Byzantine failure, etc. In all these node failure settings, the two fundamental problems of agreement and leader election are studied extensively [1,4,10,13,16,21, 24,30,31,34]. The main challenge in a faulty setting is to ensure that all the nodes have the same view across the network. Faulty nodes may lead to an increase in the number of rounds or messages or both to ensure the correctness of the protocol. Thus, in our work, we focus on developing an algorithm to minimize the time complexity and the message complexity, simultaneously, in the Congested Clique with faulty nodes.

**Our Contributions and Comparison with Previous Work.** In [35], Kumar et al. showed that if the nodes have knowledge of their neighboring nodes (known as $KT_1$ model) then at least $\Omega(n^2)$ messages and $\Omega(f)$ rounds are required for any deterministic fault-tolerant graph realization algorithm, where $f$ is the number of faults such that $f < n$. The lower bound is essentially tight, simultaneously. They provided a deterministic algorithm for graph realization that uses only $O(n^2)$ messages and $O(f)$ rounds. Overall, their algorithm works in the following way. In the first phase, which consists of only two rounds, each node sends the message containing its ID and input degree twice (in two consecutive rounds). Based on the frequency of the received messages, each node locally creates an initial faulty list and a degree sequence. In the second phase, through sharing the information received in the first phase the same final degree sequence is generated, efficiently, for all non-crashed nodes. The non-crashed nodes then realize the overlay graph using a final degree sequence via the Havel-Hakimi algorithm. This method is not readily applicable to the $KT_0$ model in which the IDs of the neighboring nodes are unknown. In real-life applications, the nodes may not be aware of the neighbors initially. In this paper, we study the graph realization problem in $KT_0$ model and answer one of the questions asked in the paper [35] - "is it possible to remove the assumption of known IDs, that is, would it be possible to achieve optimal round and message complexity if the IDs of the clique nodes are not known?" We show that even if the model is $KT_0$, $O(n^2)$ messages and $O(f)$ rounds bound is still achievable. We also prove that these bounds are optimal, simultaneously. While our algorithm is simple and the idea is similar to [35], the analysis is involved. In particular, we show the following main results.

**Theorem 1.** *Consider an n-node Congested Clique with $KT_0$ model, in which $f < n$ nodes may crash arbitrarily at any time. Given an arbitrary n-length degree sequence $D = (d_1, d_2, \ldots, d_n)$ as an input such that each $d_i$ is only known to one node in the clique. Then there exists a deterministic algorithm that solves the fault-tolerant graph realization problem in $O(f)$ rounds and using $O(n^2)$ messages such that $f$ is unknown to the network.*

**Theorem 2.** *Any algorithm that solves the distributed graph realization with $f$ crash failures in an n-node Congested Clique requires $\Omega(f)$ rounds in some admissible execution.*

**Theorem 3.** *In the CONGEST model, any algorithm that solves the distributed graph realization problem of n node network (with or without faults) requires $\Omega(n^2)$ messages in some admissible execution.*

**Paper Organization:** Section 2 presents the model and definitions. In Sect. 3, we provide a brief description of the sequential Havel-Hakimi solution for the graph realization problem. Section 4 contains the fault-tolerant graph realization problem and matching lower bounds. Finally, we conclude with some interesting problems in Sect. 5.

## 2    Model and Definitions

We consider the distributed computing model used in [35] except the network is anonymous in which a node does not know the IDs of its neighbors initially. This model is known as $KT_0$ (knowledge till 0 hop). On the other hand, [35] used the $KT_1$ (knowledge till 1 hop) model, in which nodes know their neighbors [41]. The underlying network is a Congested Clique [33,37]. The Congested Clique consists of $n$ nodes which possess a unique ID of size $O(\log n)$. These nodes are directly connected to each other via a communication link and communicate with each other by sending the message of size $O(\log n)$. This is known as CONGEST communication model [41]. Communication among the nodes occurs in synchronous rounds.

A network is called $f$-resilient if there are at most $f$ nodes which may fail by crashing. We assume that up to $f < n$ nodes may fail by crashing and $f$ is unknown to the network. A faulty node may crash in any round during the execution of the algorithm. If a node crashes in a round, then an arbitrary subset of its messages may not reach the destination in that particular round, which is determined by an adversary. The crashed node does not participate in further communication throughout the execution of the algorithm. If a node does not crash in a particular round then all the messages sent in that round are delivered to the destination. We consider the *adaptive* adversary which controls the faulty nodes and selects the faulty nodes during the execution of the algorithm. The adversary decides when and how the nodes would crash.[1]

The message complexity of the algorithm is the total number of messages exchanged in the network throughout the execution. The round complexity is the number of rounds taken by the algorithm to execute.

**Definition 1 (Distributed Graph Realization [3]).** *Let us have a set of nodes $V = \{v_1, \ldots, v_n\}$ in the network such that each $v_i$ only knows the corresponding degree $d_i$ for the input degree sequence $D = (d_1, d_2, \ldots, d_n)$. The distributed degree realization problem holds if the set of nodes $V$ constructs the degree sequence $D$ of $G$ same across the network and each $v_i$ possess the corresponding degree $d_i$, $\forall i \in \{1, \ldots, n\}$. Thus, a solution outputs the degree sequence $D$ if the graph is a realizable degree sequence; otherwise, output* unrealizable.

**Definition 2 (Distributed Graph Realization with Faults [35]).** *Let $V = \{v_1, \ldots, v_n\}$ be the set of nodes in the network such that each $v_i$ only knows the corresponding degree $d_i$ for the input degree sequence $D = (d_1, d_2, \ldots, d_n)$. The arbitrary subset of faulty nodes ($F \subset V$) in the network is such that $|F| < n$. $D'$ is the modified degree sequence after losing the degrees of some faulty nodes and $G'$ is the corresponding overlay graph over $D'$. This problem requires those* non-faulty *nodes in $V$ to construct a graph realization of $D'$ such that the resulting overlay graph $G'$ holds the following conditions:*

1. *$D' \subseteq D$.*
2. *$D' \geq n - |F|$.*
3. *$D - D'$ is the set of degrees of the nodes which are crashed, and their degree is unknown to the non-faulty nodes.*
4. *For any edge $e = (u, v) \in G'$, both $u$ and $v$ know the existence of $e$.*

*The required output is an overlay graph $G'$ if $D'$ is realizable; otherwise, the output is* unrealizable.

## 3 Preliminary: The Havel-Hakimi Algorithm

We concisely state the sequential Graph Realization problems that inspired our fault-tolerant distributed version of the same problem. The basic premises of the problem

---

[1] In contrast, a non-adaptive (static) adversary selects the nodes to be faulty before the execution of the algorithm starts.

are as follows: given a particular degree sequence $D = (d_1, d_2 \ldots d_n)$, is it possible to generate a graph $G$ whose degree sequence is $D$? The most well-known characterization is given independently by Havel [27] and Hakimi [26], which can be stated briefly as follows.

**Theorem 4 (Based on [26] and [27]).**  *A non-increasing sequence $D = (d_1, d_2, ..., d_n)$ is graphic (i.e., graph is realizable) if and only if the sequence $D' = (d'_2, ..., d'_n)$ is graphic, where $d'_j = d_j - 1$, for $j \in [2, d_1 + 1]$, and $d'_j = d_j$, for $j \in [d_1 + 2, n]$*

This characterization directly implies a $O\left(\sum_{i=1}^{n} d_i\right)$ time (in terms of number of edges) sequential algorithm, known as the Havel-Hakimi algorithm, for constructing a realizing graph $G = (V, E)$ where $V = \{v_1, ...v_n\}$ and $d(v_i) = d_i$, or deciding that no such graph exists. The algorithm works as follows. Initialize $G = (V, E)$ to be an empty graph on $V$. For $i = 1$ to $n$, in step $i$ do the following:

1. Sort the remaining degree sequence in non-increasing order ($d_i \geq d_{i-1} \geq \ldots d_n$).
2. Remove $d_i$ from $D$, and set $d_j = d_j - 1$ for all $j \in [i + 1, d_i + i + 1]$.
3. Set the neighborhood of the node $v_i$ to be $\{v_{i+1}, v_{i+2}, ...v_{i+1+d_i}\}$.

At any step, if $D$ contains a negative entry then the sequence is not realizable.

# 4   Graph Realization with Faults

Recall that we are given an $n$-node Congested Clique anonymous network, i.e., nodes do not know each other's communication link with respect to their IDs, in which at most $f < n$ nodes may crash arbitrarily at any time. A degree-sequence $(d_1, d_2, \ldots, d_n)$ is also given to all the nodes in such a way that each $d_i$ is known to only one node. In this section, we present an algorithm that guarantees that (i) all the non-crashed nodes learn and recreate a list whose size is at least $n - f$, and (ii) this list is the same for all the nodes. This allows the non-crashed nodes to locally realize the overlay graph with the help of Havel-and-Hakimi's algorithm, described in Sect. 3. Our algorithm does not require any knowledge of the number of faulty nodes and IDs of neighboring nodes in the CONGEST model. We use only a few numbers of nodes to propagate the information about the crashed nodes to the other nodes in the network which help to minimize round and message complexity.

## 4.1   Algorithm

The challenging part in designing an efficient algorithm is handling the faulty nodes. A faulty node may crash in some rounds and its message may not reach all the destination nodes in that round. Thus, there might be two sets of nodes with different degree-sequence. This may lead to an incorrect graph realization whose realization is not the same throughout the network. One of the simple ways to solve this problem is as follows: every node $u_i$ sends its ID-degree pair $\langle u_i, d(u_i) \rangle$ and whatever new degree-ID pair received as a message from other nodes to all the other nodes. Since there are

$n$ nodes, therefore, each node sends $n$ messages to the $n$ nodes. In this way, we have message complexity $O(n^3)$, and round complexity $O(n)$ when $f$ is unknown to the network. Our main aim is to get the optimal round complexity, i.e., $O(f)$ by keeping the message complexity as small as possible.

The idea is to run the algorithm in three phases, the initialization phase, performance phase, and finalization phase. In the initialization phase which consists of only two rounds, in which, each node sends the message as its ID-degree pair twice to other nodes based on the received frequency of messages which can be one or two. The receiver node maintains three lists, faulty-list, final-list, and non-faulty-list. In the performance phase, by sharing the degree of the faulty-list and final-list nodes update the non-faulty-list, at last non-faulty-list becomes the same across the network (for all non-crashed nodes). In the finalization phase, the network realizes the overlay network based on the received degree-sequence by Havel-and-Havkimi's algorithm.

We will now explain the algorithm in detail whose complete pseudocode is given in Algorithm 1 and terminologies are summarized along with their definition in Table 1.

**Initialization Phase:** This phase consists of two rounds. For the first two rounds, each node $u_i$ broadcasts a message which contains its ID along with its degree, i.e., $\langle u_i, d(u_i) \rangle$ to all the nodes. Each node $u_i$ maintains three lists in sorted order (ascending, based on ID), faulty-list ($F_{u_i}$), final-list ($L_{u_i}$) and non-faulty-list ($D'_{u_i}$) based on the frequency of received messages. Faulty-list ($F_{u_i}$) maintains the messages (ID and corresponding degree) which were received once, i.e., a list of those nodes which have surely crashed during the initialization phase. On the other hand, final-list ($L_{u_i}$) maintains the messages (ID and corresponding degree) which were received twice, i.e., a list of those nodes which have not crashed according to $u_i$. Notice that as $u_i$ knows the corresponding link of the received messages included in the final-list, therefore, $u_i$ can communicate to those particular nodes (if required). Final-list is final in the sense that no new message will be included in the final-list throughout the execution of the algorithm. Non-faulty-list ($D'_{u_i}$) possesses the messages (ID-degree pair) which were received twice, the degree of the list will be used for the graph realization problem, at last, by using Havel-Hakimi's algorithm. This non-faulty-list will be updated by only including more messages. Each node $u_i$ keeps its message in $L_{u_i}$ and $D'_{u_i}$ since it has not crashed for the first two rounds. Notice that no message will be removed from the non-faulty-list during the execution of the algorithm. Only faulty-list perform the operation include/remove during the execution of the algorithm. Each $u_i$ performs sorting operations in all the three lists, after an update, in their respective list w.r.t. ID in ascending order.

**Performance Phase:** In this phase, nodes send/receive the messages based on the received messages and the status of the $F_u$, $L_u$, and $D'_u$. We study the algorithm from the perspective of some node $u_i$. We consider two states of the node $u_i$: (i) active state and (ii) standby state. In the case of an active state, node $u_i$ which possesses the minimum ID among the non-crashed nodes broadcast the message to all other nodes. While in the case of standby state, non-crashed node $u_i$ sends the messages to a particular node based on the condition that arises. This is done by performing the following steps iteratively.

---

**Algorithm 1** FAULT-TOLERANT $KT_0$ AGREEMENT: CODE FOR A NODE $u_i$.

---

**Input:** A complete $n$ nodes $U = \{u_1, u_2, \ldots, u_n\}$ anonymous network with unique ID. Each node $u_i$ possess a degree of $d(u_i)$ of size $O(\log n)$.

**Output:** A corresponding graph realization that satisfies the condition of distributed graph realization with faults.

  INITIALIZATION PHASE:
1: For the first two rounds, each node $u_i$ broadcasts a message which contains its ID along with its degree, i.e., $\langle u_i, d(u_i) \rangle$ to all the nodes.
2: Based on the frequency (once or twice) of received messages, each node $u_i$ maintain the faulty-list $F_{u_i}$, final-list $L_{u_i}$ and non-faulty-list $D'_{u_i}$ with received IDs and the corresponding degrees $d(ID)$. If $u_i$ received the message once, then moves the message into $F_{u_i}$ otherwise moves into $L_{u_i}$ and $D'_{u_i}$.                    ▷ Message contains $\langle ID, d(ID) \rangle$
3: Each node $u_i$ keeps its message in $L_{u_i}$ and $D'_{u_i}$.
4: For a message $s$, $s^{ID}$ is the ID of the message $s$. $min(L_{u_i})^{ID}$ and $min(L_{u_i})^{d(ID)}$ is the minimum ID and corresponding degree in the list $L_{u_i}$, respectively. $F_{u_i}$, $L_{u_i}$ and $D'_{u_i}$ remain sorted whenever an include/remove operation takes place (in the sorted order w.r.t. their ID).
  PERFORMANCE PHASE:
5: Perform performance-phase $(u_i)$.
  FINALIZATION PHASE:
6: Every non-crashed node $u_i$'s non-faulty-list $D'_{u_i}$ have the same view of the degree sequence $D'$. Therefore, graph realized by all the non-crashed nodes remain same by using the Havel-Hakimi's algorithm.

---

Active State: Node $u_i$ reaches in active state if $u_i$ has received the message from some node $u_j$ whose minimum ID's entry in the $L_{u_j}$ is $u_i$ or $L_{u_i}$ has the minimum ID as $u_i$ then $u_i$ checks its $F_{u_i}$. There can be two conditions with $F_{u_i}$ either $F_{u_i}$ is empty or not. In case, if $F_{u_i}$ is empty then $u_i$ asks all other nodes to send messages from their respective faulty-list (if any). If $F_{u_i}$ is empty (did not receive the message in the next round) then $u_i$ sends the message "agreed" to all other nodes and moves to the finalization phase. On the other hand, if $u_i$ received some messages from this call then $u_i$ includes those messages in the $F_{u_i}$. On the assumption, If node $u_i$ has not moved to the finalization phase then there might have arisen a second condition, i.e., $F_{u_i}$ is non-empty. In that event, $u_i$ sends the message from $F_{u_i}$ twice and asks whether they have any message in their faulty-list whose ID is less than what they just received. If other nodes have such messages in their respective faulty-list then they send all these messages one by one to $u_i$. $u_i$ includes all these coming messages into $F_{u_i}$ and sends twice to all other nodes. In between, all other nodes send their appropriate messages from faulty-list (messages whose IDs are less than what they just received, only once to $u_i$) only if they are receiving the messages from $u_i$. In case, $u_i$ is not sending the message in some rounds which signifies that $u_i$ has crashed, therefore, there is no need to send the message to $u_i$. If $F_{u_i}$ becomes empty at last and $u_i$ has not crashed then $u_i$ sends the "agreed" message to all other nodes and moves to the finalization phase.

---

**Algorithm 2** PERFORMANCE-PHASE $(u_i)$

---

1: **while** $L_{u_i} \neq \phi$ **do**

 ACTIVE STATE:

2:    **if** $min(L_{u_i})^{ID} = u_i$, or $\exists u_j$ such that $min(L_{u_j})^{ID} = u_i$ **then**

3:       **if** $F_{u_i} = \phi$ **then**

4:          $u_i$ sends $F_{u_i} = \phi$ to all other nodes.

5:          **if** $u_i$ receives some message, say $s$, after sending $F_{u_i} = \phi$ **then**

6:             $u_i$ includes the unique $s$ (w.r.t. their ID) into $F_{u_i}$.

7:          **end if**

8:       **end if**

9:       **while** $F_{u_i} \neq \phi$ **do**

10:          $u_i$ broadcasts the message $\langle min(F_{u_i})^{u_i}, min(F_{u_i})^{d(u_i)} \rangle$ twice to all other nodes. $u_i$ removes the sent message from $F_{u_i}$ and moves the message into $D'_{u_i}$.

11:          **if** $u_i$ receives the message $s$ from $u_j$ such that $s^{ID} < min(F_{u_i})^{ID}$ **then**

12:             $u_i$ includes those messages in the $F_{u_i}$.

13:          **end if**

14:       **end while**

15:       $u_i$ sends the "agreed" message to all other nodes and moves to *finalization phase*.

16:    **end if**

 STANDBY STATE:

17:    **if** $min(L_{u_i})^{ID} = u_j$ and $u_i$ did not receive any message in last round **then**

18:       $u_i$ sends $min(L_{u_i})^{ID}$ to $u_j$ and removes the $u_j$ from $L_{u_i}$.

19:    **end if**

20:    **if** $u_i$ received the message from $u_j$ such that $F_{u_j} = \phi$ and $F_{u_i} \neq \phi$ **then**

21:       $u_i$ sends all messages from $F_{u_i}$ to $u_j$, one-by-one, till receives the messages from $u_j$.                                          ▷ $u_i$ does not remove any ID from $F_{u_i}$

22:    **end if**

23:    **if** $u_i$ received the message from $u_j$, say $s$, such that $s^{ID} > min(F_{u_i})^{ID}$ **then**

24:       $u_i$ sends all messages whose $s^{ID} > min(F_{u_i})^{ID}$ to $u_j$, one-by-one, till receives the messages from $u_j$.                                ▷ $u_i$ does not remove any ID from $F_{u_i}$

25:    **end if**

26:    **if** $u_i$ receives the message from $u_j$ such that $u_j > min(L_{u_i})^{ID}$ **then**

27:       $u_i$ removes all such $min(L_{u_i})^{ID}$, iteratively.

28:    **end if**

29:    **if** $u_i$ receives the message from $u_j$ once, say $s$, and $s \notin F_{u_i}$ **then**

30:       $u_i$ includes $s$ into $F_{u_i}$.

31:    **end if**

32:    **if** $u_i$ receives the message from $u_j$ twice, say $s$, and $s \in F_{u_i}$ **then**

33:       $u_i$ removes $s$ from $F_{u_i}$.

34:       **if** $s \notin D'_{u_i}$ **then**

35:          $u_i$ includes $s$ in $D'_{u_i}$.

36:       **end if**

37:    **end if**

38:    **if** $u_i$ receives the message "agreed" from any node $u_j$ **then**

39:       $u_i$ sends the message "agreed" to all other nodes and moves to *finalization phase*.

40:    **end if**

41: **end while**

---

**Table 1.** Terminology and their definition used throughout the Algorithms 1

Terminology at a Glance for a node $u_i$

| Terminology | Definition |
| --- | --- |
| Faulty-list ($F_{u_i}$) | List of known faulty IDs (and their corresponding degrees) at $u_i$ |
| Final-list ($L_{u_i}$) | List of IDs (and their corresponding degrees) which sent their messages in first two rounds at $u_i$ and have not crashed as per $u_i$ |
| Non-faulty-list ($D_{u_i}$) | List of IDs (and their corresponding degrees) which were received/sent twice from/to $u_i$. In *finalization phase*, contains the final degree-sequence used for graph realization at $u_i$ |
| Standby State | A node is in the Standby state if it is waiting for its turn (to broadcast entries from its faulty-list) or to receive an "agreed" message |
| Active State | A node is in the active state if it is broadcasting from its faulty-list |
| "agreed" | Node $u_i$ received "agreed" message then $u_i$ becomes agree for the *finalization phase*. $u_i$ broadcasts the "agreed" message and moves to the *finalization phase* |

Standby State: In this state, node $u_i$ communicates to the particular node, say $u_j$. For a node $u_i$ when minimum ID of $L_{u_i}$ is $u_j$ and $u_i$ did not receive any message in the previous round then $u_i$ asks $u_j$ for initiation and removes the $u_j$ from $L_{u_i}$. There is a possibility that $u_j$ has not crashed since it sent the message twice in the initialization phase. In case, $F_{u_i} \neq \phi$ and $u_i$ receives the messages such that $F_{u_j} = \phi$ then $u_i$ sends all messages from ($F_{u_i}$) to $u_j$, one-by-one, till receives the message from $u_j$. Also, if $u_i$ received the message from $u_j$ whose ID is greater than minimum of $F_{u_i}$'s ID or then $u_i$ sends all those message whose ID is greater than ($F_{u_i}$) to $u_j$, one-by-one, till receives the message from $u_j$. So that $u_j$ can convey those messages (which were not received by all the nodes) to all other nodes. Notice that $u_i$ does not remove any of these IDs from $F_{u_i}$ since there might be the case that $u_j$ crashes without conveying the message. Therefore, $u_i$ removes only those messages from $F_{u_i}$ which are received twice by the $u_i$. On the other hand, if $u_i$ receives the message from $u_j$ where $u_j$ is greater than the minimum of $L_{u_i}$ then $u_i$ removes all such messages from $L_{u_i}$, iteratively. Since those nodes have already crashed (otherwise $u_j$ has not been active to send the message) and there is no need to communicate with them in coming rounds (in case, $u_j$ crashed). Supposing $u_i$ receives the message from $u_j$ once, say $s$, and $s \notin F_{u_i}$ then $u_i$ includes $s$ into $F_{u_i}$. In case, if $u_i$ receives the message from $u_j$ twice and $s \in F_{u_i}$ then $u_i$ removes $s$ from $F_{u_i}$. If $s \notin D'_{u_i}$ and received twice then $u_i$ includes $s$ in $D'_{u_i}$. Node $u_i$ moves to the finalization phase of the algorithm if $u_i$ has received the "agreed" message from any node $u_j$. In that situation, $u_i$ sends the message "agreed" to all the nodes and moves to the finalization phase of the algorithm. The "agreed" message conveys the information to other nodes that there does not exist any node which possesses a non-empty faulty-list or a different view of non-faulty-list.

**Finalization Phase:** In this phase, each non-crashed node $u_i$'s non-faulty-list $D'_{u_i}$ has the same view of degree sequence ($D'$) throughout the network. Therefore, the graph realized by all the non-crashed nodes remains the same by using the Havel-Hakimi's Algorithm 3.

The above three phases are performed till the nodes receive the message "agreed" and the algorithm terminates. In the end, all the non-faulty nodes have the same degree sequence in their non-faulty-list, which they realize by Havel-Hakimi's algorithm.

We will now show the correctness of the algorithm with the help of Lemma 1 that the final degree sequence $D'$ of all the non-crashed nodes are the same. Thus, the algorithm correctly solves the distributed graph realization with faults in the Congested Clique in the $KT_0$ model. Further, we analyze the time and the message complexity by using the Lemma 2, Lemma 3 and Lemma 4.

**Lemma 1.** *If there exists some non-faulty nodes $u_i$ and $u_j$ such that $D'_{u_i} - D'_{u_j} \neq \phi$ at some point, then in the finalization phase of the algorithm there exist $D'_{u_i} - D'_{u_j} = \phi$.*

*Proof.* Let us suppose at some point there exists some message $s$ such that $D'_{u_i} - D'_{u_j} = s$. This implies that the sender of $s$ crashed in the second round such that $u_i$ received the message twice but not $u_j$. Now, if non-faulty node $u_j$ (or some node which has $s$ in faulty-list) becomes the minimum ID in its $L_{u_j}$ or some other node's final-list (becomes active node[2]) then $u_j$ will eventually send the $s$ to all other nodes twice. Therefore, $s$ will be part of $D'$ across the network. Similarly, if some node $u_i$ (or some node which does not possess $s$ in faulty-list) becomes active node then $u_i$ might send higher ID message to $u_j$ or send $F_{u_i} = \phi$ to $u_j$ and asks about faulty value. In that case $u_j$ sends the $s$ and $u_i$ broadcasts the $s$ to all other nodes. Therefore, $s$ becomes the part of $D'$ across the network. Hence, the lemma.    □

**Lemma 2.** *In a non-faulty setup[3], round complexity is $O(1)$ and message complexity is $O(n^2)$.*

*Proof.* In the *initialization phase*, all the nodes send their respective ID and corresponding degree twice successfully. Therefore, the faulty-list across the network remains empty. In the *performance phase*, the node with minimum ID, say $u_i$, asks all the nodes about the status of their faulty-list and waits for a round. In parallel, other nodes are also asking $u_i$ to be active. Further, after waiting for a round $u_i$ broadcasts the "agreed" message and reaches in the *finalization phase*. In the very next round, other nodes also reach *finalization phase*. In this scenario, the algorithm is executed in constant rounds and each round takes $O(n^2)$ a message. Hence, the lemma.    □

**Lemma 3.** *All the faulty-nodes $f$ cost $O(nf)$ messages and $O(f)$ rounds extra as compared to non-faulty setup.*

*Proof.* A faulty node $u_i$ may not follow the protocol during the crash. It may deviate in, mainly, two phases: (i) *initialization phase* or (ii) *performance phase*. In the initialization phase, node $u_i$ may crash during the broadcasts of the message. If it crashed in the first round or second round then some nodes possess node $u_i$ in their faulty-list. During the *performance phase*, some non-faulty nodes broadcast this value twice from

---

[2] Node which is in active state (standby state) considered as active node (standby node).

[3] A non-faulty setup is the model in which all the nodes are non-faulty.

its faulty-list. Therefore, this faulty-node may cause 2 extra rounds as compared to non-faulty setup (Lemma 2). Further, if a faulty node crash in *performance phase* during the broadcasts of the message then there might be some nodes that do not receive the message twice which would be broadcasted by some non-faulty node again. Therefore, this faulty-node may also cause 2 extra rounds as compared to non-faulty setup (Lemma 2). From the above discussion, we can see that one faulty node can cause an extra cost of $O(1)$ rounds and $O(n)$ messages (due to broadcast). Therefore, the overall extra messages and rounds cost of the algorithm for $f$ faulty nodes are $O(nf)$ and $O(f)$, respectively. □

**Lemma 4.** *The time and message complexity of the Algorithm 1 is $O(f)$ and $O(n^2)$, respectively.*

*Proof.* From the Lemma 2, in non-faulty setup, we have the round complexity $O(1)$ and message complexity $O(n^2)$. On the other hand, from the Lemma 3, we have the extra cost compared to non-faulty setup is $O(f)$ rounds and $O(nf)$ message. Therefore, we can conclude the round complexity is $O(1) + O(f) = O(f)$ and message complexity is $O(n^2) + O(nf) = O(n^2)$, as $f < n$. □

Thus, we get the following main result of fault-tolerant graph realization.

**Theorem 5.** *Consider an $n$-node Congested Clique with $KT_0$ model, in which $f < n$ nodes may crash arbitrarily at any time. Given an arbitrary $n$-length degree sequence $D = (d_1, d_2, \ldots, d_n)$ as an input such that each $d_i$ is only known to one node in the clique. Then there exists a deterministic algorithm that solves the fault-tolerant graph realization problem in $O(f)$ rounds and using $O(n^2)$ messages such that $f$ is unknown to the network.*

### 4.2  Lower Bound

Recall that our network is anonymous, i.e., a node does not know the IDs of its neighbors initially. This model is known as $KT_0$; on the other hand, in the $KT_1$ model, nodes know their neighbors. Thus, $KT_0$ model is a weaker model than $KT_1$ model, i.e., $KT_1$ model has some extra information regarding the neighbors' IDs as compared to $KT_0$. Therefore, the algorithm which can solve the graph realization problem in a Congested Clique in $KT_0$ model will also solve the problem in $KT_1$ model with matching complexity. By using the same line of argument, in the case of lower bound $KT_1$'s lower bound (shown in [35]) is the trivial lower bound for $KT_0$ model. Therefore, we also have the following results in $KT_0$ model.

**Theorem 6.** *Any algorithm that solves the distributed graph realization with $f$ crash failures in an $n$-node Congested Clique requires $\Omega(f)$ rounds in some admissible execution.*

**Theorem 7.** *In the CONGEST model, any algorithm that solves the distributed graph realization problem of $n$ node network (with or without faults) requires $\Omega(n^2)$ messages in some admissible execution.*

Notice that Algorithm 1 is, simultaneously, tight with respect to the time complexity and message complexity.

## 5    Conclusion and Future Work

In this paper, we studied the round and message complexity of the graph realization problem in the Congested Clique with faults in $KT_0$ model and provided an efficient algorithm for realizing overlays for a given degree sequence. Our algorithm is simultaneously optimal in both the round and the message complexity. This extends the results of Kumar et al. [35] who showed the same bounds and complexity for the $KT_1$ model. Given the relevance of graph realization techniques in overlay construction and the presence of faulty nodes in peer-to-peer networks, we believe there can be several gripping questions to explore in the future. Such as:

(1) Does the message and round complexity $O(n^2)$ and $O(f)$, respectively, hold for the omission failure? Also, what would be the nontrivial lower bound for the omission failure?
(2) It would be entrancing to define and analyze the graph realization problem in the presence of Byzantine faults. Since a Byzantine node can behave arbitrarily like sending the wrong message, sending a message to some nodes, or not sending the message in some rounds. Therefore, the graph realization problem needs to be defined carefully.
(3) Our paper and existing work on distributed graph realization consider a clique network [3,35]. It would be interesting to study the problem in general networks.

## References

1. Abraham, I., et al.: Communication complexity of Byzantine agreement, revisited. In: PODC, pp. 317–326 (2019)
2. Aspnes, J., Shah, G.: Skip graphs. ACM Trans. Algorithms **3**(4), 37–es (2007)
3. Augustine, J., Choudhary, K., Cohen, A., Peleg, D., Sivasubramaniam, S., Sourav, S.: Distributed graph realizations. IEEE Trans. Parallel Distrib. Syst. **33**(6), 1321–1337 (2022)
4. Augustine, J., Molla, A.R., Pandurangan, G.: Sublinear message bounds for randomized agreement. In: Proceedings of the ACM Symposium on Principles of Distributed Computing (PODC), pp. 315–324 (2018)
5. Augustine, J., Sivasubramaniam, S.: Spartan: a framework for sparse robust addressable networks. In: 2018 IEEE International Parallel and Distributed Processing Symposium (IPDPS), pp. 1060–1069. IEEE (2018)
6. Bagchi, A., Bhargava, A., Chaudhary, A., Eppstein, D., Scheideler, C.: The effect of faults on network expansion. Theory Comput. Syst. **39**(6), 903–928 (2006)
7. Barenboim, L., Khazanov, V.: Distributed symmetry-breaking algorithms for congested cliques. In: Fomin, F.V., Podolskii, V.V. (eds.) CSR 2018. LNCS, vol. 10846, pp. 41–52. Springer, Cham (2018). https://doi.org/10.1007/978-3-319-90530-3_5
8. Becker, F., Montealegre, P., Rapaport, I., Todinca, I.: The impact of locality on the detection of cycles in the broadcast congested clique model. In: Bender, M.A., Farach-Colton, M., Mosteiro, M.A. (eds.) LATIN 2018. LNCS, vol. 10807, pp. 134–145. Springer, Cham (2018). https://doi.org/10.1007/978-3-319-77404-6_11
9. Behzad, M., Simpson, J.E.: Eccentric sequences and eccentric sets in graphs. Discret. Math. **16**(3), 187–193 (1976)
10. Ben-Or, M., Pavlov, E., Vaikuntanathan, V.: Byzantine agreement in the full-information model in o(log n) rounds. In: Proceedings of the 38th Annual ACM Symposium on Theory of Computing, pp. 179–186. ACM (2006)

11. Censor-Hillel, K., Dory, M., Korhonen, J.H., Leitersdorf, D.: Fast approximate shortest paths in the congested clique. Distrib. Comput. **34**(6), 463–487 (2021)
12. Dolev, D., Lenzen, C., Peled, S.: "Tri, tri again": finding triangles and small subgraphs in a distributed setting. In: Aguilera, M.K. (ed.) DISC 2012. LNCS, vol. 7611, pp. 195–209. Springer, Heidelberg (2012). https://doi.org/10.1007/978-3-642-33651-5_14
13. Dolev, D., Strong, H.R.: Requirements for agreement in a distributed system. In: Proceedings of the Second International Symposium on Distributed Data Bases, pp. 115–129. North-Holland Publishing Company (1982)
14. Drucker, A., Kuhn, F., Oshman, R.: On the power of the congested clique model. In: Proceedings of the 2014 ACM Symposium on Principles of Distributed Computing, pp. 367–376 (2014)
15. Erdös, P., Gallai, T.: Graphs with prescribed degrees of vertices (in Hungarian). Mat. Lapok (N.S.) **11**, 264–274 (1960)
16. Feldman, P., Micali, S.: An optimal probabilistic protocol for synchronous Byzantine agreement. SIAM J. Comput. **26**(4), 873–933 (1997)
17. Fiat, A., Saia, J.: Censorship resistant peer-to-peer content addressable networks. In: Proceedings of the Thirteenth Annual ACM-SIAM Symposium on Discrete Algorithms, pp. 94–103. Society for Industrial and Applied Mathematics (2002)
18. Frank, A.: Augmenting graphs to meet edge-connectivity requirements. SIAM J. Discrete Math. **5**, 25–43 (1992)
19. Frank, A.: Connectivity augmentation problems in network design. In: Mathematical Programming: State of the Art, pp. 34–63. Univ. Michigan (1994)
20. Frank, H., Chou, W.: Connectivity considerations in the design of survivable networks. IEEE Trans. Circuit Theory **17**(4), 486–490 (1970)
21. Galil, Z., Mayer, A.J., Yung, M.: Resolving message complexity of Byzantine agreement and beyond. In: 36th Annual Symposium on Foundations of Computer Science, pp. 724–733. IEEE Computer Society (1995)
22. Ghaffari, M., Nowicki, K.: Congested clique algorithms for the minimum cut problem. In: Proceedings of the 2018 ACM Symposium on Principles of Distributed Computing (PODC), pp. 357–366 (2018)
23. Ghaffari, M., Parter, M.: MST in log-star rounds of congested clique. In: Proceedings of the 2016 ACM Symposium on Principles of Distributed Computing (PODC), pp. 19–28 (2016)
24. Gilbert, S., Kowalski, D.R.: Distributed agreement with optimal communication complexity. In: SODA, pp. 965–977 (2010)
25. Gomory, R., Hu, T.: Multi-terminal network flows. J. Soc. Ind. Appl. Math. **9**(4), 551–570 (1961)
26. Hakimi, S.L.: On realizability of a set of integers as degrees of the vertices of a linear graph - I. SIAM J. Appl. Math. **10**(3), 496–506 (1962)
27. Havel, V.: A remark on the existence of finite graphs. Casopis Pest. Mat. **80**, 477–480 (1955)
28. Jurdziński, T., Nowicki, K.: Connectivity and minimum cut approximation in the broadcast congested clique. In: Lotker, Z., Patt-Shamir, B. (eds.) SIROCCO 2018. LNCS, vol. 11085, pp. 331–344. Springer, Cham (2018). https://doi.org/10.1007/978-3-030-01325-7_28
29. Jurdziński, T., Nowicki, K.: MST in $O(1)$ rounds of congested clique. In: Proceedings of the Twenty-Ninth Annual ACM-SIAM Symposium on Discrete Algorithms (SODA), pp. 2620–2632 (2018)
30. Kapron, B.M., Kempe, D., King, V., Saia, J., Sanwalani, V.: Fast asynchronous Byzantine agreement and leader election with full information. In: SODA, pp. 1038–1047. SIAM (2008)
31. King, V., Saia, J.: Byzantine agreement in expected polynomial time. J. ACM **63**(2), 1–21 (2016)

32. Konrad, C.: MIS in the congested clique model in $\log \log \Delta$ rounds. arXiv preprint arXiv:1802.07647 (2018)
33. Korhonen, J.H., Suomela, J.: Towards a complexity theory for the congested clique. In: Proceedings of the 30th on Symposium on Parallelism in Algorithms and Architectures, pp. 163–172. SPAA 2018, ACM, New York (2018)
34. Kumar, M., Molla, A.R.: Brief announcement: on the message complexity of fault-tolerant computation: leader election and agreement. In: PODC 2021, pp. 259–262. ACM (2021)
35. Kumar, M., Molla, A.R., Sivasubramaniam, S.: Fault-tolerant graph realizations in the congested clique. In: ALGOSENSRS (2022). https://doi.org/10.48550/arXiv.2208.10135
36. Lesniak, L.: Eccentric sequences in graphs. Period. Math. Hung. **6**(4), 287–293 (1975). https://doi.org/10.1007/BF02017925
37. Lotker, Z., Patt-Shamir, B., Pavlov, E., Peleg, D.: Minimum-weight spanning tree construction in $o(\log \log n)$ communication rounds. SIAM J. Comput. **35**(1), 120–131 (2005)
38. Lua, E.K., Crowcroft, J., Pias, M., Sharma, R., Lim, S.: A survey and comparison of peer-to-peer overlay network schemes. IEEE Commun. Surv. Tutor. **7**(2), 72–93 (2005)
39. Malatras, A.: State-of-the-art survey on P2P overlay networks in pervasive computing environments. J. Netw. Comput. Appl. **55**, 1–23 (2015)
40. Patt-Shamir, B., Teplitsky, M.: The round complexity of distributed sorting. In: Proceedings of the 30th Annual ACM SIGACT-SIGOPS Symposium on Principles of Distributed Computing, pp. 249–256 (2011)
41. Peleg, D.: Distributed Computing: A Locality Sensitive Approach. SIAM (2000)
42. Ratnasamy, S., Francis, P., Handley, M., Karp, R.M., Shenker, S.: A scalable content-addressable network. In: SIGCOMM, pp. 161–172. ACM (2001)
43. Stoica, I., et al.: Chord: a scalable peer-to-peer lookup protocol for internet applications. IEEE ACM Trans. Netw. **11**(1), 17–32 (2003)
44. Upfal, E.: Tolerating linear number of faults in networks of bounded degree. In: Proceedings of the Eleventh Annual ACM Symposium on Principles of Distributed Computing, pp. 83–89. PODC 1992, ACM, New York (1992)

# A Perspective of IP Lookup Approach Using Graphical Processing Unit (GPU)

Veeramani Sonai[1]([✉]), Indira Bharathi[1], and Sk. Noor Mahammad[2]

[1] Department of Computer Science Engineering, School of Computing,
Amrita Vishwa Vidyapeetham, Chennai, India
{s_veeramani,b_indira}@ch.amrita.edu
[2] Department of CSE, IIITDM Kanceepuram, Chennai 600127, Tamil Nadu, India
noor@iiitdm.ac.in

**Abstract.** Due to increases in communication link capacity and growth of the Internet traffic, packet processing like IP address lookup and classification becomes a major concern in the network. The packet processing performed at Switch/Router does not cope up with the growing link speed. Since Graphics processing unit (GPU) has high parallelism and more flexibility for the programmers, it can be used for solving IP address lookup problem in the Switch/Router devices. This paper proposes a variant of trie based approach to find a solution for longest prefix match (LPM) problem using GPU. In this paper, IP address database is partitioned into a different table based on first $k$ bits of IP address, then a variant of trie approach is proposed to find the next hop. The proposed lookup approach shows 64.46% and 94.32% improvement than binary trie and BST implementations.

**Keywords:** IP lookup · GPU · CUDA · Trie · Multibit-trie

## 1 Introduction

As the packet arrival rate increases drastically, IP address lookup in the network Router/Switch become so complex. One of the main objective of Switch/Router is to find an appropriate next hop information by comparing the incoming packet destination IP address against the prefix information stored in the forwarding table. The current network architecture follows Classless Inter-domain Routing (CIDR [11]) due to the depletion of address spaces in the class-based addressing scheme. Most of the prefix length varies from 16 to 24 in the case of IPv4 [1]. Since the destination IP address of the incoming packet does not carry the length information, it has to find the bit pattern as well as the length. In such a scenario, there might be a possibility of multiple matches for a given destination IP address. In order to find the best match, a method called Longest Prefix Match (LPM) is utilized.

There are two approaches can be used for solving the IP address lookup problems namely hardware based and software based approach. Since the hardware-based [6] approach suffers from cost, power consumption, and flexibility, many

A. R. Molla et al. (Eds.): ICDCIT 2023, LNCS 13776, pp. 98–103, 2023.
https://doi.org/10.1007/978-3-031-24848-1_7

kinds of research focusing on software-based approaches. Though several variants of trie based approach are proposed to solve the problem of finding LPM [14,15,19] to further improve the speed up a GPU is utilized. GPU can offer a large number of Arithmetic Logic Unit (ALU), many threads can be executed simultaneously and each thread can run in a short duration. Because of massive parallelism in GPU, multiple lookups can be done in a short period of time. Compute Unified Device Architecture (CUDA [3]) is platform can achieve parallelism in GPU environment. This paper proposes a GPU based IP lookup approach which uses a variant of a trie-based data structure to perform IP lookup operation efficiently. The proposed approach uses independent memory for different table partitions. The proposed approach reduces the lookup time by choosing appropriate stride value.

The rest of the paper is organized as follows Sect. 2 discuss related works, Sect. 3 brief about proposed solution for GPU based IP lookup, Sect. 4 results and discussion and followed by Sect. 5 conclusions.

## 2   Related Works

A binary trie based parallel IP address lookup approach is proposed in [13]. In this paper, IP address is partitioned into different groups based on its prefix length. Since the level can be searched independently it can be provisioned separately. A shape-shifting trie based IP lookup is proposed in [10] with the worst case time complexity of $O(n)$. The author [20,21] and [22] proposes a software based approaches which uses a combination of trie and hashing approach to perform efficient IP lookup. A level-compressed trie based IP lookup approach is proposed in [8]. This approach suffers when there was no balanced structure found in the structure. An efficient binary search approach for the IP lookup proposed in [23] which takes the worst case time complexity of $O(\log n)$. The author [18] proposes an efficient virtual compression approach to compress the flow table. Recently a GPU based implementation of IP lookup is proposed in many of the research [4,7] and [17]. A radix tree based IP lookup algorithm using GPU is proposed in [7]. A packetshader which utilizes GPU multi-core facilities [4]. Amar Shekhar et al., proposes a GPU based lookup which uses binary search tree (BST) data structure to perform IP lookup operation. Though it is a low-cost solution, it suffers from update problem when the forwarding table is very large. The author proposes a trie based IP lookup approach to be used in the forwarding table [5]. It suffers from the worst case time complexity of $O(w)$, where $w$ is maximum length of the IP address. The IP lookup problems are useful in packet/traffic classification and intrusion detection in the network [9,16]

## 3   Proposed Solution for GPU Based IP Lookup

The proposed approach consists of two phases namely partitioning and lookup. The forwarding table in the data plane of the switch can be partitioned based on the first $k$ bits of IP prefix. The Fig. 1 shows the overview of IP lookup operation takes place in the GPU platform. Since the tables $\{T_0, T_1, \ldots T_{2^k-1}\}$ are copied into the global memory of GPU, it can be accessed by different threads easily. The Algorithm 1 brief about partitioning of forwarding table into different groups of table $\{T_0, T_1, \ldots T_{2^k-1}\}$. For each entry available in each of the tables $\{T_0, T_1, \ldots T_{2^k-1}\}$ trie can constructed using Algorihtm 2. Algorithm 2 construct trie with specified stride value and insert values into the level tables according to the active path from root node to the leaf node. The Algorithm 3 perform insert operation for each entry in the routing table. In order to update an entry in the multi-bit trie, in the worst case it will take $O(w/k)$ where $k$ represents the chosen stride value. Every prefix (IP address) will have different length $l$ (Mask) value. Now, divide this database into multiple groups based on the leftmost $k$ bits of the prefix. There are a different groups of prefixes table based on the leftmost $k$ bits. Construct multi-bit trie [15] with stride value for each of the groups separately. In order to search an entry in the trie, find the middle of the level using a binary search. Then extract the leftmost *middle* number of bits from the destination IP and apply hashing at this level to check the extracted bits present at this level. If a match is not found, apply binary search recursively till match found. Write GPU kernel code which performs a longest match operation (LPM) on each of the group simultaneously. The Algorithm 1 performs IP lookup based on given destination IP.

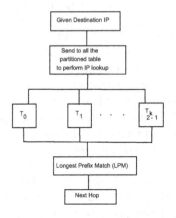

**Fig. 1.** Parallel search in the partitioned tables

**Lemma 1.** *The height of the complete binary trie* $\log n$.

**Theorem 1.** *Search complexity of complete binary trie* $O(\log \log n)$.

*Proof.* Since there are $\log n$ level table in the complete binary trie by Lemma 1. In order to search a given IP address, a binary search mechanism applied over the height $h$ of the trie. The binary search takes the effort of $\log h$. By the Lemma 1, apply substitution in place of $h = \log n$, which give the complexity of $O(\log \log n)$.

**Algorithm 1.** Table Partitioning

**Require:** The forwarding table $T$ and k
**Ensure:** Partitioned table $\{T_0, T_1, \ldots T_{2^k-1}\}$
Let $T$ has $n$ number of IP prefix entires
**for** $i = 0$ to $n$ **do**
    Based on leftmost $k$ bits insert the IP prefixes along with next hop into $T_i$, $0 \leq$ i
    $\leq 2^k - 1$
**end for**

**Algorithm 2.** Trie Construction

**Require:** Partitioned table $\{T_0, T_1, \ldots T_{2^k-1}\}$
**for** $i = 0$ to $2^k - 1$ **do**
    for each $T_i$ call *insert_trie* $(T_i)$
**end for**

**Algorithm 3.** *insert_trie* $(P)$

**Require:** stride size s
c=$\|P\|$
**for** $i = 0$ to $c$ **do**
    for each entry $c_i$ in P, construct multibit trie based on specified s
    for each level in the trie store active paths in a separate table (L)
**end for**

**Theorem 2.** *Search complexity of complete binary trie* $O((\log \log n)/s)$

*Proof.* Based on the stride value $s$, the number of level reduced by $s$. By the Theorem 1, which give the complexity of $\log \log n$. So, the proposed approach takes the lookup complexity of $O((\log \log n)/s)$

## 4    Results and Discussion

In this section, performance of the proposed GPU-based approach is compared against the existing approaches like binary trie, BST on GPU. The proposed approach is implemented on NVIDIA TESLA C2075 [2]. The TESLA card C2075 has 14 stream processor (SM) and each SM has 32 cores. All the programs are compiled with CUDA release 2.0 platforms. The dataset considered here, consists of IPv4 addressing. Initially, all the experiment are performed with various RIPE dataset [12] ranging from $RRC01$ to $RRC15$. Figure 2 shows IP address lookup time for tire implementation in GPU, BST implementation in GPU and the proposed approach with $s = 3$ and $k = 5$. One can see that the proposed IP lookup approach perform better than the other existing approaches.

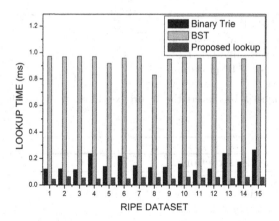

**Fig. 2.** Look up time for the forwarding table using the binary trie [13], BST on GPU [17] and the proposed GPU approach

## 5    Conclusion

The proposed approach utilizes multiple cores in GPU and divides IP lookup into a different group based on first $k$ bits of IP addresses. Using the proposed approach one can enhance the performance of IP lookup operation in Switch/Router by adopting massive data level parallelism. The proposed technique is compared with the existing GPU-based approaches where the proposed GPU implementation gives 64% and 94% improvement than GPU based binary tire and BST implementations. The future work is to enhance the proposed approach with IPv6 datasets.

## References

1. https://www.caida.org/research/traffic-analysis/
2. (2016). http://www.nvidia.com/object/cuda_get.html
3. CUDA, Compute Unified Device Architecture. http://nvidia.com/. Accessed 2016
4. Han, S., Jang, K., Park, K.S., Moon, S.: Packetshader: a GPU-accelerated software router. ACM SIGCOMM Comput. Commun. Rev. **40**(4), 195–206 (2010)
5. Indira, B., Valarmathi, K., Devaraj, D.: A trie based IP lookup for high performance router/switch. IEEE International Conference on Intelligent Techniques in Control, Optimization and Signal Processing, pp. 1–6 (2019)
6. Kasnavi, S., Gaudet, V.C., Berube, P., Amaral, J.N.: A hardware-based longest prefix matching scheme for TCAMs. IEEE International Symposium on Circuits and Systems, ISCAS, pp. 3339–3342 (2005)
7. Mu, S., Zhang, X., Zhang, N., Lu, J., Deng, Y.S., Zhang, S.: IP routing processing with graphic processors. In: Proceedings of the Conference on Design, Automation and Test in Europe, pp. 93–98 (2010)
8. Nilsson, S., Karlsson, G.: IP Address lookup using LC-tries. IEEE J. Sel. Areas Commun. **17**(6), 1083–1092 (1999)

9. Padmashani, R., Sathyadevan, S., Dath, D.: BSnort IPS better snort intrusion detection/prevention system. In: 2012 12th International Conference on Intelligent Systems Design and Applications (ISDA), pp. 46–51 (2012)
10. Pan, M., Haibin, L.: Build shape-shifting tries for fast IP lookup in o(n) time. Comput. Commun. **30**(18), 3787–3795 (2007)
11. Rekhter, Y., Li, T.: An architecture for IP address allocation with CIDR. RFC 1518 (1993)
12. RIS. Routing information service (2015)
13. Rojas-Cessa, R., Ramesh, L., Dong, Z., Cai, L., Ansari, N.: Parallel search trie-based scheme for fast IP lookup. In: IEEE Global Telecommunications Conference, GLOBECOM 2007, pp. 210–214 (2007)
14. Ruiz-Sanchez, M.A., Biersack, E.W., Dabbous, W.: Survey and taxonomy of IP address lookup algorithms. IEEE Netw. **15**(2), 8–23 (2001)
15. Sahni, S., Kim, K.S.: Efficient construction of multibit tries for IP lookup. IEEE/ACM Trans. Netw. (TON) **11**(4), 650–662 (2003)
16. Sajeev, G.P., Nair, L.M.: LASER: a novel hybrid peer to peer network traffic classification technique. In: 2016 International Conference on Advances in Computing, Communications and Informatics (ICACCI), pp. 1364–1370 (2016)
17. Shekhar, A., Goyal, J.: Parallel binary search trees for rapid IP lookup using graphic processors. In: 2nd International Conference on Information Management in the Knowledge Economy (IMKE), pp. 176–179 (2013)
18. Veeramani, S., Mahammad, N.: Minimization of flow table for TCAM based openflow switches by virtual compression approach. In: IEEE International Conference on Advanced Networks and Telecommunication Systems (ANTS), pp. 1–4 (2013)
19. Srinivasan, V., Varghese, G.: Faster IP lookups using controlled prefix expansion. ACM SIGMETRICS Perform. Eval. Rev. **26**(1), 1–10 (1998)
20. Veeramani, S., Mahammad, N.: Hybrid trie based partitioning of TCAM based openflow switches. In: IEEE IEEE International Conference on Advanced Networks and Telecommunication Systems (ANTS), pp. 1–5 (2013)
21. Veeramani, S., Mahammad, N.: Efficient IP lookup using hybrid trie-based partitioning of TCAM-based Openflow switches. Photonic Netw. Commun. **28**(2), 135–145 (2014)
22. Veeramani, S., Rahul Sharma, S., Noor Mahammad, SK.: Constructing scalable hierarchical switched openflow network using adaptive replacement of flow table management. In: 2013 IEEE International Conference on Advanced Networks and Telecommunications Systems (ANTS), pp. 1–3 (2013)
23. Yim, C., Lee, B., Lim, H.: Efficient binary search for IP address lookup. IEEE Commun. Lett. **9**(7), 652–654 (2005)

# Intelligent Technology

# MCMARS: Hybrid Multi-criteria Decision-Making Algorithm for Recommender Systems of Mobile Applications

S. Tejaswi[1,2]([✉]) [iD], V. N. Sastry[1], and S. Durga Bhavani[2] [iD]

[1] Center for Mobile Banking, IDRBT, Castle Hills, Hyderabad 500040,
Telangana, India
tejaswivij@gmail.com
[2] SCIS, University of Hyderabad, Hyderabad 500046, Telangana, India

**Abstract.** A mobile application (app) recommender system needs to support both developers and users. Existing recommender systems in the literature are based on single-criterion analysis, which is insufficient for producing better recommendations. Moreover, recommendations do not reflect the user's perspectives. To address these issues, in this paper, we present a Multi-Criteria Mobile App Recommender System (MCMARS) that assists developers in improving their apps and recommends the top-performing apps to users. We define the performance score of an app based on four criteria attributes: risk assessment score, functionality score, user rating, and the app's memory size. We define the risk assessment score for each app using multi-perspective analysis and the functionality score by assigning preference weights to the services of apps in the same category. We evaluate optimal weights of the criteria by integrating the entropy method and the extended Best-Worst method (BWM) using Hesitant-Triangular-Fuzzy information with group-decisions. Finally, the TOPSIS uses these weights to assess the app's performance. To validate our MCMARS, we prepared a dataset of 124 government-approved COVID-19 Android apps from 80 countries and made it available on GitHub for the research community. Finally, we perform a fine-grained analysis of the app's performance based on the criteria attributes that help the developers to improve their apps. The experimental results show that two independent attributes, "risk assessment score" and "functionality score", significantly measure the app's performance. According to our findings, only 12.5% of the apps in the experimental dataset provide high-performance, high-functionality, and low-risk.

**Keywords:** Mobile app recommender system · MCDM · OWASP

## 1 Introduction

With the rapid development of mobile technology, mobile apps (apps) of special categories like education, health, mobile payments, etc., play a vital role

A. R. Molla et al. (Eds.): ICDCIT 2023, LNCS 13776, pp. 107–124, 2023.
https://doi.org/10.1007/978-3-031-24848-1_8

in human life by providing online personal services. Generally, app developers upgrade their app effectiveness depending on the number of downloads and user satisfaction with app usage. User satisfaction is primarily associated with many personal preferences and experiences [5]. Generally, app developers regularly follow user feedback (subjective reviews) to ensure timely upgrades and enhancements to meet user expectations. However, it is challenging to understand user reviews. Moreover, challenges faced by the app developers while developing apps like fragmentation, testing, reuse, lack of expertise and quality assurance practices, etc., reduces app efficiency by allowing data insecurity and privacy threats [1]. If an ordinary user loses emotion and trust in an app's usage, they may discontinue it. So, the app developers need to consider user preferences while developing the apps, and also check their apps' relative performance with other apps in the same category in order to upgrade them. On the user side, many options are made available to choose an app in any category in the Google Play store. It causes an information overload problem and makes the decision-making process more complex. The app recommender systems performs a filtering process to solve the information overload problem by suggesting apps to specific users according to their preferences. Generally, users rely on a single-criterion rating (overall rating) as a primary source for the recommendation process. Authors in [21] recommended apps to users based on the app's risk and popularity. We found that this risk score evaluated based on developer-defined app permission risk categories that are different from the user-concerned risk categories [9]. Hence, incorporating user concerns into the development of the app is essential. Moreover, single-criterion analysis is not sufficient to provide accurate recommendations [10].

The essential issues discussed above motivate us to work towards providing a solution. This paper presents a hybrid Multi-Criteria Decision making Mobile App Recommender System i.e., MCMARS is built to perform these tasks : (i) it helps with decision support for app developers in order to upgrade the apps based on their relative performance and (ii) it facilitates recommendations to users by recommending the top-performing apps. MCMARS evaluates the apps based on multiple criteria such as security risk, functionality, memory size, and user assessment (rating) scores. We chose these criteria attributes based on the suggestions from the studies [7,13].

MCMARS recommends mobile apps based on their relative performance scores, which are found with the help of criteria weights and each app's criteria values. As a part of the recommendation process, MCMARS generates values for criteria, such as risk score, functionality score, etc., for each app, as the required data is not available on the internet to use directly. Moreover, MCMARS finds the criteria weights optimally by combining their subjective and objective weights, which are evaluated using BWM [15] with hesitant fuzzy information and entropy methods, respectively. Finally, the apps are sorted according to their relative performance scores to suggest the top-performing apps to users. Next, the apps in the dataset are clustered based on the overall performance and criteria values to find the impact of the criteria on app performance. We identify recommendations to the app developers based on the findings, which aid in decision-making to improve the apps. These are the highlights that emphasize the novelty of this work. We validate our MCMARS through a case study on the COVID-19 Android apps. The COVID-19 apps have various functionalities like contact tracing and syndromic surveillance to trace the origins of disease transmission and control the disease [6]. Moreover, they have proven helpful in disease monitoring and control. Furthermore, the observations made in the case study may be helpful for app developers in dealing future pandemics. Our approach is also helpful to generate recommendations to the apps in the categories of education, health and fitness, and financial services.

The main contributions in this paper are:

1. We present the framework and implementation of MCMARS, which performs dual tasks: (i) a decision support system for app developers and (ii) a recommender system for users. This proposed system highly considers the user's requirements and perspectives on app downloads as criteria attributes.
2. We define quantitative scores for criteria attributes; (i) risk score of an app using multi-perspective analysis such as code privilege analysis and static analysis, emphasising the user's risk intentions refer Table 2. Moreover, we analysed the consistency of apps with the OWASP mobile top 10 mobile security risks. (ii) an app's functionality score by assigning weights to the services it provides. As per our knowledge, this is the first work to define and use the app's criteria attributes scores in an app performance evaluation.
3. We prepare a real time dataset of COVID-19 mobile apps worldwide by collecting information from various web sources and making it available at GitHub for the research community for further exploration.
4. We introduce a modified approach to using FBWM with group decision-making in MCMARS. Our proposed approach improves execution time than the state-of-the-art by minimising the number of iterations in computing criteria weights.
5. We analyze the criteria impact on app efficiency and identify findings that assist developers in making decisions to upgrade their apps.

The rest of this paper is organized as follows: Sect. 2 discusses the related work. Section 3 presents the design of MCMARS which consists of criterion generation and recommendation methodology. Section 4 explains in detail our experimental study on a case COVID-19 Android apps. Section 5 presents the analysis of experimental results. Section 6, presents the conclusion and future work.

## 2    Related Work

In this section, we discussed the existing literature on app recommendations and limitations as below.

Many studies in the past have suggested various approaches about general app recommendations. For instance, authors of [4] have proposed a recommender system (RS) which is modelled based on user reviews and permission configurations. The authors of [16] have built a recommender system to recommend secure

**Table 1.** Comparative analysis on criteria the considered in various mobile app recommendation systems in the literature with the proposed system.

| [Ref. No] | Risk evaluation parameter(s) | Services | Memory size | Rating |
|---|---|---|---|---|
| [4] | User reviews, Permission profile | X | X | X |
| [16] | Permission profile, API calls | X | X | X |
| [8] | Permission profile | X | X | ✓ |
| [14] | Permission profile | X | X | ✓ |
| [21] | Permission profile | X | X | ✓ |
| [11] | Permission profile API calls | ✓ | X | X |
| MCMARS (proposed work) | Permission profile, Static Analysis | ✓ | ✓ | ✓ |

apps based on app permissions list and API calls. In [8], authors have proposed a recommendation framework that uses user ratings and permission configurations to derive a security score thereby recommend secure apps. In [14] have generate recommendations by framing the relationship between user interests in terms of app ratings, and risk score evaluated on the permission list of apps. [21] have implemented a RS named SPAR which provides ranks to the apps in three dimensions: security (using permission list), popularity (ratings), and hybrid (both). Through [11], authors have suggested a model which assessed the user preference for the new app by relating the user interests, security preferences, and functionalities using latent factorization techniques.

Table 1 shows the comparison between the various models from previous works, which highlights that the models are dependent extensively on only data access permissions or consumer reviews, which seem inadequate in terms of the overall evaluation possibility of the mobile apps. We have considered versatile sets of essential dimensions to address this limitation for an effective and efficient evaluation of mobile apps. These dimensions include a systematic code vetting for security (static testing) and a multi-criteria assessment for risk score, functionality score, app size and usability score, thereby recommending the best app for the user and the developers with a single performance score.

# 3   Proposed Mobile App Recommender System (MCMARS) Design

Figure 1 shows the framework of the proposed recommender system, MCMARS design which consists of three phases and five modules can be explained as follows,

## 3.1   Data Extraction and Classification

**Module 1: Data Collection.** In this module, using URL of an app, we collect the app's metadata or profile including the app category, country name, size,

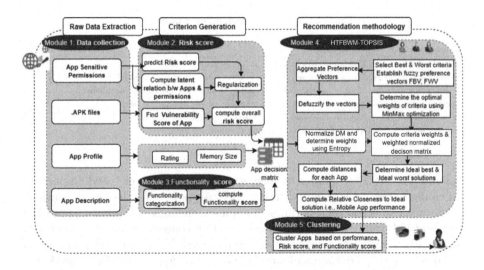

**Fig. 1.** Frame work of MCMARS

version, rating, price, developer, genres and permission configuration by scraping it from various web sources, such as the Play store, Wikipedia, etc., and create a real-time dataset with them. For some experiments, we divide the dataset into three sets: One contains the app metadata, the second set contains permission configuration, and the third set contains the functionalities of each app.

### 3.2 Criterion Generation

In this phase, we evaluate the values of two essential decision criteria, the risk score and the functionality score, as discussed below.

**Module 2: Evaluation of Risk Score.** In this module, the app risk assessment is carried out through multi-perspective analysis: codes' privileges analysis (which considers sensitive data access permissions of apps); and static analysis (which identifies security vulnerabilities that may cause information theft by the outside world). An added advantage of using this analysis is that if we remove a dangerous permission or a vulnerability from an app, the risk score will automatically get reduced. Furthermore, we employ user as well as developer defined risk intentions in the risk score evaluation Table 2, which is an essential characteristic novelty of this model. According to this approach, let $A = \{a_1, a_2, a_3, \ldots, a_M\}$ be a set of apps, $P = \{p_1, p_2, p_3, \ldots, p_N\}$ be a set of permissions, the risk score of a mobile app $(a_i)$ is formulated as,

$$Risk_i^a = RCP_i^a + RSA_i^a; i \in \{1, 2, \ldots, M\} \tag{1}$$

where $RCP_i^a$ and $RSA_i^a$ are the risk scores of the $i^{th}$ app derived from the codes' privilege analysis and static analysis, respectively.

**Table 2.** Risk categories at user point [9]

| Risk category (C) | Description | Weight ($w_j^C$) |
|---|---|---|
| Personal-privacy | Leakage of private data e.g, location, contacts, etc. | 6 |
| Personal-data Integrity | Refers to accuracy & data validity rather than data security | 5 |
| Eavesdropping | Theft of confidential information through a network by smartphone | 4 |
| Monetary risks | Accidental purchases, unauthorized access of financial info | 2 |
| Device-instability | Risks on hardware functionality of devices | 3 |
| General risks | Risks which are not harming the sensitive data | 1 |

*Evaluation of $RCP_i^a$:* In order to generate optimal risk scores based on permission configuration, it is essential to consider two points: (i) the latent relationship between apps and their permission requests, ensuring that similar apps have equivalent risk value. (ii) define the risk function concerning the permissions of some categories of apps. For example, even though location-based permissions are dangerous, they are commonly used in navigation apps. Therefore, first derive a risk score function based on app's permissions, and then regularize the scores to satisfy the above points as discussed in following steps.

– **Step 1.** To derive risk score evaluation function, we leverage the rarity based Risk Score with Scaling (RSS) method proposed in [3], which strictly depends on the fraction of apps accessing specific permission, and results in higher risk score to apps that make access more sensitive permissions. To assess the risk score more accurately, it is necessary to consider the presence of number of permissions in the same risk category. That means, an app with more number of dangerous permissions have high risk score than app with less permissions in the same category.

Therefore, we generalized RSS to incorporate the effect of presence/absence of permission within a category.

$$\tilde{R}_j^p = w_j^C * Pr(p_j^C) * ln(M/c_j)$$
$$\tilde{R}_i^a = \sum_{j=1}^{N} x_{i,j} * w_j^C * Pr(p_j^C) * ln(M/c_j) \qquad (2)$$

where $x_{ij}$ is a binary value equals to 1 when app $a_i$ requests permission $p_j$, 0 otherwise. M is the number of apps, and $c_j = \sum_{i=1}^{M} x_{ij}$. $Pr(p_j^C)$ is the probability of occurrence of permission $p_j$ in the risk category C, and $w_j^C$ is the category weight of $j^{th}$ permission according to the user intended weights as defined in [9] shown in Table 2.

– **Step 2.** *Regularization:* In this step, we use a prevalent and effective regularization technique proposed in [21] to generate optimal risk scores.

*Evaluation of $RSA_i^a$:* The $RSA_i^a$ incorporates Common Vulnerability Security Scoring (CVSS), which is the industry standard for determining the severity of a vulnerability on an app, developed by the Forum of Incident Response and

Security Teams (FIRST) [2]. It is used by the National Vulnerability Database (NVD) for scoring vulnerabilities. And, it is a logical scoring evaluation process. In this paper, we use automatic testing tool i.e., "ImmuniWeb® MobileSuite"[1] that can examine the source code by following predefined rules. It takes .apk files of apps as input, and produces a CVSS for each vulnerability of the app ($CVSS_{vul}$) as output. A software may contain multiple vulnerabilities. Hence, we sum up the $CVSS_{vul}$ values of all the vulnerabilities identified in app's software, and it is considered as $RSA_i^a$. Finally, we get the optimal risk score of $i^{th}$ app, $Risk_i^a$ by substituting the results of $RSA_i^a$ and $RCP_i^a$ in Eq. (1).

**Module 3: Evaluation of Functionality Score.** The functionality score is a pivotal factor in estimating the app's performance. Also, it is important to categorise the functionalities and assign relative weights based on their necessity.

We aim at coming up with a functionality score $f(a_i)$ for mobile apps based on their services. Let $i^{th}$ app in the dataset be represented by $a_i = [x_{i,1}, \ldots, x_{i,N}]$ where N is the number of functionalities provided by the app, and $x_{i,j} \in \{1,0\}$ indicates whether $i^{th}$ app ($a_i$) provides $j^{th}$ functionality. The functionality score of an app is determined by the number of functionalities it provides in each category. That means if an app has a high number of high-category functionalities, it will have a high functionality score. The functionality score $f(a_i)$ is defined as follows,

$$f(a_i) = \sum_{j=1}^{N} x_{ij} * P(x_{ij}^C) * w_j^C \qquad (3)$$

where $P(x_{ij}^C)$ is the probability of occurrence of $j^{th}$ functionality of $i^{th}$ app in the category C, and $w_j^C$ is the normalized categorical weight of the $j^{th}$ functionality.

### 3.3    Recommendation Methodology

In this section, we present proposed model of app recommendation using hybrid HTFBWM-Entropy-TOPSIS as follows:

**Preliminaries: Hesitant Triangular Fuzzy Sets.** The mathematical model of hesitant fuzzy set(HFS) [19] is given as, a HFS is defined as a mapping HFS:X→[0,1] and is given by:

$$A = \{< x, h_A(x) > \backslash x \in X\} \qquad (4)$$

where $h_A(x)$, is called as hesitant fuzzy element (HFE), is a set of discrete values in [0,1], denoting the membership degree. Let X be a reference set, a HTFS E on X defined as a function $f_E(x)$ as: $E = \{< x, f_E(x) >: x \in X\}$, where $f_E(x)$ is a set of various triangular fuzzy values is called as Hesitant triangular fuzzy numbers (HTFS), which are used to express the possible fuzzy membership degree of an element $x \in X$ to E.

---

[1] URL: https://www.immuniweb.com/.

**Algorithm 1.** module 5: HTFBWM-Entropy-TOPSIS

---

**Input:** Matrix $A_{M \times N}$ M: number of mobile apps, N: number of criterion
W: weight list of criteria, BC: beneficial criteria list, NBC: Non beneficial criteria list
    **Output:** Performance score of each mobile app

1: **for** each expert k $\in$ [1,r]: **do**
2:     read the fuzzy preference values of $\tilde{P}_B^k$, $\tilde{P}_W^k$ using Eq. (5) and Eq. (6), respectively.
3: **end for**
4: compute aggregate fuzzy preference values of $\tilde{AP}_B$, $\tilde{AP}_W$ using max-min with mean method.
5: defuzzify $\tilde{AP}_B$, $\tilde{AP}_W$ as $AP_B$, $AP_W$,respectively by Eq. (7)
6: determine subjective weights of the criteria using Eq. (9)
7: determine objective weights of the criteria using Eq. (12)
8: determine optimal weights of the criteria from steps 6, 7 using Eq. (13)
9: compute normalized R of A using Eq. (11)
10: compute weighted normalized V of A using Eq. (14)
11: **for** each criterion $j \in [1, N]$: **do**
12:     compute Ideal Best($A^+$) using Eq. (15)
13:     compute Ideal Worst($A^-$) using Eq. (16)
14: **end for**
15: **for** each alternative $i \in [1, M]$: **do**
16:     **for** each criterion $j \in [1, N]$: **do**
17:         compute $d_i^+$, $d_i^-$ using Eq. (17), Eq. (18), respectively.
18:     **end for**
19: **end for**
20: **for** each alternative $i \in [1, M]$: **do**
21:     compute $S_i$ using Eq. (19)
22: **end for**

---

**Module 4: Proposed Approach of HTFBWM-Entropy-TOPSIS with Group Decisions:** The proposed method that combines entropy, HTFBWM and TOPSIS to find the relative performances of the apps. The entropy, HTF-BWM are using for computing criteria weights, and the TOPSIS uses these weights to assess the app's performance.

For mobile app multi-criteria recommendation with group decision making, let A = $\{a_1, a_2, a_3, \ldots, a_M\}$ be a set of mobile apps, C = $\{C_1, C_2, C_3, \ldots, C_N\}$ be a set of performance criteria, and DM = $\{dm_1, dm_2, dm_3, \ldots, dm_r\}$ be the r experts. The goal of HTFBWM-Entropy-TOPSIS procedure is to assign a relative performance score $(S_i)$ for each app $(a_i)$ based on the preferences of a group of decision makers (DM). The HTFBWM-Entropy-TOPSIS is presented in Algorithm 1 that is explained in Steps 1–7.

- **Step 1.** In this step, we should determine the criteria set C that affects the decision-making process in mobile app recommendation.
- **Step 2. Criteria weights evaluation (HTFBWM-Entropy):** In this step, we evaluate and combine the subjective and objective weights of the criteria attributes to find the final optimal weights.

**Table 3.** Fuzzy scale for weight comparison: The linguistic terms and corresponding membership functions of Hesitant Triangular Fuzzy Numbers (TFNs) used in this work

| Linguistic terms | HTFN |
|---|---|
| Most important (MI) | (0.9, 1, 1) |
| Very important (VI) | (0.7, 0.9, 1) |
| Fairly important (FI) | (0.5, 0.7, 0.9) |
| Slightly more important (SMI) | (0.5, 0.5, 0.7) |
| Equivalent (EI) | (0.5, 0.5, 0.5) |

- **Subjective weights evaluation (HTFBWM):** The $dm_k \in$ DM, $k = 1, 2, \ldots, r$ should select the most beneficial (Best) criterion, $C_B^k$ and the least beneficial (Worst) criterion $C_W^k$. And, give the hesitant fuzzy preferences of the $C_B^k$ over the remaining criteria using Table 3 which gives fuzzy best vector, FBV.

$$\tilde{P}_B^k = [\tilde{P}_{B1}^k, \tilde{P}_{B2}^k, \ldots, \tilde{P}_{BN}^k], k = 1, 2, \ldots, r \quad (5)$$

where $\tilde{P}_{Bj}^k$ implies the fuzzy preference of the best criterion $C_B^k$ over the criterion $j$, and $\tilde{P}_{BB}^k = (0.5, 0.5, 0.5)$. Also determine the hesitant fuzzy preferences of the remaining criteria over $C_W^k$ using Table 3 which gives fuzzy worst vector FWV.

$$\tilde{P}_W^k = [\tilde{P}_{1W}^k, \tilde{P}_{2W}^k, \ldots, \tilde{P}_{NW}^k]^T, k = 1, 2, \ldots, r \quad (6)$$

where $\tilde{P}_{jW}^k$ is implies the fuzzy preference of the criterion j over $C_W^k$, and $\tilde{P}_{WW}^k = (0.5, 0.5, 0.5)$.

* An expert's judgement purely depends on his opinion on particular item (apps) that could be varying from other experts. So in order to get an accurate judgement, it is necessary to aggregate expert opinion. Here, we use min-max with mean method [12] that expands the range of aggregate value by including best and worst judgement.

Finally, we get aggregated fuzzy preferences of $C_B$ over remaining criteria as $\tilde{AP}_B = [\tilde{AP}_{B1}, \tilde{AP}_{B2}, \ldots, \tilde{AP}_{BN}]$, and fuzzy preferences of remaining criteria over $C_W$ as, $\tilde{AP}_W = [\tilde{AP}_{1W}, \tilde{AP}_{2W}, \ldots, \tilde{AP}_{NW}]^T$, where $\tilde{AP}_W = (\tilde{AP}_{jW}^l, \tilde{AP}_{jW}^m, \tilde{AP}_{jW}^h)$ be the aggregated fuzzy value if $j^{th}$ criterion over best criterion.

* Next defuzzification can be carried out to convert fuzzy preference values into equivalent crisp preference values as follows, Let $\tilde{AP}_{ij} = (\tilde{AP}_{ij}^l, \tilde{AP}_{ij}^m, \tilde{AP}_{ij}^h)$ be the fuzzy preference value of $i^{th}$ criterion over $j^{th}$ criterion can be defuzzyfied as [12],

$$AP_{ij} = \frac{(\tilde{AP}_{ij}^l + 4 * \tilde{AP}_{ij}^m + \tilde{AP}_{ij}^h)}{6}; i, j \in 1, 2, \ldots, N. \quad (7)$$

As results, the crisp preference value of $C_B$ over remaining criteria as $AP_B = [AP_{B1}, AP_{B2}, \ldots, AP_{BN}]$, and the crisp preference value of remaining criterion over $C_W$ as $AP_W = [AP_{1W}, AP_{2W}, \ldots, AP_{NW}]^T$.

* The optimal subjective weights for criteria $C_j$ is the one that satisfies the condition, $\mathcal{W}_B/\mathcal{W}_j = AP_{Bj}$ and $\mathcal{W}_j/\mathcal{W}_W = AP_{jW}$, where $\mathcal{W}_B, \mathcal{W}_W, \mathcal{W}_j$ be the weights of best, worst, $j^{th}$ criterion, respectively. In order to find the solution to satisfy these conditions to all j, there is a necessity to minimize the maximum absolute gaps $|\mathcal{W}_B/\mathcal{W}_j - AP_{Bj}|$ and $|\mathcal{W}_j/\mathcal{W}_W - AP_{jW}| \forall$ j is minimized. To satisfy this condition for all j, it should be solve the following nonlinear programming problem:

$$min \ max_j \left( \left| \frac{\mathcal{W}_B}{\mathcal{W}_j} = AP_{Bj} \right|, \left| \frac{\mathcal{W}_j}{\mathcal{W}_W} = AP_{jW} \right| \right) s.t \begin{cases} \sum_{j=1}^{N} \mathcal{W}_j = 1 \\ \mathcal{W}_j \geq 0, \forall j \end{cases}$$

(8)

According to author in [15], problem (8) can be transformed as follows,

$$min \ \eth \ s.t \begin{cases} \mathcal{W}_B - \eth \leq AP_{Bj}\mathcal{W}_j, \ \forall j \\ \mathcal{W}_B - \eth \geq AP_{Bj}\mathcal{W}_j, \ \forall j \\ \mathcal{W}_j - \eth \leq AP_{jW}\mathcal{W}_W, \ \forall j \\ \mathcal{W}_j - \eth \geq AP_{jW}\mathcal{W}_W, \ \forall j \\ \sum_{j=1}^{N} \mathcal{W}_j = 1 \\ \mathcal{W}_j \geq 0, \ \forall j \end{cases}$$

(9)

Solving Problem (9), the optimal subjective weights $(\mathcal{W}_1^*, \mathcal{W}_2^*, \ldots, \mathcal{W}_N^*)$, and the optimal value of $\eth$ as $\eth^*$ can be obtained. After assessing the optimal weights, the consistency has been checked through computing the consistency ratio (CR) as following:

$$ConsistencyRatio(CR) = \eth^*/ConsistencyIndex \qquad (10)$$

the Consistency Index (CI) can be evaluated by quadratic equation [15]

$$\eth^2 - (1 - 2U)\eth + (U^2 - U) = 0$$

where U is the upper value of HTFN of corresponding linguistic value.

• **Evaluation of objective weights of criteria:** The evaluation of objective weights of criteria can be carried out using Shannon entropy weighing method based on criteria values of apps as follows:

* **Construct normalized decision matrix:** For decision matrix $A_{M \times N}$ which consists of M apps with N criteria values, construct normalized decision matrix $R_{M \times N}$ as follows,

$$R_{ij} = A_{ij}/(\sqrt{\sum_{i=1}^{M} A_{ij}^2}), \quad i = 1, 2, \ldots, M; \quad j = 1, 2, \ldots, N. \qquad (11)$$

where $R_{ij}$ is the normalized value of $j^{th}$ criterion of $i^{th}$ mobile app.

\* Defining the objective weight as follows:

$$W_j^o = 1 - Ej/(\sum_{j=1}^{N} 1 - E_j); E_j = -k \sum_{i=1}^{M} P_{ij} ln P_{ij}, \qquad (12)$$

in which k= 1/ ln(M)
- **Integrating weights of criteria:** The subjective weights computed from Eq. (9) and the objective weights from Eq. (12) can be integrated to determine the overall weights of the criteria as follows:

$$W_j^* = (W_j^* * W_j^o)/(\sum_{j=1}^{N} W_j^* * W_j^o) \qquad (13)$$

- **Step 3. TOPSIS: Construct weighted normalized decision matrix:** To construct weighted normalized decision matrix, by multiply each criterion of normalized decision matrix with associated weight as shown in Eq. (14):

$$V_{ij} = W_j * R_{ij}, \quad i = 1, 2, \ldots, M; \quad j = 1, 2, \ldots, N. \qquad (14)$$

where $W_j$ is weight of criterion j.
- **Step 4. Calculate ideal best and ideal worst solutions:** Ideal best solution

$$A^+ = \{v_1^+, v_2^+, \ldots, v_N^+\}, v_j^+ = \begin{cases} max_{i=1,2,\ldots,M}(V_{ij}), & if \ j \in BC \\ min_{i=1,2,\ldots,M}(V_{ij}), & if \ j \in NBC \end{cases} \qquad (15)$$

Ideal worst solution.

$$A^- = \{v_1^-, v_2^-, \ldots, v_N^-\}, v_j^- = \begin{cases} min_{i=1,2,\ldots,M}(V_{ij}), & if \ j \in BC \\ max_{i=1,2,\ldots,M}(V_{ij}), & if \ j \in NBC \end{cases} \qquad (16)$$

- **Step 5. Compute distance measures for each alternative:** The distances from the ideal best and ideal worst solutions using Eq. (17) and Eq. (18) respectively.

$$d_i^+ = \left\{ \sum_{j=1}^{N} (v_{ij} - v_j^+)^2 \right\}^{\frac{1}{2}}, \quad i = 1, 2, \ldots, M. \qquad (17)$$

$$d_i^- = \left\{ \sum_{j=1}^{N} (v_{ij} - v_j^-)^2 \right\}^{\frac{1}{2}}, \quad i = 1, 2, \ldots, M. \qquad (18)$$

- **Step 6. Calculate the relative closeness to the ideal solution using:**

$$S_i = \frac{d_i^-}{(d_i^+ + d_i^-)}, \quad i = 1, 2, \ldots, M. \ S_i \in [0, 1] \qquad (19)$$

where $S_i$ denotes the performance score of $i^{th}$ app.
- **Step 7.** Finally, sort the apps according to their performance score to recommend top high performances apps to users.

**Module 5: Cluster Analysis.** Each app $a_i, i = 1, \ldots, M$ is represented as a 4-tuple (Risk score, functionality score, rating, and memory size). In this module, we perform fine-grained analysis to know the impact of criteria on app performance by clustering the apps based on their performance score, risk score, and functionality score using the K-means clustering algorithm [8], and validate the consistency of cluster data by Silhouette score.

## 4    Validation of MCMARS with COVID-19 Mobile Apps

Here, we present our work on the COVID-19 worldwide Android apps. For this study, we have prepared a real-world dataset (see Subsect. 3.1) consisting of information about 124 COVID-19 Android apps from 80 countries, collected from various web sources during the period between June, 2020 and November, 2020, and made it available to the research community at "https://github.com/tejaswivij/COVID-19"'. In the criteria generation phase, for risk score evaluation, we have manually divided standard permission risk levels (dangerous, signature, normal) into six categories (see Table 2) and assigned weights to them from 1 to 6, depending on the severity of the risk level. For regularisation, the parameters $\lambda = 0.5$ and $\mu = 1$ are set. We obtained an optimised $RCP_i^a$. For static security testing, we have used the tool "ImmuniWeb® MobileSuite" to find the overall vulnerability score ($RSA_i^a$) of each app. Finally, we got the overall risk score of each app by combining these two risk values. For the functionality score, we have manually divided the functionality list into three categories (i.e., HIGH, MEDIUM, LOW) based on their service necessity in the pandemic period. We identified 12 functionalities such as contact tracing, syndrome surveillance, self-diagnosis, exposure detection, quarantine enforcement, isolation, 24/7 online doctor, General Information, Medical Information, Geo-fencing, Medical Reporting and Training, etc. in COVID-19 Android apps. We classified four of them as high: contact tracing, syndrome surveillance, self-diagnosis, and exposure detection. The three functionalities such as quarantine enforcement, isolation, and 24/7 online doctor are medium, and the remaining are as low category functionalities. We assigned weights to these categories to show the relative difference, such as HIGH = 5, MEDIUM = 3, LOW = 2, and used normalized weights in the evaluation. Finally, we got a real number score for each app by applying Eq. (3).

In the Recommendation Methodology phase, MCMARS did the process in two folds. In the first, it evaluated criteria weights optimally by integrating objective weights and subjective weights of the criteria attributes, viz., c1 = risk assessment score, c2 = rating, c3 = memory size, and c4 = functionality score of a mobile app. Here, we randomly chose three people as experts who know about app usage, security, and follow regular communication on COVID-19 for judgements while evaluating weights for the decision criteria. The entire process of steps 1 & 2 in Sect. 3.3 is shown in Table 4. By solving the model in Eq. (9), we obtained the subjective weight vector of the decision criteria $\mathcal{W}^*$ as [0.33, 0.21, 0.16, 0.3], and $\eth^* = 0.010$. In this process we check the consistency ratio using Eq. (10). By Eq. (12) we found objective weight vector as $\mathcal{W}^o$ as [0.23, 0.26,

Table 4. Decision criteria weights by experts

| Expert-FBV | BEST | Risk | Rating | Size | Functionality |
|---|---|---|---|---|---|
| dm1 | Risk | EI | VI | VI | SMI |
| dm2 | Risk | EI | FI | MI | SMI |
| dm3 | Risk | EI | VI | MI | FI |
| **Expert-FWV** | **WORST** | **Risk** | **Rating** | **Size** | **Functionality** |
| dm1 | Size | MI | FI | EI | MI |
| dm2 | Size | VI | FI | EI | MI |
| dm3 | Size | MI | VI | EI | FI |
| Aggregate | $\tilde{AP}_B$ | (0.5,0.5,0.5) | (0.63,0.83,0.97) | (0.83,0.97,1) | (0.5,0.56,0.76) |
| opinion | $\tilde{AP}_W$ | (0.79,0.95,1) | (0.5,0.7,0.9) | (0.5,0.5,0.5) | (0.9,1,1) |
| Crisp | $AP_B$ | 0.5 | 0.8 | 0.95 | 0.58 |
| opinion | $AP_W$ | 0.93 | 0.7 | 0.5 | 0.98 |

Table 5. Relative performance score (S) evaluation for a sample of 5 COVID-19 apps

| Name | Weighted normalized matrix | | | | $d^+$ | $d^-$ | S |
|---|---|---|---|---|---|---|---|
| | **Rating** | **Size** | **Risk** | **Functionality** | | | |
| AAROGYA SETU | 0.023 | 0.001 | 0.044 | 0.049 | 0.04 | 0.11 | 0.71 |
| Open WHO | 0.022 | 0.006 | 0.046 | 0.011 | 0.07 | 0.09 | 0.58 |
| COVID-19 AP | 0.022 | 0.012 | 0.019 | 0.046 | 0.03 | 0.10 | 0.79 |
| Odisha COVID | 0.023 | 0.006 | 0.039 | 0.047 | 0.04 | 0.10 | 0.72 |
| Protect Scotland | 0.020 | 0.034 | 0.036 | 0.021 | 0.06 | 0.07 | 0.52 |

0.26, 0.25]. Finally, we obtained consistent optimal weights by model Eq. (13) as [0.34, 0.22, 0.14, 0.3]. Next, these weights are used by TOPSIS (steps 3–7 in Sect. 3.3) to evaluate the apps' performance score by considering rating and functionality are as beneficial, size and risk score of the apps are taken as non-beneficial/cost criteria. The results shown in Table 5. **Note:** In order to make top-N app recommendations, all apps are sorted according to their performance score(S). In this study, we do not sort the apps because every country has their own government-issued COVID-19 Android apps, and they work in the specified geographic region only.

## 5   Results and Discussion

In this section, we analyse and present the relative performances of COVID- 19 Android apps in relation with multiple criteria, and also discuss the comparison of proposed approach of FTFBWM with other MCDM techniques briefly. The various notable results and recommendations are discussed below.

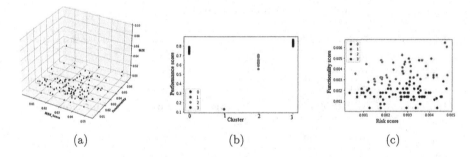

(a)                    (b)                    (c)

**Fig. 2.** (a) 3D plot of the mobile apps' risk score (X-axis), functionality score (Y-axis), and size (z-axis). (b) Clusters of mobile apps based on their Performance score, show that cluster 3 is desirable cluster which consists of mobile apps with high performance. (c) Mobile app clusters based on risk and functionality scores show that most of the mobile apps in cluster 2 are at desirable level with low risk and high functionality.

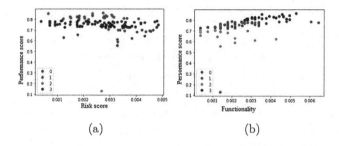

(a)                    (b)

**Fig. 3.** Performance of mobile apps w.r.t. Risk and w.r.t. Functionality; (a) shows most of the mobile apps in cluster 1 are at desirable level (b) shows cluster 3 is the desirable cluster of mobile apps with high performance and high functionality.

## 5.1   Criteria Analysis

1. **Analysis on the overall performance of the apps:** The app's performance score has been derived based on four necessary criteria; risk score, functionality score, memory size, and rating. The apps are classified into four performance categories: High, Medium, Low and Very low. Figure 2 (a) shows the 3D plot of risk score, functionality score, and memory size, which are the main values because most of the apps have high rating scores. We observed that only 13.7% of apps are developed with high functionality, low memory, and low-risk score. The results shown in Fig. 2 (b), indicate that the desirable cluster (cluster 3) contains 21.77% of the apps. 63.71% of apps perform at a medium level, 13.71% of apps perform at a low level, while the rest are very low.

2. **Risk Vs functionalities of apps:** It is a general assumption that the security risk score of the app may increase with increased functionality. However, as per the result shown in Fig. 2, nearly 17% of overall apps (cluster 2) have a desirable level of functionality with a low-risk score. The apps in cluster 3

(20.97% of overall apps) have a low-security risk score with low functionality, and cluster 1(21% of overall apps) has high functionality with a high-risk score. These results show that there are possibilities to improve the apps in clusters 1 and 3 by enhancing the functionality with security measures.

3. **Risk Vs Performance:** Here, we divided the risk score into two regions as low risk and high risk. By observing the Fig. 3, we can say nearly, 22% of mobile apps are at high performance with low risk. Nearly, 63% of the mobile apps(cluster 3) are at medium performance, among which 57% apps are at medium performance with low risk.

4. **Functionality Vs Performance:** As discussed in previous sections, the functionalities like contact tracing, syndromic surveillance, self-diagnosis, and exposure detection are considered high priority functionalities. The functionality score is divided into two categories: low(0-0.29) and high(0.03-above). The results are shown in Fig. 3 reveal that only small amount(nearly 22.5%) of apps have high performance score with high functionality score. Our further analysis reveals that 62.09% of apps provide only one high priority functionality. In our study, it also observed that approximately 59.68% of the apps work with high priority functionality but not reached in the high-performance list. From both the Fig. 3, it is noticed that without impacting the risk score, if the additional functionalities can be added to the apps of cluster 0 in Fig. 3b, and cluster 3 in Fig. 3a, then the performance value may be improved.

From the above, we can identify three typical trends in app development to yield low-performance, they are: (i) Enhancement of functionalities which occupy larger memory (ii) Implementing features without adequate security measures (iii) Building functionalities of elevated security impact. Hence, the study recommends that developers focus on functionality enhancement by adhering to security measures and reducing the app's memory size in order to improve app performance.

5. **Static security testing analysis:** We have conducted static analysis testing over COVID-19 apps in the dataset. As a result, we found 17 vulnerabilities of which eight (47.8%) vulnerabilities are identified as low level, and five (29.4%) are at medium level vulnerabilities, and remaining are at high level. There are no critical level vulnerabilities. In most of the apps (above 80%) "missing tap jacking protection" and "hard coded data" which are considered as low-level vulnerabilities, have been identified. More than half of the apps contained one high-level (but not critical) vulnerability, "Clear text SQLite Database" (55.83%), and one medium-level vulnerability, "Predictable Random number generator" (65% of apps). These are related to data storage and cryptography methods. Finally, we observed that the majority of the apps in the dataset are secure as we found no critical vulnerabilities.

## 5.2 Comparison of Proposed Method with Other MCDM Techniques

The benefits of the proposed methodology are demonstrated in this section by choosing FBWM and FAHP models for comparison. As the validity of both

**Table 6.** Comparison of FBWM in MCMARS with other MCDM techniques

| Criteria | FAHP [18] | EFAHP [17] | FBWM [20] | Proposed method |
|---|---|---|---|---|
| C1 (Risk) | 0.319 | 0.342 | 0.346 | 0.34 |
| C2 (Rating) | 0.183 | 0.158 | 0.205 | 0.22 |
| C3 (Size) | 0.173 | 0.172 | 0.136 | 0.140 |
| C4 (Functionality) | 0.325 | 0.328 | 0.313 | 0.30 |
| Comparisons | n(n−1)/2 | n(n−1)/2 | 2n−3 | 2n−3 |
| Computational complexity | $O(n^2)$ | $O(n^2)$ | O(n) | O(n) |
| Avg. execution time | 9.62 ms | 40.5 ms | 24.5 ms | 10.5 ms |

techniques are predicated on meeting the constraints of mathematical transitivity and pairwise comparison of criteria. In comparison with the above subjective model, to find $n$ criteria coefficients, the HTFBWM requires $2n-3$ pairwise comparisons whereas the FAHP method requires $n(n-1)/2$ comparisons. An increase in the number of criteria in FAHP models produces an exponential increment in pairwise comparisons, which highly increases the computational complexity. For instance, for determining the relative weights of 9 criteria, FBWM requires 15 comparisons, where as FAHP requires 36 pairwise comparisons. By using the methods, FAHP with geometric mean (FAHP-G) [18], FAHP with extent analysis (EFAHP) [17], the method of FBWM in [20] on the dataset in this paper, it has been shown that our proposed application of HTFBWM gives better computational complexity and execution time (shown in Table 6). Because, unlike conventional FBWM, our MCMARS first aggregates the relative preferences of experts and then follows the remaining steps to find optimal weights for the criteria. That means, in the case of $r$ number of experts participating in FBWM with group decision making, $r$ number of iterations will take place for aggregation in conventional FBWM, whereas in our case, only one iteration is carried out for finding the optimal weights.

## 6   Conclusion and Future Work

In this paper, we proposed a novel multi-criteria decision-making based recommender system that helps developers and end users to take decisions on apps. Through MCMARS, evaluation of the relative performance of mobile apps is carried out based on the essential criteria, viz. security risk, services provided, memory size and usability rating, which are necessary to consider while downloading a app. Most of the papers in the literature report the app security testing based either on insecure data-access permissions or static security analysis testing or user reviews. In this paper, we perform both permission-based and static security testing to find the risk score by emphasizing the users' preferences. Furthermore, functionality score based on services offered by the apps has been proposed. MCMARS combines the MCDM techniques such as HTFBWM with group decision-making to obtain realistic judgements and TOPSIS while evaluating apps' relative performance scores. Our approach of HTFBWM with group decision-making ends with better computational complexity and execution time than traditional FBWM. Our case study reveals the efficacy and effectiveness of MCMARS in addressing some underlying issues. From this, we arrived at following conclusions and recommendations:

1. Through the relative performance of an app, MCMARS can effectively recommend developers to upgrade their apps with respect to specific aspects.. Also, can recommend high-performance apps to the consumer without going through overwhelming technicalities while selecting public domain apps like health and government service apps.
2. The findings from the case study highlights that a significant number of apps are built with high functionality but with moderate and higher risks which puts these apps at lower performance level. Developers may need to give priority to minimize risk and memory efficiency so that the app can be placed at an expected level of overall performance. Our analysis found that only 12.5% of the apps from experimental dataset provide high performance with high functionality and low risk.
3. We have analyzed consistency of COVID-19 Android apps' with the OWASP mobile top 10 security risks. From the results, we learnt that there is a potential scope for improvement to make apps free from any kind of vulnerabilities. Developers can make use of these results to improve the quality of their apps.

In future, we will concentrate on incorporating large-scale group decision-making and including social network relations among the decision-makers to generate trustworthy multi-criteria recommendations with good performance.

# References

1. Alazab, M., Shalaginov, A., Mesleh, A., Awajan, A.: Intelligent mobile malware detection using permission requests and API calls. Futur. Gener. Comput. Syst. **107**, 509–521 (2020)
2. FIRST: Cvss v3.0 specification document, first (2015). https://www.first.org/cvss/v3.0/specification-document. Accessed 26 Jan 2021
3. Gates, C.S., et al.: Generating summary risk scores for mobile applications. IEEE Trans. Dependable Secure Comput. **11**(3), 238–251 (2014)
4. Gómez, M., Rouvoy, R., Monperrus, M., Seinturier, L.: A recommender system of buggy app checkers for app store moderators. In: 2015 2nd ACM International Conference on Mobile Software Engineering and Systems, pp. 1–11. IEEE (2015)
5. Hazrati, N., Ricci, F.: Recommender systems effect on the evolution of users' choices distribution. Inf. Process. Manage. **59**(1), 102766 (2022)
6. Huang, W., Cao, B., Yang, G., Luo, N., Chao, N.: Turn to the internet first? Using online medical behavioral data to forecast COVID-19 epidemic trend. Inf. Process. Manage. **58**(3), 102486 (2021)

7. Jain, P., Sharma, A., Aggarwal, P.K.: Key attributes for a quality mobile application. In: 2020 10th International Conference on Cloud Computing, Data Science & Engineering (Confluence), pp. 50–54. IEEE (2020)

8. Jisha, R., Krishnan, R., Vikraman, V.: Mobile applications recommendation based on user ratings and permissions. In: 2018 International Conference on Advances in Computing, Communications and Informatics (ICACCI), pp. 1000–1005. IEEE (2018). https://doi.org/10.1109/ICACCI.2018.8554691

9. Jorgensen, Z., Chen, J., Gates, C.S., Li, N., Proctor, R.W., Yu, T.: Dimensions of risk in mobile applications: a user study. In: Proceedings of the 5th ACM Conference on Data and Application Security and Privacy, pp. 49–60. ACM (2015). https://doi.org/10.1145/2699026.2699108

10. Kumar, G., Parimala, N.: A sensitivity analysis on weight sum method MCDM approach for product recommendation. In: Fahrnberger, G., Gopinathan, S., Parida, L. (eds.) ICDCIT 2019. LNCS, vol. 11319, pp. 185–193. Springer, Cham (2019). https://doi.org/10.1007/978-3-030-05366-6_15

11. Liu, B., Kong, D., Cen, L., Gong, N.Z., Jin, H., Xiong, H.: Personalized mobile app recommendation: reconciling app functionality and user privacy preference. In: Proceedings of the Eighth ACM International Conference on Web Search and Data Mining, pp. 315–324. ACM (2015). https://doi.org/10.1145/2684822.2685322

12. Liu, Y., Eckert, C.M., Earl, C.: A review of fuzzy AHP methods for decision-making with subjective judgements. Expert Syst. Appl. **161**, 113738 (2020). https://doi.org/10.1016/j.eswa.2020.113738

13. Ma, Y., Liu, X., Liu, Y., Liu, Y., Huang, G.: A tale of two fashions: an empirical study on the performance of native apps and web apps on android. IEEE Trans. Mob. Comput. **17**(5), 990–1003 (2017)

14. Peng, M., Zeng, G., Sun, Z., Huang, J., Wang, H., Tian, G.: Personalized app recommendation based on app permissions. World Wide Web **21**(1), 89–104 (2017). https://doi.org/10.1007/s11280-017-0456-y

15. Rezaei, J.: Best-worst multi-criteria decision-making method. Omega **53**, 49–57 (2015)

16. Rocha, T., Souto, E., El-Khatib, K.: Functionality-based mobile application recommendation system with security and privacy awareness. Comput. Secur. **97**, 101972 (2020). https://doi.org/10.1016/j.cose.2020.101972

17. Samanlioglu, F., Taskaya, Y.E., Gulen, U.C., Cokcan, O.: A fuzzy AHP-TOPSIS-based group decision-making approach to IT personnel selection. Int. J. Fuzzy Syst. **20**(5), 1576–1591 (2018). https://doi.org/10.1007/s40815-018-0474-7

18. Thapar, S.S., Sarangal, H.: Quantifying reusability of software components using hybrid fuzzy analytical hierarchy process (FAHP)-metrics approach. Appl. Soft Comput. **88**, 105997 (2020)

19. Torra, V.: Hesitant fuzzy sets. Int. J. Intell. Syst. **25**(6), 529–539 (2010)

20. Xu, Y., Zhu, X., Wen, X., Herrera-Viedma, E.: Fuzzy best-worst method and its application in initial water rights allocation. Appl. Soft Comput. **101**, 107007 (2021)

21. Zhu, H., Xiong, H., Ge, Y., Chen, E.: Mobile app recommendations with security and privacy awareness. In: Proceedings of the 20th ACM SIGKDD International Conference on Knowledge Discovery and Data Mining, pp. 951–960. ACM (2014). https://doi.org/10.1145/2623330.2623705

# Opinion Maximization in Signed Social Networks Using Centrality Measures and Clustering Techniques

Leela Srija Alla and Anjeneya Swami Kare[✉]

School of Computer and Information Sciences, University of Hyderabad,
Hyderabad, India
{17mcme18,askcs}@uohyd.ac.in

**Abstract.** In a social network, OPINION MAXIMIZATION is a problem that targets spreading the desired opinion across the social network. In the real world, every user/individual has their own opinion. The opinion of an individual can be favorable or unfavorable. Each individual can affect the others positively or negatively. In this paper, we consider the OPINION MAXIMIZATION problem for signed, weighted and directed social networks with Multi-Stage Linear Threshold Model for information propagation. The seed nodes are responsible for spreading the desired opinion in the network. The OPINION MAXIMIZATION problem asks to compute the minimum number of seed nodes such that the overall opinion of the network is maximized.

In this paper, we proposed three heuristics to select the seed nodes. The proposed methods use centrality measures and clustering of the social network. The proposed methods are tested on real-world as well as synthetic data sets.

**Keywords:** Opinion maximization · Signed social networks · Dynamic opinions

## 1 Introduction

Social network analysis has lately garnered attention due to the quick expansion of social media. It can offer valuable insights in various fields like economics [1], healthcare [2], and social studies [3–5]. Influence Maximization (IM) [6–15] mainly used for product promotions, and rumor control [16–19] is one of the typical applications of social network analysis. A practical extension of IM is Opinion maximization (OM). IM assumes all the users are potential customers, which is not the case in the real world. In contrast, OM considers that users/consumers can have variety of opinions, which makes it unlikely for some people to purchase a specific product or hold the desired opinion.

There are many instances where a business or organization's interest would be to sway the opinions of its clients, customers, or just a group of individuals. The organisation would benefit greatly from the capacity to create an

A. R. Molla et al. (Eds.): ICDCIT 2023, LNCS 13776, pp. 125–140, 2023.
https://doi.org/10.1007/978-3-031-24848-1_9

algorithm that can simulate the behavioural changes in people. In real world, every user/individual have their own opinion. Opinion of an individual can be favourable or unfavourable. OM is a way to manipulate users opinions using his/her connections in the network.

For Example - During the COVID 19 pandemic, a section of India's population hesitated to take the vaccines provided by the government. Although, eventually government used different methods to assure people the safety of vaccines in various ways, using actors, influencers and politicians that eventually led to a sizable population taking the vaccines. India's citizens serve as the network's nodes in this scenario, while the connections between them serve as its edges. The government needed to change people's perceptions of vaccinations, and to do this, it turned to the network's powerful or influential nodes- actors and politicians.

The addition of negative opinions may alter the characteristics of the objective function that is used to represent propagation of opinion for the OM problem. Thus, objective of OM is to spread the desired opinion using the influential nodes and maximize its desired opinion propagation within a certain time.

We consider a directed, weighted signed social network $G = (V, E, W)$, where $V = \{v_1, v_2, \cdots v_n\}$ and $E = \{e_1, e_2, \cdots, e_m\}$ are the vertex and edge sets respectively, and $W = \{w_{e_i} | e_i \in E\}$ represent the weights of the edges. A node can influence neighbors positively or negatively. If the sign of edge $e_i = (u, v)$ is positive then the node $u$ has positive influence on the node $v$, otherwise the influence is negative. The weight of the edge represents the quantum of the influence.

For the OPINION MAXIMIZATION problem apart from the input the information diffusion model is also important. In this paper, we consider the Multi-stage Linear Threshold model [20] for information diffusion. The OPINION MAXIMIZATION problem asks to compute the minimum number of seed nodes such that the overall opinion is maximized. In this paper, we propose few approaches to compute the seed set.

## 2    Related Work

Influence Maximization (IM) has been a problem of interest for several years now. It was first formulated as an optimization problem by Kempe et al. [15]. Zhu et al. [8] proposed a Balanced Influence and Profit (BIP) maximization problem. They considered PR-L (Price related Linear Threshold) and PR-I (Price realted Independent cascading) models and proposed a greedy method for the BIP problem. A class of impact models with an empirical motivation was defined by Aral et al. [10] and their effects on the IM were examined. They proposed a greedy algorithm to solve the IM problem based on the monotonicity and submodularity of the influence spread function. Liang et al. [12] proposed R (Restricted)-Greedy and Linear Threshold in Signed networks (LT-S) to formulate an RLP (R-Greedy with Live-edge and Propagation-path) model. He et al. [13] made a two-stage iterative framework for the influence maximization using an overlapping influence. Yao et al. [6] proposed a model where the rumour and the truth are spreading simultaneously. The paper concentrates on how to choose boosting-blocking users to regulate the spread. The study developed a Multi Hop Neighbour Boosting (MHNB) approach to address it.

Gionis et al. [21] discussed a *CAMPAIGN* problem, considering opinion cannot have a binary state - active and inactive. The problem is based on the presumption that opinions can have continuous values over an interval and that each user can have both an internal and an expressed opinion that is constantly changing. A discrete cascade model is combined with this to spread opinion. Lue et al. [4] proposed an *AcTive Opinion Maximization* that uses an active learning framework aCtive OpinioN Estimator (CONE) that uses both dynamic opinions and Multi Stage Linear Threshold model to diffuse the opinion in unsigned undirected graphs.

He et al. [22] put up a model that took the network's users' unfavourable opinions into account in an unsigned undirected network. A parameter known as the quality factor was introduced to an IC-N (Independent Cascading - Negative) model that simulates users' typical behaviour. Another model proposed by them was a *Multi-Stage Opinion Maximization Scheme (MOMS)*, which uses Multi Stage IC model combined with dynamic opinion formation in unsigned directed networks. A *Multi-Stage Heuristic Algorithm (MHA)* [23] was also developed by them for similar kind of networks blended with *Dynamic Opinion Maximization Framework (DOMF)* using Q-learning for convergence. *Reinforcement learning based Opinion maximization (RLOM)* [1] is also a Q-learning based algorithm developed for unsigned directed networks in a competetive network where a user may not be a potential customer. An *Active opinion maximization framework (AOMF)* [20] was also introduced in particular for signed directed networks. This model also simultaneously addresses dynamic opinion development and node activation.

Nayak et al. [3] modelled interactions as a *Dynamic Bayesian network (DBN)*. Unlike most other papers where OM focuses on changing the opinion of the users, this model poses OM as an information spreading problem. The study proposes a centralised and a de-centralised algorithm considering the intractability and scalability issues respectively.

Cao et al. [24] proposed Optimal Allocation in Social NETworks (OASNET) which is the first community based approach for the IM problem. Zhang et al. [25] developed an algorithm to select influential nodes from the network using community detection. An information transfer probability matrix was used with $k$-medoid clustering algorithm to select seed nodes. Most of the community detection algorithms uses modularity and assume networks with positive edges. Tragg et al. [26] used Potts [27] model to consider both positive and negative links in the network.

Guo et al. [28] proposed an algorithm for unweighted/weighted, un-directed/ directed and signed networks by developing a method that uses dissimilarity distance matrix and Affinity propagation to find the cluster centers. Androulidakis [29] introduced two kinds of matrices co-reference and co-citation matrices for signed directed networks to form a similarity matrix, which can further be used as inputs for $K$-means clustering and Affinity Propagation Algorithms. Kessler [30] first developed a co-reference matrix to count the number of papers that are commonly cited by two documents. Co-citation matrix which was first given by Small [31], is the number of documents that cite both documents as part of a bibliometric study. Table 1 compares few relevant algorithms.

**Table 1.** Comparison between existing models

| Model | Network type | Dynamic opinions | Community detection |
|-------|--------------|------------------|---------------------|
| AOMF  | Signed, Directed | ✓ | |
| MOMS  | Directed | ✓ | |
| CAOM  | Unsigned, Undirected | ✓ | ✓ |
| MHA   | Directed | ✓ | |
| RLOM  | Signed, Directed | ✓ | |
| DBN   | Directed | | |
| ATOM  | Unsigned, Undirected | ✓ | |
| CIM   | Unsigned, Undirected | ✓ | ✓ |
| CoFIM | Unsigned, Undirected | | ✓ |

In this paper, we consider Multi-Stage Linear threshold Model and dynamic opinion propagation. First we compute a set of seed nodes called candidate seed nodes and from the candidate seed nodes, the seed nodes are selected greedily [20]. We proposed the following approaches to solve the OM problem.

1. Selecting the candidate seed nodes using centrality measures of signed social networks.
2. Selecting the candidate seed nodes using communities in the signed social network.
3. Selecting the candidate seed nodes based on both centrality measures and communities of the signed social networks.

We have done an extensive experimentation on both real world and synthetic datasets. The proposed methods are compared with themselves and $AOMF$ [20].

The rest of the paper is structured as follows. Section 3 discusses the proposed approaches. The experimental results are presented in Sect. 4. Finally, the paper is concluded in Sect. 5.

## 3    Proposed Approaches of OM

In this section, first we discuss the mathematical formulation of the OM problem and then we discuss the proposed approaches. The opinion maximization is mathematically described as $\Gamma(S)$. We need to maximise positive nodes and reduce negative nodes.

$$\Gamma(S) = \Sigma_{v_i \epsilon C^+} o_i + \Sigma_{v_j \epsilon C^-} o_j \qquad (1)$$

Here, $C^+$ is the set of nodes with positive opinions and are activated and $C^-$ is the set of activated nodes with negative opinions. $S$ represents the seed node set.

## 3.1   Diffusion Model

Multi-Stage Linear Threshold (MSLT) model is used to spread the desired opinion across the network. According to MSLT, a node $v_i$ is said to be active at time $t$, if $\Sigma_{v_j \epsilon C^{(t-1)}} w_{ji} > \theta$, where $C^{(t-1)}$ is the set of active users before stage $t$ starts and node $v_j$ is in-neighbor of node $v_i$. $\theta$ is the threshold to be reached by the node to be activated.

Each user has a static opinion and dynamic opinion. The dynamic opinion is updated using the Eq. 2. To find the updated opinion at every stage a variant of De-Groot model that considers both static and dynamic opinions was introduced. The opinion at time $t + 1$ is given by

$$o_i^{(t+1)} = \frac{s_i + \Sigma_{j \epsilon N_i^{in}} I_{ji} w_{ji} s_j}{1 + \Sigma_{j \epsilon N_i^{in}} I_{ji} w_{ji} o_j^t} \qquad (2)$$

where $s_i$ is the static opinion of node $v_i$ and $I_{ji}$ is $+1$ or $-1$ depending on the sign of edge between node $v_j$ and $v_i$. $w_{ji}$ is the weight on the edge between node $v_j$ and $v_i$.

## 3.2   Centrality Measures Based Approach

Two centrality measures given by Liu et al. [32] specially made for signed directed networks are used to select candidate seeds. The two measures $\eta$ and $\mu$, represent the total effect of the node on the network and total effect of the network received by the node respectively.

**Deriving $\eta$ and $\mu$.** It starts with calculating the effect of nodes on each other. Consider an signed directed graph, the effect of node $v_i$ on node $v_j$ (which means there is an edge coming from $i$ to $j$) is defined as $a_{ij}$

$$a_{ij} = I_{ij} \times \frac{1}{D_j} \qquad (3)$$

where $D_j$ is the degree of node $v_j$ and $I_{ij}$ is the sign of the edge between nodes $v_i$ and $v_j$. Consider $v_k$ is a neighbour of $v_j$ and $v_j$ is a neighbour of $v_i$, the effect of node $v_i$ on $v_k$ through node $v_j$ is given by $a_{ik}$

$$a_{ik} = a_{ij} \times a_{jk} = (I_{ij} * \frac{1}{D_j}) \times (I_{jk} * \frac{1}{D_k}) \qquad (4)$$

where $D_k$ is the degree of node $v_k$. So, to calculate effect of node $v_i$ on $v_k$, we should consider all the nodes that are in the path $v_i \rightarrow v_j$.

So, the total effect of the node $v_i$ on node $v_j$ in $n$ steps is represented by $TE_{ij,n}$. If there are $p$ $n$-step pathways from node $v_i$ to $v_j$ then $TE_{ij,n}$ represents the sum of those $p$ effects. Cumulative effect of node $v_i$ on node $v_j$ is given by

$$CTE_{ij,n} = g(1) * TE_{ij,1} + g(2) * TE_{ij,2} + g(3) * TE_{ij,3} + \dots g(n)TE_{ij,n} \quad (5)$$

$g(n)$ is a weight function in Eq. 5 and it is defined as

$$g(p) = \frac{1}{p} \quad (6)$$

An $N \times N$ matrix is generated for the network where $N$ is the total number of nodes in the network. Sum of all the values in row $i$ gives the total effect of node $v_i$ on the whole network($\eta_i$).

$$\eta_{i,n} = \Sigma_{j=1}^{N} CTE_{ij,n} \quad (7)$$

Sum of all the values in column $i$ gives the total effect received by node $v_i$ from the network ($\mu_i$).

$$\mu_{i,n} = \Sigma_{i=1}^{N} CTE_{ij,n} \quad (8)$$

Using the centrality measures derived, a heuristic to calculate influence score that combines static and dynamic opinions of the node is developed. The first $l * k$ nodes with highest influence score are selected.

To calculate influence score, we developed a heuristic that measures the influence of a node in the network. The influence score of node $v_i$ is given by

$$IS_i = s_i + \eta_i + \kappa * (\mu_i) \quad (9)$$

where $s_i$ is the static opinion of node $v_i$, $\kappa$ is a weight parameter, $\eta_i$ is the total effect of the node $v_i$ on the network and $\mu_i$ is the total effect received by $v_i$ from the network. Algorithm 1 shows the selection of candidate seed nodes using Eq. 9.

---

**Algorithm 1.** Selection of candidate seed nodes using centrality measures

---

**Input**: Network $G = (V, E, W)$ number of seed nodes $k$, parameter $l$, parameter $\kappa$
**Output**: Candidate seed node set

1: **procedure** SELECTCANDIDATE($G$, $l$, $k$)
2:     $CD \leftarrow \phi$                    ▷ Initialize candidate seed node set
3:     **for each** $v_i \in V$ **do**
4:         $IS_i = s_i + \eta_i + \kappa * (\mu_i)$
5:     **end for**
6: Rank the nodes and according to Influence score($IS_i$) and select the highest influence score $l * k$ nodes
7: **end procedure**

---

To select seed nodes, we add the node to the seed node set and check the increase in potential opinions. The change in opinions of the seed node set can be defined as follows

$$PO_v = \Gamma(S \cup v) - \Gamma(S) \tag{10}$$

As we are trying to maximize the opinions, we are selecting node with highest $PO_v$ value. That is, the seed nodes are selected using greedy method.

In each stage of MSLT, the opinions are dynamically updated using Eq. 2, nodes are activated and new nodes are added to seed node set similar to that used in *AOMF*. Algorithm 3 shows the complete algorithm with SELECTCANDIDATE$(G, l, k)$ from Algorithm 1.

### 3.3 Community Detection Based Approach

He et al. [29] proposed a community based approach for unsigned undirected networks. We will now apply a similar algorithm for signed weighted directed network combined with AOMF [33].

The algorithm uses two communication detection algorithms introduced by Androulidakis [29] for signed directed networks. A similarity matrix is created by representing the network as a collection of data clusters and using the network's topological qualities to determine how similar the nodes are to one another. Two different kinds of similarity matrices are developed. One to measure incoming similarity, $S_{in}(i, j)$ is the number of nodes that effects both $v_i$ and $v_j$ either positively or negatively, other outgoing similarity, $S_{out}(i, j)$ is the number of nodes that are being positively or negatively influenced by both $v_i$ and $v_j$. The final similarity is denoted by $Similarity[i, j]$.

$$Similarity[i, j] = \frac{S_{in}(i, j) + S_{out}(i, j)}{max(D(i), D(j))} \tag{11}$$

where $D(i)$, $D(j)$ denotes the total number of edges involving $v_i$ and $v_j$ respectively.

The similarity matrix will then be fed to Affinity propagation or $K$-means clustering to divide into communities. After detecting communities, to further reduce complexity, significant communities in the network are selected. A heuristic for the same was developed by He at al. [33].

$$sig_t = \frac{\Sigma_{i=1}^{p} n_i}{k} \tag{12}$$

where $sig_t$ represents the threshold, the community should reach for it to be significant community, $n_i$ represents the number of nodes in that community and $p$ is the total number of communities. So, if the number of nodes in each community is greater than or equal to $sig_t$, it is considered a significant community.

It is more efficient to select candidate seed nodes depending on the significance of the community, instead of selecting equal number of candidates from each community. To determine the number of candidate seed nodes to be selected from each community we use

$$CS_i = \frac{\lambda * k * n_i}{\Sigma_{h=1}^{g} n_h} + 1 \qquad (13)$$

where $CS_i$ represents number of candidate seed nodes to be selected from significant community $i$. $\lambda$ represents weight parameter, $k$ is the number of seed nodes to be determined in the end, $n_i$ is the number of nodes in community $i$ and $g$ represents the number of significant communities. Algorithm 2 summarizes the selection of candidate seed nodes from the communities.

---

**Algorithm 2.** Selection of candidate seed nodes

---

**Input**: Network $G = (V, E, W)$, number of seed nodes $k$, parameters $l$, $\lambda$, $\gamma$, $\delta$
**Output**: Candidate seed node set CD

 1: **procedure** SELECTCANDIDATE($G$, $l$, $k$)
 2:     Divide G into communities using Kmeans or Affinity
 3:     Calculate $sig_t$ using Equation 12
 4:     Consider $C = \{C_1, C_2, C_3, ...C_d\}$          ▷ $d$ being the number of communities
 5:     $C_s \leftarrow \phi$                              ▷ Initialize significant community set
 6:     **for** each $C_i$ in $C$ **do**
 7:         $n_i \leftarrow |C_i|$                        ▷ $|C_i|$ is the number of nodes in $C_i$
 8:         **if** $n_i \geq sig_t$ **then**
 9:             $C_s \leftarrow C_s \cup C_i$
10:         **else**
11:             $C_i$ is insignificant community
12:             continue
13:         **end if**
14:     **end for**
15:     Initialize candidate seed node set $CD \leftarrow \phi$
16:     **for** each $C_i$ in $C_s$ **do**
17:         Calculate $CS_i$ using Equation 13
18:         **for** $v_i \in C_i$ **do**
19:             Calculate $IS_i$ using Equation 14
20:         **end for**
21:         Rank the nodes and according to Influence score and select $CS_i$ highest influence score nodes
22:     **end for**
23:     Rank all the nodes from significant communities and select $l * k$ nodes with highest influence score
24: **end procedure**

---

Once significant communities are chosen, we coupled heuristics developed by He et al. [20] that employs $O_i^{out}$ - neighbour sets of node $v_i$ beyond the community and $O_i^{in}$ neighbour sets within the community [33] with influence score in that uses the static opinions of nodes [20] to determine the most influential nodes.

$$IS_i = |O_i^{in}| + \gamma|O_i^{out}| + \delta(s_i) \tag{14}$$

The algorithm for selecting seed nodes from the candidate seed nodes obtained is same as the one used in Centrality Measures Based Approach. Algorithm 3 shows the complete approach with SELECTCANDIDATE$(G, l, k)$ from Algorithm 2.

### 3.4  Centrality Measures and Community Structure Based Approach

Here, we use community detection techniques used in Community Detection Based Approach and change the formula used for calculating influence score. In this method, we calculate influence score after community detection, using centrality measures proposed in Centrality Measures Based Approach.

After detecting significant communities, the following heuristic is used to calculate influence score.

$$IS_i = \eta_i * (s_i) + |O_i^{in}| + \gamma|O_i^{out}| \tag{15}$$

where $\eta_i$ is the total effect of the node on the network, $s_i$ is the static opinion of the node. Hence, in Algorithm 2 we use Eq. 15 instead of Eq. 14.

### 3.5  Example Network

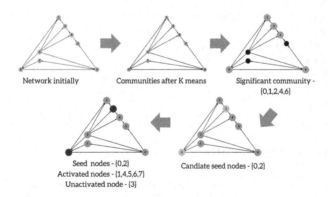

**Fig. 1.** Opinion maximization using $K$-Means and centrality measures

The network in Fig. 1 contains 8 nodes with opinions - (node 1 - opinion 0.8, 2 - 0, 3 - -0.5, 4 - 1, 5 - 0.6, 7 - 0.3, 8 - 0.2) with edges [(1,2), (3,2), (4,1), (4,3), (4,5), (4,7), (7,6), (7,8), (8,1) ] having sign +1 and [(1,7), (3,5), (4,2), (5,6)] having sign -1. All the edge weights are considered 1. Nodes 1,4,5,6,7 are activated on implementing Centrality Measures and Community Structure Based Approach using $K$-means clustering. The method for selecting seed nodes and opinion propagation is the same as the above algorithms. Algorithm 3 shows the complete algorithm.

**Algorithm 3.** OM using Community Detection and MSLT

**Input**: Network G = (V,E,W), number of seed nodes k, threshold $\theta$,parameter l
**Output**: Seed node set S.

1: $CD \leftarrow$ SELECTCANDIDATE$(G, l, k)$                    ▷ Obtain candiate seed node set
2: $AC_{-1} \leftarrow \phi$ ▷ $AC$ represents activated node set and initially consider no nodes are activated
3: **for** $r = 1 : R$ **do**
4:    $AC_{(r)} \leftarrow \phi$                    ▷ Initialize activated seed node set at that stage
5:      **for** $t = 1 : k/R$ **do**
6:        **for** $v \in CD/S$ **do**
7:            Optimal potential opinion $v^* \leftarrow argmax_{v \in CD/S}\Gamma(S \cup v) - \Gamma(S)$
8:        **end for**
9:        $S^{(r)} \leftarrow S^{(r)} \cup \{v*\}$
10:        $CD \leftarrow CD\backslash\{v*\}$
11:        $AC_{(r-1)} = AC_{(r-1)} \cup S$
12:        **for** each $v_i \in V$ **do**
13:            **if** $\Sigma_{v_j \in AC_{(r-1)}} w_{ji} > \theta$ **then**
14:                Node $v_i$ is activated
15:            **else**
16:                Continue
17:            **end if**
18:            $AC_{(r-1)} \leftarrow AC_{(r)} \cup v_i$
19:            $V \leftarrow V/v_i$
20:        **end for**
21:      **end for**
22:      **for** each $v_j \in AC_{(r-1)}$ **do**
23:        **for** $t = 1 : t$ **do**
24:            Calculate $o_j^t$ using Equation 2
25:        **end for**
26:      **end for**
27: **end for**
28: Obtain seed node set

## 4  Experimental Results

In this section, we will give more information about datasets used and discuss the results obtained by implementing AOMF and three proposed methods.

### 4.1  Datasets

In this paper, nine social networks are used as part of experimentation. Five datasets are randomly generated using number of nodes and connectivity of the network as input. For all five networks a connectivity of 0.75 was used. Real-world datasets Bitcoin Alpha, Bitcoin OTC [34,35], Wiki Admin [36] and Wiki Elec [37] are used. All the weights are generated randomly in the range $(0,1)$. Sign of the edge is either randomly generated or scaled to $-1$ to $+1$ depending on the dataset.

## 4.2  Parameters

In this section, we discuss the parameters that should be defined for the algorithms.

- All three methods need $R$ - number of stages, $l$ - proportion parameter and $k$ - number of seed nodes as input initially and $\theta$ threshold of MSLT.
- In Centrality Measures Based Approach, $\kappa$ is a weight parameter for centrality measure $\mu$.
- In Community Detection Based Approach, to calculate influence score for selection of candidate seed nodes parameters - $\gamma, \delta$ are weight parameters for nodes outside community and static opinion respectively. As $K$-means requires number of clusters as input, we divide the network into $\sqrt{\frac{n}{2}}$ where $n$ is the number of nodes in the network.
- In Centrality Measures and Community Structure Based Approach, to calculate influence score, the parameter $\gamma$ is the same as the one used on Community Detection Based Approach.

## 4.3  Experiments

- For selecting candidate seed nodes we experimented using different values of weight parameters, values of $R$ ranging from 1 to 5, number of seed nodes $k$ ranged from 1 to 10, $l$ value was from range $(1, 2)$ and $l > 1$. Value of $\theta$ is set to 0.01 in all cases.
- $p$ value in Centrality Measures Based Approach that represents number of steps to calculate the centrality measures is set to 2 in our case for all datasets and $\kappa$ value is taken as 0.1.
- We experimented with $\gamma$ values in the range $(1, 5)$ and $\delta$ value is set to 1.

## 4.4  Results

Table 2 shows the values of initial positive node ratio and total opinion of the network. The first column in the table is the dataset used and second column represents positive ratio and third represents sum of opinions of all nodes in the network. The positive ratio of a network is defined as:

$$P_N = \frac{\text{Number of nodes with positive opinion}}{\text{Total number of nodes in the network}} \qquad (16)$$

Table 3 shows the $P_N$ values and sum of opinions of all nodes in the network after OM using $AOMF$[1] and methods proposed in the paper. The $R$, $l$, $k$ and $\gamma$ values are changed and experimented with for all algorithms.

---

[1] We implemented AOMF [20].

**Table 2.** Initial values

| Dataset | $P_N$ initially | Sum initially |
|---------|-----------------|---------------|
| 100 nodes | 0.47 | −4.27 |
| 500 nodes | 0.484 | −14.87 |
| 1000 nodes | 0.499 | −22.54 |
| 2000 nodes | 0.499 | −17.53 |
| 3000 nodes | 0.488 | −22.04 |
| Bitcoin Alpha | 0.497 | −16.37 |
| Bitcoin OTC | 0.505 | 47.72 |
| Wiki Admin | 0.498 | −22.10 |
| Wiki Elec | 0.499 | −3.41 |

**Table 3.** $P_N$ values and Sum of opinions of all nodes in the network on implementing AOMF, Method 1, Method 2 and Method 3

| Datasets | AOMF | | Method 1 | | Method 2 | | | | Method 3 | | | |
|----------|------|-----|----------|-----|----------|-----|----------|-----|----------|-----|----------|-----|
| | | | | | Affinity | | K means | | Affinity | | K means | |
| | $P_N$ value | Sum | $P_N$ value | Sum | $P_N$ value | Sum | $P_N$ value | Sum | $P_N$ value | Sum | $P_N$ Value | Sum |
| 100 nodes | 0.53 | −3.41 | 0.53 | −3.41 | 0.52 | 51.28 | 0.52 | 51.289 | 0.52 | 51.289 | 0.52 | 51.289 |
| 500 nodes | 0.49 | 85.522 | 0.484 | 4.90 | 0.49 | 85.522 | 0.49 | 85.522 | 0.484 | 4.904 | 0.484 | 4.904 |
| 1000 nodes | 0.5 | 96.917 | 0.5 | 96.917 | 0.5 | 96.917 | 0.5 | 96.917 | 0.5 | 96.917 | 0.5 | 96.917 |
| 2000 nodes | 0.530 | 171.556 | 0.4945 | 328.312 | 0.538 | 274.078 | 0.538 | 274.078 | 0.507 | 234.480 | 0.507 | 234.480 |
| 3000 nodes | 0.553 | 294.209 | 0.488 | 201.234 | 0.5 | 790.995 | 0.5 | 790.995 | 0.488 | 201.234 | 0.488 | 201.234 |
| Bitcoin Alpha | 0.516 | 618.136 | 0.507 | 53.049 | 0.516 | 618.136 | 0.516 | 618.136 | 0.507 | 53.049 | 0.507 | 53.049 |
| Bitcoin OTC | 0.519 | 185.980 | 0.519 | 185.980 | 0.519 | 185.980 | 0.519 | 185.980 | 0.519 | 185.980 | 0.519 | 185.980 |
| Wiki Admin | 0.512 | 1453.451 | 0.504 | 545.503 | 0.512 | 1453.451 | 0.512 | 1453.451 | 0.504 | 545.503 | 0.504 | 545.503 |
| Wiki Elec | 0.511 | 3425.3296 | 0.503 | 332.746 | 0.511 | 3425.3296 | 0.511 | 3425.3296 | 0.503 | 335.746 | 0.503 | 335.746 |

### 4.5  Inferences

Figure 2 shows a comparison in $P_N$ values initially, with AOMF and three methods proposed. Method 2 has given relatively better results than the other two algorithms in most datasets.

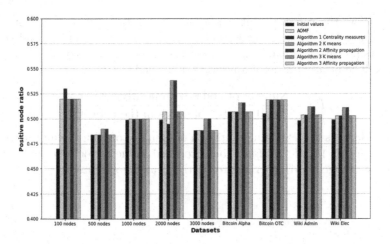

**Fig. 2.** Bar graph representing positive node ratio before and after OM in each dataset

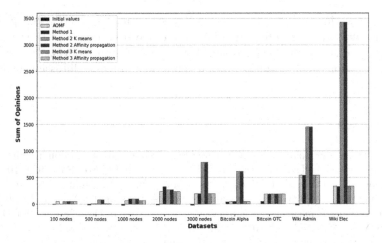

**Fig. 3.** Bar graph representing sum of all opinions before and after OM in each dataset

Figure 3 shows a comparison in initial and final opinions in all datasets using AOMF and the three proposed approaches. In this case too, method 2 increased the potential opinions of nodes in network better than the other two methods and AOMF.

- The final $P_N$ values are almost same for all three methods.
- In Method 2 and 3, $K$-means and Affinity propagation methods, despite detecting slightly different communities sometimes, give same $P_N$ and opinions sum in the end.
- In case of dataset, with 2000 nodes, there is an increase of 1.5% in $P_N$ value.
- For Wiki elec dataset, the difference in the values of sum of opinions in the end is greater than 3000.
- For datasets with less connectivity there is a chance the community detection algorithms used may not converge.

## 5   Conclusions

In this paper, we study OM as an optimization problem in signed directed weighted social networks. We propose three approaches to solve for OM. In the first approach, we proposed a heuristic to select candidate seed nodes using centrality measures. In the second approach, we divide the network into communities and a heuristic was developed to calculate influence of nodes in the communities to select candidate seed nodes. In the third approach, we combine community detection with a heuristic developed using centrality measures. We further determine the seed nodes from the candidate seed nodes in all the methods using an approach similar to that of AOMF. The three algorithms are experimented on nine datasets - 5 randomly generated signed social networks and 4 real world datasets. The community structure based approach provides better results than prevailing algorithms and the ones proposed in the paper. In future, we study the OM problem for other diffusion models. An extensive experimentation on the proposed approaches over different diffussion models is interesting.

# References

1. He, Q., et al.: Reinforcement-learning-based competitive opinion maximization approach in signed social networks. IEEE Trans. Comput. Soc. Syst. **9**, 1–10 (2021)
2. Abebe, R., Kleinberg, J., Parkes, D., Tsourakakis, C.E.: Opinion dynamics with varying susceptibility to persuasion. In: Proceedings of the 24th ACM SIGKDD International Conference on Knowledge Discovery amp; Data Mining. KDD 2018, New York, NY, USA, pp. 1089–1098. Association for Computing Machinery (2018)
3. Nayak, A., Hosseinalipour, S., Dai, H.: Smart information spreading for opinion maximization in social networks. In: IEEE INFOCOM 2019 - IEEE Conference on Computer Communications, pp. 2251–2259 (2019)
4. Liu, X., Kong, X., Yu, P.S.: Active opinion maximization in social networks. In: Proceedings of the 24th ACM SIGKDD International Conference on Knowledge Discovery and Data Mining. KDD 2018, New York, NY, USA, pp. 1840–1849. Association for Computing Machinery (2018)
5. Chen, X., Deng, L., Zhao, Y., Zhou, X., Zheng, K.: Community-based influence maximization in location-based social network. World Wide Web **24**(6), 1903–1928 (2021). https://doi.org/10.1007/s11280-021-00935-x
6. Yao, X., Gao, N., Gu, C., Huang, H.: Enhance rumor controlling algorithms based on boosting and blocking users in social networks. IEEE Trans. Comput. Soc. Syst. (2022)
7. Li, Y., Chen, W., Wang, Y., Zhang, Z.L.: Voter model on signed social networks. Internet Math. **11**(2), 93–133 (2015). A preliminary version appears as "Influence diffusion dynamics and influence maximization in social networks with friend and foe relationships", WSDM'2013
8. Zhu, Y., Li, D., Yan, R., Wu, W., Bi, Y.: Maximizing the influence and profit in social networks. IEEE Trans. Comput. Soc. Syst. **4**, 1–11 (2017)
9. Quach, T.-T., Wendt, J.D.: A diffusion model for maximizing influence spread in large networks. In: Spiro, E., Ahn, Y.-Y. (eds.) SocInfo 2016. LNCS, vol. 10046, pp. 110–124. Springer, Cham (2016). https://doi.org/10.1007/978-3-319-47880-7_7
10. Aral, S., Dhillon, P.: Social influence maximization under empirical influence models. Nat. Hum. Behav. **2**, 375–382 (2018)
11. Topîrceanu, A.: Benchmarking cost-effective opinion injection strategies in complex networks. Mathematics **10**(12), 2067 (2022)
12. Liang, W., Shen, C., Li, X., Nishide, R., Piumarta, I., Takada, H.: Influence maximization in signed social networks with opinion formation. IEEE Access **7**, 68837–68852 (2019)
13. He, Q., Wang, X., Lei, Z., Huang, M., Cai, Y., Ma, L.: TIFIM: a two-stage iterative framework for influence maximization in social networks. Appl. Math. Comput. **354**, 338–352 (2019)
14. Cai, T., Li, J., Mian, A.S., Li, R.H., Sellis, T.K., Yu, J.X.: Target-aware holistic influence maximization in spatial social networks. IEEE Trans. Knowl. Data Eng. **34**, 1993–2007 (2022)

15. Kempe, D., Kleinberg, J., Tardos, E.: Maximizing the spread of influence through a social network. In: Proceedings of the ACM SIGKDD International Conference on Knowledge Discovery and Data Mining, pp. 137–146 (2003)
16. Zhu, J., Ghosh, S., Wu, W.: Robust rumor blocking problem with uncertain rumor sources in social networks. World Wide Web 24, 229–247 (2021). https://doi.org/10.1007/s11280-020-00841-8
17. Ni, Q., Guo, J., Huang, C., Wu, W.: Community-based rumor blocking maximization in social networks: algorithms and analysis. Theoret. Comput. Sci. 840, 257–269 (2020)
18. Guo, J., Chen, T., Wu, W.: A multi-feature diffusion model: rumor blocking in social networks. IEEE/ACM Trans. Networking 29(1), 386–397 (2021)
19. He, Q., Lv, Y., Wang, X., Huang, M., Cai, Y.: Reinforcement learning-based rumor blocking approach in directed social networks. IEEE Syst. J. 16, 1–11 (2022)
20. He, Q., et al.: Positive opinion maximization in signed social networks. Inf. Sci. 558, 34–49 (2021)
21. Gionis, A., Terzi, E., Tsaparas, P.: Opinion maximization in social networks. In: IEEE Transactions on Knowledge and Data Engineering (2013)
22. Chen, W., et al.: Influence maximization in social networks when negative opinions may emerge and propagate, pp. 379–390. SIAM/Omnipress (2011)
23. He, Q., Fang, H., Zhang, J., Wang, X.: Dynamic opinion maximization in social networks. IEEE Trans. Knowl. Data Eng. 35, 350–361 (2021)
24. Cao, T., Wu, X., Wang, S., Hu, X.: OASNET: an optimal allocation approach to influence maximization in modular social networks. In: Proceedings of the 2010 ACM Symposium on Applied Computing, pp. 1088–1094 (2010)
25. Zhang, X., Zhu, J., Wang, Q., Zhao, H.: Identifying influential nodes in complex networks with community structure. Knowl.-Based Syst. 42, 74–84 (2013)
26. Traag, V., Bruggeman, J.: Community detection in networks with positive and negative links. Phys. Rev. E, Stat. Nonlin. Soft Matter Phys. 80, 036115 (2009)
27. Reichardt, J., Bornholdt, S.: Statistical mechanics of community detection. Phys. Rev. E 74, 016110 (2006)
28. Guo, W.F., Zhang, S.W.: A general method of community detection by identifying community centers with affinity propagation. Physica A 447, 508–519 (2016)
29. Androulidakis, M.A.: Community Detection in Signed Directed Graphs. PhD thesis, University of Piraeus (2021)
30. Kessler, M.M.: Bibliographic coupling between scientific papers. Am. Doc. 14, 10–25 (1963)
31. Small, H.: Co-citation in the scientific literature: a new measure of the relationship between two documents. J. Am. Soc. Inf. Sci. 24, 265–269 (1973)
32. Liu, W.C., Huang, L.C., Liu, C., Jordán, F.: A simple approach for quantifying node centrality in signed and directed social networks. Appl. Netw. Sci. 5, 46 (2020)
33. He, Q., et al.: CAOM: a community-based approach to tackle opinion maximization for social networks. Inf. Sci. 513, 252–269 (2020)
34. Kumar, S., Spezzano, F., Subrahmanian, V., Faloutsos, C.: Edge weight prediction in weighted signed networks. In: Data Mining (ICDM), 2016 IEEE 16th International Conference on, pp. 221–230. IEEE (2016)
35. Kumar, S., Hooi, B., Makhija, D., Kumar, M., Faloutsos, C., Subrahmanian, V.: REV2: fraudulent user prediction in rating platforms. In: Proceedings of the Eleventh ACM International Conference on Web Search and Data Mining, pp. 333–341. ACM (2018)

36. West, R., Paskov, H.S., Leskovec, J., Potts, C.: Exploiting social network structure for person-to-person sentiment analysis. Trans. Assoc. Comput. Linguist. **2**, 297–310 (2014)
37. Leskovec, J., Huttenlocher, D., Kleinberg, J.: Signed networks in social media. In: Proceedings of the SIGCHI Conference on Human Factors in Computing Systems. CHI 2010, New York, NY, USA, pp. 1361–1370. Association for Computing Machinery (2010)

# Detection of Object-Based Forgery in Surveillance Videos Utilizing Motion Residual and Deep Learning

Mrinal Raj and Jamimamul Bakas[(✉)] [iD]

School of Computer Engineering, Kalinga Institute of Industrial Technology,
Bhubaneswar 751024, India
mrinalraj860@gmail.com, jamimamul.bakasfcs@kiit.ac.in

**Abstract.** In recent years, video surveillance plays an important role in security applications as well as legal sectors. For example, CCTV footage can be submitted as evidence of a crime scene in a courtroom. However, the recorded video footage can be manipulated, for example, an object can be removed from a video through low-cost and easily available video editing tools. Any sensitive surveillance video should be verified before being accepted as a piece of evidence of events/circumstances. In this paper, we present a motion residual and Deep Learning based forensic approach to identify object-based forged frames in surveillance videos. Firstly, the motion residuals are computed from the videos sequences to extract the video forgery footprints, which are generated due to the manipulation operations. Secondly, the computed motion residuals are fed to VGG-16 network for detecting authentic and forged frames in a surveillance video.

**Keywords:** Intra-frame forgery · Object-based forgery · Surveillance video · Video forensics

## 1 Introduction

In the modern era, the use of cameras is massively increasing for day-to-day uses. Surveillance video has become the most important tool for everyone's life regardless of their positions. It is being used in legal sectors, roads, offices, etc. Even in a courtroom, a CCTV video can be used as evidence in the courtroom. However in today's world manipulating the video is no big deal, it can be done by low-cost editing software and sometimes the editing becomes much good that no one can detect it with the naked eye. Hence, it is very much required to authenticate the video before taking decisions. Video forgery can be categorise into inter-frame forgery and intra-frame forgery based on the operation of forgery [16,18]. Manipulation of a video by deleting, adding, or duplicating some frames inside the video sequence is known as inter-frame forgery. In the other hand, when a manipulation is performed within the selected frames in a video sequence is

A. R. Molla et al. (Eds.): ICDCIT 2023, LNCS 13776, pp. 141–148, 2023.
https://doi.org/10.1007/978-3-031-24848-1_10

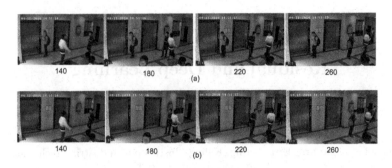

**Fig. 1.** A sample of authentic and forged video sequence [6]: (a) Authentic frames, and (b) Corresponding forged frames.

called as intra-frame forgery [18], which is also known as *object-based forgery in videos*. An example of object-based forgery is shown in Fig. 1, where a person (near to lift) is removed from the video sequence. Due of the greater similarity between manipulated and genuine locations, it is exceedingly challenging to identify forgery.

This paper investigates an object-based forgery in surveillance videos/footage. Due to the limitation of storage, the surveillance videos are always saved in compressed form, and the following steps are performed for creating the object-based forgery [5,6] in a video:

1. Decompress a compressed video into frames.
2. Select some consecutive targeted frames and manipulated those frames by adding/ removing object(s).
3. Re-compress all frames (both authentic and forged) to generate forged a video.

Hence, a forged video contains both authentic (untouched) and forged (manipulated) frames. In other words, all frames in an authentic video are authentic and single-compressed. Whereas, all frames in a forged video are double compressed, but some frames are forged (manipulated) and the remaining are authentic (untouched). Based on this property, we categorize the frames of a forged video into authentic and forged frames. Authentic frames mean those frames are double-compressed but untouched, and forged frames mean those are double-compressed and manipulated. Hand-crafted steganalysis features [9–11] can be used to detect object-based forgery in videos. Recently, Capsule Network based Deep Learning model [5] is utilized to identify the authentic and forged frames in videos. However, the forged frames detection rate of this method is poor (maximum accuracy is only 84.97%), and needs to be improved.

Here, we build a VGG-16 [17] network for detection of object-based video forgery. At first, the motion residuals are computed from the videos, then computed motion residuals are fed into the VGG-16 [17] network for finding the forged frames in a video. We present the efficiency of the presented model for detection of forged frames through a number of experiments on static surveillance videos, with varying video resolution.

The rest paper is as follows: Sect. 2 outlines relevant efforts towards object-based forgery detection in videos. A detailed description of the proposed forensic method for object-based forgery detection is presented in Sect. 3. Section 4 presents and discusses the experiment results. Finally, we concluded the paper by pointing out the future research directions in Sect. 5.

## 2    Related Works

In recent years, many notable works have been found in the field of video forensics [1,4,6,8,12,14], and few of them are for object-based video forgeries [2,5,6]. Object insertion and object removal video forgery are the two forms of object-based video forgery. Blue screen composition is one example of object insertion video forgery. To detect such kind of blue screen compositing, the correlation between blurring features [3] or edge features [20] can be used. The Changes correlation in patterns between these features are exploited to detect forgery. If the background of a video is Blue or Green, the correlation based approaches fails to detect object-based video forgery. Some other methods, such as [21,22], employ local characteristics, such as DCT coefficients [21] or brightness and contrast [22], to find the similarity in between foreground and background in a video. The inconsistency in computed similarity index is helps to identify the video forgery. However, these methods suffers for a low bit rate videos. D'Avino et al. [7] proposed a Deep Learning based forensic technique to inpainting type object-based video forgery.

For detecting the object removal based video forgery, Zhang et al. [23] exploits inconsistencies between trajectory of moving foreground and foreground mosaic by using the ghost shadow artifact. Richao et al. [15] proposed a machine learning based method for detection of moving erased object(s) in a static background video. The authors extracted object contour features and fed to a support vector machine (SVM) to identify the forged videos. In 2016, Chen et al. [6] proposed a machine learning based forensic approach for object-based forgery detection in videos. They extracted handcrafted steganalysis features [10] from computed motion residuals of a video sequence and fed them to an ensemble classifier. Recently, Bakas et al. [5] proposed a Deep Learning based technique to detect the object-based video forgery. The authors computed motion residuals from the video sequence and fed them to a capsule network [13] to identify single compressed authentic, double compressed authentic and double compressed forged frames.

## 3    Proposed Method

This section presents a forensic approach for identifying object-based forgery in surveillance footage. We employ *VGG-16* [17] as a Deep Learning network to extract the spatial information from the frames of a video clip. Due to the fact that object-based forgeries are executed in the spatio-temporal domain, simply spatial domain characteristics are not adequate for object-based forgery

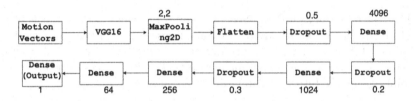

**Fig. 2.** The operational pipeline of the proposed video forensic model.

detection in videos. In order to store temporal information of a video sequence, we first compute the motion residuals [6] and the computed motion residuals are then fed as input to the proposed Deep Learning network to detect the forged frames from the original ones. Figure 2 presents an operational pipeline of the proposed method. The following sections present the proposed method in detail.

### 3.1   Motion Residual Extraction

Since, a video is constituting a series of successive frames, there is a significant correlation between each pair of adjacent frames. Hence, a local temporal frame segment can be considered a fixed component and a moving component. The moving component denotes a displacement of the moving scene relative to its reference frame, whereas the stationary part denotes an exact similarity to its preceding frames. The computed motion residuals store the displacement of the moving scenes respective to their reference frames. In other words, the temporal information of a video sequence is stored in motion residuals.

We can consider each fixed-size GOP as a local temporal segment for MPEG-1 and MPEG-2 videos, where I-frame may act as a reference frame. But, most advanced video encoding scheme, such as MPEG-4, H.264, uses variable GOP size in a video. In such cases, we cannot consider each GOP as a local temporal segment. To overcome such limitation, the proposed method uses a fixed size (say 21) local temporal segment and employs a collusion operator [19] for computing the motion residuals over the temporal segment. We follow the below process to compute the motion residuals [6] for a video sequence.

1. Perform the pixel-wise collusion procedure in a temporal segment of a video sequence $V'$ and generate a colluded result $R^i$ of frame $I^{(i)}$ as below:

$$R^i = (R_{p,q}^{(i)}) = \mathfrak{C}[(I_{p,q}^{(i-l_h)}), ...(I_{p,q}^{(i)}), ...(I_{p,q}^{(i+l_h)})] \tag{1}$$

where a collusion operator $\mathfrak{C}$ aggregates the pixels of every frame of a temporal segment of size $L$. Where $L = 2 \times l_h + 1$ and $l_h$ indicates number of immediate preceding/ succeeding neighbours frames of $I^{(i)}$). In this work, a min collusion operation is performed as below:

$$\mathfrak{C} = min_{l \in [-l_h, l_h]} \{I_{p,q}^{i+l}\} \tag{2}$$

2. Compute motion residual $M^{(i)}$ of frame $I^{(i)}$ using the following equation.

$$M^{(i)} = |I^{(i)} - R^{(i)}|$$
$$= (M_{p,q}^{(i)}) = (|I_{p,q}^{(i)} - R_{p,q}^{(i)}|) \tag{3}$$

From Eqs. 1–3, it is found that the value range of $M^{(i)}$ is within $[0, 255]$. Hence, the resulting $M^{(i)}$ can be treated as a gray image. The following section briefly presents the proposed Deep Learning model for object-based video forgery detection.

### 3.2 Deep Learning Model

In this work, we use the VGG-16 [17] network, excluding top fully connected layers, and develop a Deep Learning model to detect object-based forgery in surveillance videos. After fifth *pooling* layer of VGG-16 network, we use three *dropout*, four *fully connected* layers and one *output* layer, as shown in Fig. 2.

The portion of the VGG-16 [17] network is utilised as a features extractor from motion residuals of a video sequence (up to the fifth pooling layer). To distinguish manufactured frames from real ones, a succession of dropout and fully connected layers are fed the extracted VGG-16 characteristics.

## 4   Experimental Results and Discussion

We use 100 forged videos from SYSU-OBJFORG [6] dataset, which are recorded using static surveillance cameras. The forged object(s) in this dataset is both static and movable. H.264 video encoding library is used to compress the recorded videos, with bit rate 3 Mbps, resolution $1280 \times 720$ and frame rate 25.

We also re-encoded the forged videos with 1.5 Mbps to test the robustness of the proposed technique against low-quality videos. We create total six sub-datasets by re-encoding the forged videos with resolution $256 \times 256$, $400 \times 400$ and $512 \times 512$ for both 1.5 and 3 Mbps bit rate videos. It can be noted that we do not use the videos with resolution $1280 \times 720$ in our experiments, due to limitations of our system configuration (low GPU memory). Empirically, we set learning rate $= 0.01$ to train the proposed network.

### 4.1   Evaluation Parameters and Experimental Results

In this experiment, we use *Precision* $(P)$, *Recall* $(R)$, and *Accuracy* $(A)$ to measure the effectiveness and robustness of our proposed approach. The above evaluation parameters are defined as follows:

$$P = \frac{a}{a + b} \times 100\% \tag{4}$$

$$R = \frac{a}{a + c} \times 100\% \tag{5}$$

$$A = \frac{a + d}{a + d + b + c} \times 100\% \tag{6}$$

where $TP$ and $FP$ represent the numbers of correct and false detected forged frames respectively. $TN$ and $FN$ denote the number of correct and false detected authentic frames respectively.

For conducting the experiments, we randomly select the training, validation and testing datasets in 50:20:30 ratios for authentic as well as forged frames. Empirically, we select $l_h = 10$ for extracting the motion residuals from the video sequences.

In this section, we demonstrate the performance of the proposed approach for different degrees of video compression (bit rate) and resolution. The experimental results for low compressed (bit rate 3 M) and high compressed (1.5 M) videos are shown in Tables 1 and 2 respectively, by varying the resolution.

**Table 1.** Forgery detection results for 3 M dataset.

| Resolution | Precision (%) | Recall (%) | Accuracy (%) |
|------------|---------------|------------|--------------|
| 256 × 256  | 90.95         | 81.16      | 90.14        |
| 400 × 400  | 91.90         | 80.86      | 90.38        |
| 512 × 512  | 90.82         | 87.43      | 92.16        |

Table 1 shows that our approach provides maximum accuracy of 92.16% for 3 M bit rate and 512 × 512 resolution. Whereas for bit rate 1.5 M and resolution 256 × 256, the proposed approach provides maximum accuracy of 96.68%. From Tables 1 and 2, it can be also noticed that precision increases for higher resolution in most cases. Because, higher frame resolution improves frame quality by reducing the noise, which reduces the number of false positives.

**Table 2.** Forgery detection results for 1.5 M dataset.

| Resolution | Precision (%) | Recall (%) | Accuracy (%) |
|------------|---------------|------------|--------------|
| 256 × 256  | 94.19         | 81.69      | 90.68        |
| 400 × 400  | 96.28         | 77.25      | 90.08        |
| 512 × 512  | 97.48         | 76.98      | 90.02        |

### 4.2   Comparison with State-of-the-Art

This section presents a comparison results of our proposed approach with a state-of-the-art scheme, proposed by Bakas et al. [5], in terms of authentic and forged frame detection accuracy. The results of this comparison are presented in Table 3 for frame resolution 360 × 360. Table 3 shows that the proposed approach outperforms the scheme [5] for both low compressed (bit rate 3M) and high compressed (bit rate 1.5M) videos.

**Table 3.** Comparison with state-of-the-art for frame resolution 360 × 360.

| Bit rate | Scheme | Accuracy (%) |
|----------|--------|--------------|
| 3 M | Bakas et al. [5] | 89.54 |
|  | Proposed | **93.59** |
| 1.5 M | Bakas et al. [5] | 85.30 |
|  | Proposed | **90.32** |

## 5  Conclusion

In this research, we proposed a forensic method based on Deep Learning to identify object-based forgery in surveillance footage. We computed motion residuals, which store the forgery footprints, from the video sequences, and we also used VGG-16 network to extract the suitable features to identify the forged frames in a video. The proposed approach outperforms the state-of-the-art in terms of forged frame detection accuracy.

However, the performance of the proposed approach is not still satisfactory. Also, the proposed method is not capable to localize the forgery within a forged frame, this is out of scope of the this work. Hence, improving the performance and finding the precise position of forgery in a forged frame could be future research in this direction.

## References

1. Abbasi Aghamaleki, J., Behrad, A.: Malicious inter-frame video tampering detection in MPEG videos using time and spatial domain analysis of quantization effects. Multimed. Tools Appl. **76**(20), 20691–20717 (2016). https://doi.org/10.1007/s11042-016-4004-z
2. Aloraini, M., Sharifzadeh, M., Schonfeld, D.: Sequential and patch analyses for object removal video forgery detection and localization. IEEE Trans. Circuits Syst. Video Technol. **31**(3), 917–930 (2020)
3. Bagiwa, M.A., Wahab, A.W.A., Idris, M.Y.I., Khan, S., Choo, K.K.R.: Chroma key background detection for digital video using statistical correlation of blurring artifact. Digit. Investig. **19**, 29–43 (2016)
4. Bakas, J., Naskar, R.: A digital forensic technique for inter–frame video forgery detection based on 3D CNN. In: Ganapathy, V., Jaeger, T., Shyamasundar, R.K. (eds.) ICISS 2018. LNCS, vol. 11281, pp. 304–317. Springer, Cham (2018). https://doi.org/10.1007/978-3-030-05171-6_16
5. Bakas, J., Naskar, R., Nappi, M., Bakshi, S.: Object-based forgery detection in surveillance video using capsule network. J. Ambient. Intell. Humaniz. Comput. 1–11 (2021). https://doi.org/10.1007/s12652-021-03511-3
6. Chen, S., Tan, S., Li, B., Huang, J.: Automatic detection of object-based forgery in advanced video. IEEE Trans. Circuits Syst. Video Technol. **26**(11), 2138–2151 (2016). https://doi.org/10.1109/TCSVT.2015.2473436

7. D'Avino, D., Cozzolino, D., Poggi, G., Verdoliva, L.: Autoencoder with recurrent neural networks for video forgery detection. Electron. Imaging **2017**(7), 92–99 (2017)
8. Johnston, P., Elyan, E.: A review of digital video tampering: from simple editing to full synthesis. Digit. Investig. **29**, 67–81 (2019)
9. Kodovskỳ, J., Fridrich, J.: Calibration revisited. In: Proceedings of the 11th ACM Workshop on Multimedia and Security, pp. 63–74 (2009)
10. Kodovsky, J., Fridrich, J.: Steganalysis of JPEG images using rich models. In: Media Watermarking, Security, and Forensics, vol. 8303, p. 83030A. International Society for Optics and Photonics (2012)
11. Kodovsky, J., Fridrich, J., Holub, V.: Ensemble classifiers for steganalysis of digital media. IEEE Trans. Inf. Forensics Secur. **7**(2), 432–444 (2011)
12. Mizher, M.A., Ang, M.C., Mazhar, A.A., Mizher, M.A.: A review of video falsifying techniques and video forgery detection techniques. Int. J. Electron. Secur. Digit. Forensics **9**(3), 191–208 (2017)
13. Nguyen, H.H., Yamagishi, J., Echizen, I.: Capsule-forensics: using capsule networks to detect forged images and videos. In: IEEE International Conference on Acoustics, Speech and Signal Processing (ICASSP), pp. 2307–2311 (2019)
14. Pandey, R.C., Singh, S.K., Shukla, K.K.: A passive forensic method for video: exposing dynamic object removal and frame duplication in the digital video using sensor noise features. J. Intell. Fuzzy Syst. **32**(5), 3339–3353 (2017)
15. Richao, C., Gaobo, Y., Ningbo, Z.: Detection of object-based manipulation by the statistical features of object contour. Forensic Sci. Int. **236**, 164–169 (2014)
16. Shanableh, T.: Detection of frame deletion for digital video forensics. Digit. Investig. **10**(4), 350–360 (2013)
17. Simonyan, K., Zisserman, A.: Very deep convolutional networks for large-scale image recognition. In: Computer Science. arXiv preprint arXiv:1409.1556 (2014)
18. Sitara, K., Mehtre, B.M.: Digital video tampering detection: an overview of passive techniques. Digit. Investig. **18**, 8–22 (2016)
19. Su, K., Kundur, D., Hatzinakos, D.: Statistical invisibility for collusion-resistant digital video watermarking. IEEE Trans. Multimedia **7**(1), 43–51 (2005)
20. Su, Y., Han, Y., Zhang, C.: Detection of blue screen based on edge features. In: 2011 6th IEEE Joint International Information Technology and Artificial Intelligence Conference, vol. 2, pp. 469–472. IEEE (2011)
21. Xu, J., Yu, Y., Su, Y., Dong, B., You, X.: Detection of blue screen special effects in videos. Phys. Procedia **33**, 1316–1322 (2012)
22. Xu, R., Li, X., Zhou, B., Loy, C.C.: Deep flow-guided video inpainting. In: Proceedings of the IEEE/CVF Conference on Computer Vision and Pattern Recognition, pp. 3723–3732 (2019)
23. Zhang, J., Su, Y., Zhang, M.: Exposing digital video forgery by ghost shadow artifact. In: Proceedings of the First ACM Workshop on Multimedia in Forensics, pp. 49–54 (2009)

# Mapped-RRT* a Sampling Based Mobile Path Planner Algorithm

Rapti Chaudhuri$^{(\boxtimes)}$ , Suman Deb , and Soma Saha

National Institute of Technology Agartala, Agartala, India
{rapti.ai,sumandeb.cse}@nita.ac.in

**Abstract.** Relative efficient computation of motion plans by a Wheeled Robot Platform (WRP) has resulted through an incremental sampling of the considered environment. Although the existing sampling-based path planning techniques hardly converge into an optimal solution and this creates a challenge for a path planner specifically in case of a complex environment occupied with dynamic obstacles. A partially unknown environment creates reasonable problems for a Point-To-Point (PTP) robot to decide certain steps which would merge to the ultimate optimum path solution. To tackle the aforesaid challenge, this concerned paper proposes a noble algorithmic approach, Mapped-RRT*, for concurrent accumulation of optical information from a concerned surrounding GPS (Global Positioning System)-denied indoor environment by combining 2D (Two-Dimensional) and 3D (Three Dimensional) sensor data followed by the application of a sampling pathfinding a strategy for obtaining near-optimum point to point navigation by a wheeled robot. The VO (Visual Odometry) is obtained from a simultaneously created map of the traversed trajectory and serves as an instant perceptible reference during the movement of the mobile robot. For near-perfect detection of every possible on-path obstacles both 2D as well as 3D sensors are used together to collect fused data based on their respective preferential identification process, and a unique algorithmic approach has been proposed for run time traversal map creation along with probabilistic optimized path plan, executable within the optimum amount of time. A numerical comparison of the proposed technique with the performance of conventional strategies with respect to taken parameters confirms the reliability of the carried-out technique. The experimental results would be a citation for future research work in justifying the constructed algorithm within the domain of clash-free ORN (Optimized Robot Navigation).

**Keywords:** Wheeled robot platform · Sampling based path planning · Mapped-RRT* · GPS-denied indoor environment · Visual odometry · Optimized robot navigation

## 1 Introduction

Accumulation of optical information specifically from a concerned indoor environment constitutes a magnificent share in the research domain of robotics and

A. R. Molla et al. (Eds.): ICDCIT 2023, LNCS 13776, pp. 149–164, 2023.
https://doi.org/10.1007/978-3-031-24848-1_11

machine vision [1]. One of the basic building blocks to perceive visual citation is VSLAM (Visual Simultaneous Localization And Mapping) [2,3]. Run time visualization of map produced by navigation of mobile robot serves as a referential structure for the mobile agent to make decision in optimized path planning executed in a sampling indoor area. This concerned research work primarily takes inspiration from VSLAM to compute optimal time point to point mobile agent locomotion. Figure 1 portrays the customized mobile robot platform as well as the 2D [4] and 3D sensors used with their characteristic features.

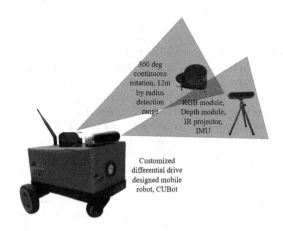

**Fig. 1.** Portrait of customized differential drive design with the sensors denoting their characteristic features.

Unlike any grid-based path searching technique [5], sampling-based algorithm has been taken into prioritization depending on the current probabilistic complex environment. Family of Rapid exploring Random Trees (RRT) cites as one of the perfect sampling-based path searching technique. The decision making module of the presented work is inspired from RRT* [6] and the execution follows its principle to a great extent. A noble approach of optimized path estimation fused with run time visual inference using multi-sensor has been proposed in this paper. Calculation of the obtained results has been done on the basis of collected measurements taken by both 2D LiDAR (Light Detection And Ranging) [7] and 3D RGB-D (Red, Green, Blue Depth) sensors [8] for near perfect minimum-time on-path obstacle capture. Primary contributions of this work have been categorized below:

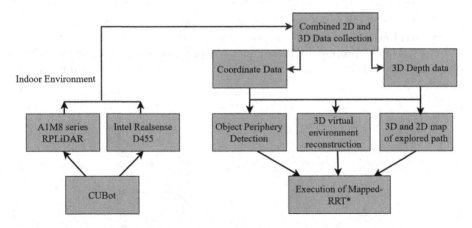

**Fig. 2.** Schematic architectural representation of the adapted methodology depicting the entire simultaneous and consequent working procedure.

– Architectural model formation of a 2-active wheeled differential drive designed mobile robot platform (CUBot) for computation of the desired experimental applications.
– Proposal of Fusing both 2D and 3D sensors data for obtaining perfect obstacle detection resulting in optimal-time mobile robot (CUBot) navigation.
– Proposal of Rapidly exploring Random Trees with Simultaneous Mapping (Mapped-RRT*) algorithm for simultaneous real-time intake of visual perception followed by optimal-time sampling-based path search.

From computational point of view, RRT* converges to optimum minima in less amount of time as compared to RRT [9] after the iteration reaches o infinity. Unlike RRT, RRT* explores every possible direction of the considered environment to find the best suitable path towards the goal. RRT* completes the coverage of every possible path towards the goal point in a relatively complex sampling structure. Mapped-RRT* would save some amount of time by citing visual reference of non-optimum explored trajectory, preventing exploration of every corner. In the sake of actual comparison, performance of the proposed methodology has been included along with parameterized action of both existing sampling-based path searching techniques, RRT and RRT* considered on the same taken indoor obstructed stature. The visual and numerically analysed presentation based on optimized working function and time, proves the supremacy of the research architecture. Figure 2 portrays the schematic architecture of the methodology adapted to carry out the experiment. Figure 3 depicts the geometrical structure of RP LiDAR and Intel Realsense[1] respectively.

---

[1] https://support.intelrealsense.com/hc/user_images/fzXb7K7btFFWhf_z3iszMQ.
png.

The outline of the paper consists of brief description about reviewed literature survey, architectural method of the working procedure adapted and module wise elaboration of the proposed strategy and comparative analytical portrait of the obtained experimental results with other existing conventional computational models.

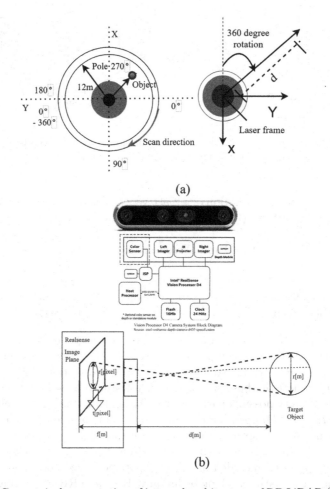

**Fig. 3.** (a) Geometrical presentation of internal architecture of RP LiDAR (b) Geometrical presentation and working procedure of used RGB-Depth sensor, Intel Realsense D455

## 2    Related Works

This section briefly depicts the reviewed related survey including the respective proposed works with research procedure. The gaps in experimentation have been noticed and the concerned paper tries to improvise the existing work.

Fragkopoulos et al. in the paper [10] portrays a technically new approach of CellBiRRT (Cell based Bi-directional Rapidly exploring Random Trees) for a manipulator to handle pose constraints and it is applicable mostly in difficult environments reducing probability of collision. The work described in the paper [11] proposes ERRT (Execution Extended RRT) algorithm for executing collision free locomotion of mobile robots mostly in dynamic environments. Gammell et al. proposes Informed RRT* in the paper [12] briefly describing the experimentation on sampling-based path planning by RRT* for achieving optimum path. The research work mentioned in the paper [13] suggests a new RRT*- inspired online informative path planning algorithm which shows its working principle by continuous expansion of a single tree of candidate trajectories along with rewiring of nodes supported with refinement of intermediate paths. Rapti Chaudhuri et al. in the paper [14] proposed a technique of combining the 3D information as visual reference followed by a comparative performance of well-known bio intelligent path planning approach with respect to a customized mobile robot platform.

The paper [15] makes a comparative study on the parameterized comparison of three variations of A-star algorithm in a perfect maze environment by a mobile robot executing rescue function. The same work has been carried out in a considered indoor obstructed lab environment. A fused application of RGB-D camera, working according to the kinect principle and an IMU (Inertial Measurement Unit), generating 3D feature points presenting the visual Odometry with aggregation of depth information into RGB color information in highly dynamic environments has been proposed in paper [16]. The work described in paper [17] evaluates LiDAR SLAM work with respect to comparative accuracy in map construction for accomplishing smooth point to point mobile robot navigation. Zingg et al. in the paper [18] presents an important work with video cameras as the main sensor for navigation and a laser sensor as auxiliary to detect and avoid obstacles in situations like intersections, left or right corridors and open doors to right or left. Labbé et al. [19] proposed first remarkable SLAM (Simultaneous Localization And Mapping) method for near-perfect detection and tracking of moving obstacles using laser range finder and presented a verified result in outdoor urban environment. Rapti Chaudhuri et al. in the paper [20] proposes a different solution for fused virtual 3D reconstruction of indoor environment using point cloud and intelligent object identification using machine learning module for saving the memory of already traversed path by customized mobile robot.

The executed research work modifies the existing aforesaid works by algorithmic incorporation of sampling based path planning technique with real-time simultaneous mapping of the traversed path by the considered mobile agent. The experimental comparative results confirm the optimality in achieving collision-free path navigation.

## 3    Methodology

This section gives a brief depiction of the schematic presentation of working procedure followed to carry out the research experiment. It is further divided into subsections mentioned below giving the step wise idea of the work done. Figure 4 denotes geometrical architecture of the methodology adapted.

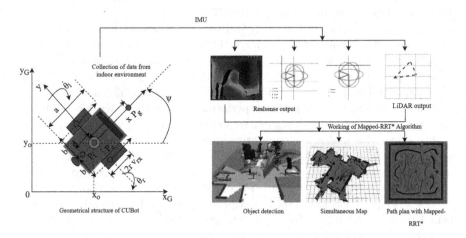

**Fig. 4.** Geometrical structure of the entire process to be adapted for achieving optimized Mapped-RRT* mediated point to point robot navigation.

### 3.1    Combined Sensor Calibration

Parameterized calibration of RPLiDAR and RGB-Depth sensor have been done for achieving correct precision in on-route obstacle detection with respect to every possible aspects. Combined calibration procedure has been numerically presented needed for visual referential output. Relative LiDAR and Realsense RGB-D sensor calibration strategy are visually presented with geometrical configurations in Fig. 5. World coordinate taken as Z has been projected to image coordinate whose numerical representation is shown in Eq. (1).

$$Z = I(JZ + t) \tag{1}$$

where I denotes $3 \times 3$ intrinsic parameter of the Realsense sensor, J depicts $3 \times 3$ Realsense sensor orientation matrix and t presents $3 \times 1$ vector relative to its position. Rigid transformation from Realsense coordinate system to LiDAR system has been denoted by $Z_f$ in Eq. (2).

$$Z_f = \sigma Z + \omega \tag{2}$$

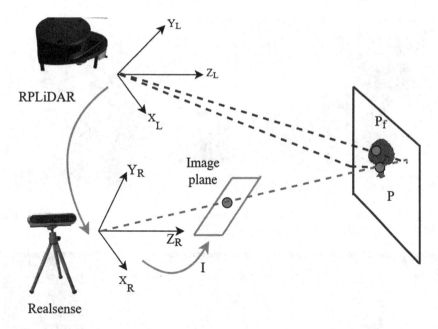

**Fig. 5.** Parameterized calibration procedure adapted to carry out the experimentation with both 2D LiDAR and 3D RGB-Depth sensor.

$\sigma$ presents $3 \times 3$ Realsense sensor orientation orthonormal matrix with respect to LiDAR and $\omega$ depicts $3 \times 1$ vector relative to its respective position. Calibrated sensors are then made to scan and map the taken obstructed indoor environment simultaneously with simulation of optimum path searching technique.

## 3.2   Simultaneous Mapping with Path Search

Individual procedural mapping has been discussed respectively with consequent description of combined structured mapping strategy with combined data input from both RPLiDAR and Intel Realsense D455.

*LiDAR Mapping.* LiDAR mapping is generated and obtained visually from Rviz visualizer of ROS environment. Run time mapping of the trajectory is obtained based on the input data captured by the RPLiDAR. Function, f(s), in the form of Laser input and a prior map are considered in LiDAR mapping. Probabilistic

**Fig. 6.** Point to point correlation obtained from odometry data visualized on RTabmap, a ROS visualizer resulted from input collected by Intel Realsense D455.

**Fig. 7.** 3D depth obtained in mapping module of the concerned procedure visualized on RTabmap, a ROS visualizer by taking input from Intel Realsense D455.

estimation of grid is followed by creation of grid map. Integer coordinates results in gradients and derivatives which are used for formation of 2D pose expression [21]. Primary concern of the aforesaid technique is to obtained optimized estimation of the rigid transformed robot pose [22]. $(\delta) = [I_x, I_y, \phi]$ from the robot to the prior map [23]. $M(R_j(\delta))$ is the value of the map at $R_j(\delta)$, world coordinate of scan end points $R_j = (S_{j,x}, X_{j,y})^T$. $M(R_j(\delta))$ obeys the function as presented in Eq. (3):

$$R_j(\delta) = \begin{bmatrix} cos\phi & -sin\phi \\ sin\phi & cos\phi \end{bmatrix} \begin{bmatrix} S_{j,x} \\ S_{j,y} \end{bmatrix} + \begin{bmatrix} I_x \\ I_y \end{bmatrix} \tag{3}$$

The optimized map is formed by computing the loss and error costs by matching the laser data with the raw map obtained.

*RGB-D Mapping.* 3D Depth data is captured from RGB-Depth sensor needed for 3D virtual reconstruction of indoor scenes [24]. Computation of rigid transformation is resulted from RGB-D mapping using RANSAC (Random Sample Consensus) which filters out the outliers [25]. Input is obtained from Intel Realsense D455 and the output is obtained in the form of loop closure detection, 3D point cloud mapping [26], 2D linear mapping odometry [27], etc. Estimated depth is achieved from internal triangulation procedure. Emission of the IR (Infra Red) structured light produces a constant pattern of speckles and the IR module in Realsense camera captures the pattern for correlation [28] with reference pattern. Occupancy grid map is obtained as output along with 3D point cloud formation presenting the 3D scene scan of the obstructed indoor environment. Figure 6 and Fig. 7 presents the result obtained from detected 3D depth collection section of respective simultaneous mapping module.

*Combined Mapping.* Combined mapping structure [29] is produced by fusion of data obtained from both LiDAR sensor as well as Realsense RGB-D sensor resulting in relatively better precise and accurate detection of on-route obstacles [30]. Real-time quick detection and simultaneous mapping of the trajectory serving as visual citation result in optimum-time path navigation by the concerned customized mobile agent. Combined data is obtained in the form of coordinate as well as 3D depth format. In case of original RGB image, LiDAR mapping is performed and for 3D depth image as input, pairwise 6D transformation is done with achievement of 3D point cloud and optimized 2D pose estimated graph of the explored trajectory. Figure 8 depicts the visual output of 3D reconstruction as well as periphery detection of the on-path obstacles followed by 3D map creation of the explored trajectory by CUBot.

### 3.3   Algorithmic Description of Mapped-RRT* (Rapidly Exploring Random Trees with Simultaneous Mapping)

The proposed technique has been illustrated with required numerical presentation in Algorithm 1. The first module of the algorithm depicts the real time simultaneous mapping of the scanned environment and it is consequently followed by Random Tree exploration based sampling path searching procedure. Figure 9 presents the executed graphical structure of Mapped-RRT* in one of the considered scene in taken indoor environment.

---

**Algorithm 1.** Mapped-RRT* for optimized path search

---

**Input:** - $f(s_1)$, $f(s_2)$, $map_{a-priori}$, Laser data, n, $X_{node}$, $X_{best\_neighbour}$, $X_{new}$, $X_{nearest\_neighbour}$ // LiDAR data, depth data, and a priori map

**Output:** - $\epsilon^*$,$a_c$,$a_p$, Occupancy Grid map, G(V,E) // Final graph obtained, Optimal Pose estimation of the robot

1: radius // r
2: counter = 0
3: num = n // number of iterations
4: G($V, E$)// Concerned graph with V vertices and E edges
5: Probabilistic estimation of a grid being occupied.
6: Division of the grid map.
7: **for** RGB image, I_m $\in map\_a - priori$ **do**
8:     $\nabla M(I\_m) = \left( \frac{\delta\ M}{\delta\ x}(I\_m), \frac{\delta\ M}{\delta\ y}(I\_m) \right)$

9: Approximation of gradient and derivatives using four closest integer coordinates $I_{00}$, $I_{01}$, $I_{10}$ and $I_{11}$.

10: $M(I_m) \approx \frac{y-y_0}{y_1-y_0}\left(\frac{x-x_0}{x_1-x_0}M(I_{11}) - \frac{x_1-x}{x_1-x_0}M(I_{01})\right) + \frac{y_1-y}{y_1-y_0}\left(\frac{x-x_0}{x_1-x_0}M(I_{10}) - \frac{x_1-x}{x_1-x_0}M(I_{10}) - \frac{x_1-x}{x_1-x_0}M(I_{00})\right)$

11: $\frac{\partial M}{\partial x}(I_m) \approx \frac{y-y_0}{y_1-y_0}M\left((I_{11}) - M(I_{01})\right) + \frac{y_1-y}{y_1-y_0}M\left((I_{10}) - M(I_{00})\right)$

12: $\frac{\partial M}{\partial y}(I_m) \approx \frac{x-x_0}{x_1-x_0}M\left((I_{11}) - M(I_{10})\right) + \frac{x_1-x}{x_1-x_0}M\left((I_{01}) - M(I_{00})\right)$

13: **for** RGB image, Depth Image **do**
14:     Pairwise 6D Transformation Estimation (RANSAC)

15: Pose expression in 2D Environment.
16: $\epsilon = (I_x, I_y, \chi)^T$.
17: Global Pose Graph Optimization ($g^2o$)
18: 3D point clouds formation
19: Achievement of optimized pose expression by matching the laser data and the map.
20: **for** $\epsilon = (Ix, Iy, \chi)^T, map_{output} \leftarrow M(rj(\epsilon))$ at $Rj(\epsilon)$ **do**
21:     $\epsilon^* = \arg\min_\epsilon \sum_{j=1}^{N}[1 - M(Rj(\epsilon))]^2$
22:     **for** $\epsilon \leftarrow \epsilon + \Delta\epsilon$ **do**
23:         $\sum_{j=1}^{N}[1 - M(Rj(\epsilon))]^2 \rightarrow 0$

24: Accuracy in obstacle detection ($a_c$)
25: Accurate explored path ($a_p$)
26: $\epsilon^* \leftarrow \epsilon$
27: **while** counter < num **do**
28:     $X_{node} = Random_{explored\_position}()$
29:     **if** $ISINObstacle(X_{node})$ == True: **then**
30:         try again
31:     $X_{nearest_{neighbour}}$ = $(G(V, E), X_{node})$
32:     Cost($X_{node}$) = Dist($X_{node}$, $X_{nearest_{neighbour}}$)
33:     $X_{best\_neighbour}$ = findNeighbours($G(V, E)$, $X_{node}$, radius)
34:     Join = Chain($X_{node}$, $X_{best\_neighbour}$)
35:     **for** $X_{new}$ in $X_{neighbours}$ **do**
36:         **if** cost($X_{node}$) + Dist($X_{node}, X_{new}$) < cost($X_{new}$) **then**
37:             Cost($X_{new}$) = cost($X_{node}$) + Dist($X_{node}$, $X_{new}$)
38:             Parent($X_{new}$) = $X_{node}$
39:             G = G + ($X_{node}$, $X_{new}$)
40:     G.append(Join)
41: Return G

---

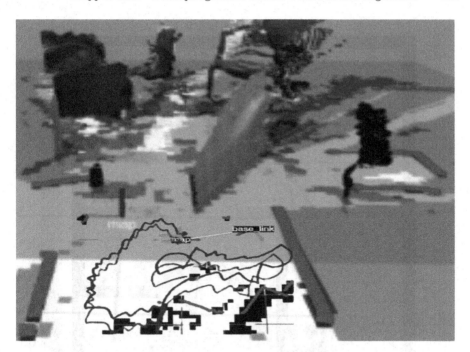

**Fig. 8.** Virtual 3D reconstruction with obstacle periphery detection and consequent map creation of the traversed path by the mobile agent.

## 4    Experiment Analysis and Results

This section presents the analytical results obtained by carrying out the experiment. This has been divided further into sub categories characterizing the functions at each level with graphical and tabular comparative results.

**Table 1.** Parameterized Characteristic features of considered 2D and 3D sensors

| Sensor used | Visual Output | Data type obtained | Data obtained |
|---|---|---|---|
| A1M8 series RPLiDAR | 2D scanned environment with explored path | Coordinate data | Minimum and Maximum angle, angle increment, time, range, intensity |
| Intel Realsense D455 | 3D voxelNet and 3D path map | 3D Depth data | RGB image, Depth image, Accelerometer and Gyro data, 3D point map |

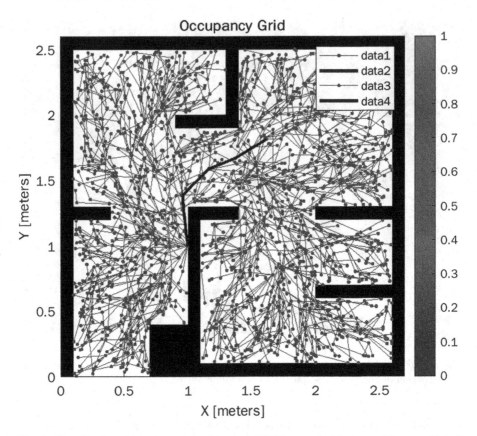

**Fig. 9.** Graphical architecture of the Mapped-RRT* executed in one of the considered scene of indoor environment.

### 4.1 Combined Visual Data Perception

The coordinate formated data is obtained from obtained from A1M8 series RPLiDAR fusing with the 3D depth formatted data achieved from Intel Realsense D455 Depth sensor. A1M8 model of 2D RPLiDAR is preferred for 2D scan of indoor surrounding by detecting periphery of the same plane obstacles using its limited triangular measurement [31]. Maximum range of scanning is approximately 12m by radius. The limited capacity of 2D LiDAR is tackled by 3D RGB-Depth sensor [32] which makes probabilistic obstacle detection comparatively more precise. Physical features of Intel Realsense D455 consist of RGB colour sensor, 3D depth module, IR projector, IMU and Vision Processor. Table 1 presents the characteristic features of used sensors for experimentation.

### 4.2 Environment Mapping Results

Both the periphery of the on-path obstacles and trajectory map of the non-optimum explored path are achieved by the application of simultaneous mapping module of the proposed algorithmic architecture. The usage of multi sensor have facilitated the customized mobile path planner to have precise optimum-time sensible perception of present objects. Use of both the concerned sensors overcome the performance limitations of each. Formation of voxelNet and occupancy grid map serve as real time visual reference to make further decision in executing sampling-based path searching procedure. The mapping module is basically inspired from the combined characteristic features of both Hector SLAM as well as RGB-D SLAM [25]. Obstacle detection in any plane is accompanied with each IMU updated visual data representation. Figure 10 denotes the visual representation of RRT execution with simultaneous mapping and Mapped-RRT* respectively in an indoor environment.

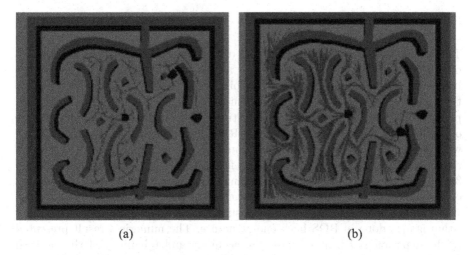

(a)                              (b)

**Fig. 10.** (a) Simulated Execution of RRT with simultaneous mapping in an indoor scene (b) Simulated Execution of Mapped-RRT* in the same indoor scene.

### 4.3   Comparative Analysis of Proposed Method with Conventional Techniques

The performance of the proposed technology has been analysed with respect to the considered complex indoor environment and a bench marking has been done along with performance of existing path searching sampling-based processes. As a matter of visual comparative study, the technique of simultaneous combined mapping has been applied and experimented both with RRT as well as RRT* and presented in Fig. 10. Table 2 denotes the numerically analysed results executed by the customized mobile agent, CUBot. The optimum-time measurement of adapted technology confirms its reliability and superiority over other executed conventional procedures evident from the result table.

**Table 2.** Comparative performance analysis of proposed algorithm

| Algorithm | Obstacle present | Detected obstacles | Accuracy of detection | Output | Path | Time |
|---|---|---|---|---|---|---|
| RRT* with RPLiDAR | 32 | 26 | 81.25% | path | 30 m | 7,735.2 s |
| RRT* with Realsense | 32 | 28 | 87.5% | path | 30 m | 7,735.2 s |
| RRT* with Multi sensor | 32 | 29 | 90.625% | path with map | 30 m | 7,051.2 s |
| Mapped-RRT* with Multi sensor | 32 | 30 | **93.75%** | Optimum path with map | 30 m | **7,051.2 s** |

## 5  Conclusion

This paper proposes a noble approach of unique architectural working procedure for achieving optimum-time path navigation. The proposed process has been carried out using a customized mobile agent, CUBot, working according to the principle of differential drive mechanism and it has been executed in a sampling complex indoor environment consisting of both static and dynamic on-route obstacles. The performance of Mapped-RRT* has been comparatively analysed with respect to conventionally used technique. The sampling-based RRT* algorithm has been considered in different forms consisting of individual 2D and 3D optical sensor as well as with combined data input respectively. Simultaneous accurate obstacle detection is executed with consequent Mapped-RRT*. The visual reference is obtained from ROS visualizer and the computation of the algorithm is also done in ROS back end structure. The numerical result presented in the experimental table ensures the accuracy and reliability of the adapted technique over other existing procedures for executing optimum point to point movement by customized mobile platform. The concerned research outputs with proposed structured path searching algorithm could be a reference citation for future researchers in the domain of Machine Vision mediated Optimized Robot Navigation.

## References

1. Azzam, R., Taha, T., Huang, S., Zweiri, Y.: Feature-based visual simultaneous localization and mapping: a survey. SN Appl. Sci. **2**(2), 1–24 (2020)
2. Yousif, K., Bab-Hadiashar, A., Hoseinnezhad, R.: An overview to visual odometry and visual slam: applications to mobile robotics. Intell. Ind. Syst. **1**(4), 289–311 (2015)
3. Merzlyakov, A., Macenski, S.: A comparison of modern general-purpose visual slam approaches. In: 2021 IEEE/RSJ International Conference on Intelligent Robots and Systems (IROS) (2021)
4. Subbanna, B. B., Choudhary, K., Singh, S., Kumar, S.: 2D material-based optical sensors: a review. ISSS J. Micro Smart Syst. **11**(1), 169–177 (2022)

5. Yap, P.: Grid-based path-finding. In: Cohen, R., Spencer, B. (eds.) AI 2002. LNCS (LNAI), vol. 2338, pp. 44–55. Springer, Heidelberg (2002). https://doi.org/10.1007/3-540-47922-8_4

6. Jeong, I.-B., Lee, S.-J., Kim, J.-H.: RRT*-quick: a motion planning algorithm with faster convergence rate. In: Kim, J.-H., Yang, W., Jo, J., Sincak, P., Myung, H. (eds.) Robot Intelligence Technology and Applications 3. AISC, vol. 345, pp. 67–76. Springer, Cham (2015). https://doi.org/10.1007/978-3-319-16841-8_7

7. Hess, W., Kohler, D., Rapp, H., Andor, D.: Real-time loop closure in 2d lidar slam. In: 2016 IEEE International Conference on Robotics and Automation (ICRA), pp. 1271–1278. IEEE (2016)

8. Li, J., Gao, W., Wu, Y., Liu, Y., Shen, Y.: High-quality indoor scene 3D reconstruction with RGB-D cameras: a brief review. Comput. Vis. Media 1–25 (2022)

9. Islam, F., Nasir, J., Malik, U., Ayaz, Y., Hasan, O.: RRT-smart: rapid convergence implementation of RRT towards optimal solution. In: 2012 IEEE International Conference on Mechatronics and Automation, pp. 1651–1656. IEEE (2012)

10. Fragkopoulos, C., Graeser, A.: A RRT based path planning algorithm for rehabilitation robots. In: ISR 2010 (41st International Symposium on Robotics) and ROBOTIK 2010 (6th German Conference on Robotics), pp. 1–8 (2010)

11. Bruce, J., Veloso, M.M.: Real-time randomized path planning for robot navigation. In: Kaminka, G.A., Lima, P.U., Rojas, R. (eds.) RoboCup 2002. LNCS (LNAI), vol. 2752, pp. 288–295. Springer, Heidelberg (2003). https://doi.org/10.1007/978-3-540-45135-8_23

12. Gammell, J.D., Srinivasa, S.S., Barfoot, T.D.: Informed RRT*: optimal sampling-based path planning focused via direct sampling of an admissible ellipsoidal heuristic. In: 2014 IEEE/RSJ International Conference on Intelligent Robots and Systems, pp. 2997–3004. IEEE (2014)

13. Schmid, L., Pantic, M., Khanna, R., Ott, L., Siegwart, R., Nieto, J.: An efficient sampling-based method for online informative path planning in unknown environments. IEEE Robot. Autom. Lett. 5(2), 1500–1507 (2020)

14. Chaudhuri, R., Deb, S., Shubham, S.: Bio inspired approaches for indoor path navigation and spatial map formation by analysing depth data. In: 2022 IEEE International Conference on Distributed Computing and Electrical Circuits and Electronics (ICDCECE), pp. 1–6 (2022)

15. Liu, X., Gong, D.: A comparative study of a-star algorithms for search and rescue in perfect maze. In: 2011 International Conference on Electric Information and Control Engineering, pp. 24–27. IEEE (2011)

16. Bloesch, M., Omari, S., Hutter, M., Siegwart, R.: Robust visual inertial odometry using a direct EKF-based approach. In: 2015 IEEE/RSJ International Conference on Intelligent Robots and Systems (IROS), pp. 298–304. IEEE (2015)

17. Fan, X., Wang, Y., Zhang, Z.: An evaluation of lidar-based 2d slam techniques with an exploration mode. In: Journal of Physics: Conference Series, vol. 1905, p. 012021. IOP Publishing (2021)

18. Zingg, S., Scaramuzza, D., Weiss, S., Siegwart, R.: Mav navigation through indoor corridors using optical flow. In: 2010 IEEE International Conference on Robotics and Automation, pp. 3361–3368. IEEE (2010)

19. Wang, C.-C., Thorpe, C., Thrun, S., Hebert, M., Durrant-Whyte, H.: Simultaneous localization, mapping and moving object tracking. Int. J. Robot. Res. 26(9), 889–916 (2007)

20. Chaudhuri, R., Deb, S.: Adversarial surround localization and robust obstacle detection with point cloud mapping. In: Das, A.K., Nayak, J., Naik, B., Vimal, S.,

Pelusi, D. (eds.) Computational Intelligence in Pattern Recognition. CIPR 2022. LNNS, vol. 480, pp. 100–109. Springer, Singapore (2022). https://doi.org/10.1007/978-981-19-3089-8_10

21. Heo, J., Savvides, M.: Gender and ethnicity specific generic elastic models from a single 2d image for novel 2d pose face synthesis and recognition. IEEE Trans. Pattern Anal. Mach. Intell. **34**(12), 2341–2350 (2011)

22. Mustafa, M., Stancu, A., Guteirrez, S.P., Codres, E.A., Jaulin, L.: Rigid transformation using interval analysis for robot motion estimation. In: 2015 20th International Conference on Control Systems and Computer Science, pp. 24–31. IEEE (2015)

23. Zhang, X., Lai, J., Xu, D., Li, H., Fu, M.: 2d lidar-based slam and path planning for indoor rescue using mobile robots. J. Adv. Transp. (2020)

24. Chen, R., Jing, X., Zhang, S.: Comparative study on 3d optical sensors for short range applications. Opt. Lasers Eng. **149**, 106763 (2022)

25. Zhang, S., Zheng, L., Tao, W.: Survey and evaluation of RGB-D slam. IEEE Access **9**, 21367–21387 (2021)

26. Guo, Y., Wang, H., Qingyong, H., Liu, H., Liu, L., Bennamoun, M.: Deep learning for 3D point clouds: a survey. IEEE Trans. Pattern Anal. Mach. Intell. **43**(12), 4338–4364 (2020)

27. Fraundorfer, F., Scaramuzza, D.: Visual odometry: Part ii: matching, robustness, optimization, and applications. IEEE Robot. Autom. Mag. **19**(2), 78–90 (2012)

28. Rukhin, A.L.: Pattern correlation matrices and their properties. Linear Algebra Appl. **327**(1–3), 105–114 (2001)

29. Glaw, X., Inder, K., Kable, A., Hazelton, M.: Visual methodologies in qualitative research: autophotography and photo elicitation applied to mental health research. Int. J. Qual. Methods **16**(1), 1609406917748215 (2017)

30. Dieterle, T., Particke, F., Patino-Studencki, L., Thielecke, J.: Sensor data fusion of lidar with stereo RGB-D camera for object tracking. In: 2017 IEEE Sensors, pp. 1–3 (2017)

31. Markom, M.A., et al.: A mapping mobile robot using RP lidar scanner. In: 2015 IEEE International Symposium on Robotics and Intelligent Sensors (IRIS), pp. 87–92 (2015)

32. Da Silva Neto, J.G., et al.: Comparison of RGB-D sensors for 3D reconstruction. In: 2020 22nd Symposium on Virtual and Augmented Reality (SVR), pp. 252–261 (2020)

# Intelligent Optimization Algorithms for Disruptive Anti-covering Location Problem

Edukondalu Chappidi[1] , Alok Singh[1(✉)] , and Rammohan Mallipeddi[2]

[1] School of Computer and Information Sciences, University of Hyderabad,
Hyderabad 500046, India
alokcs@uohyd.ernet.in
[2] School of Electronics Engineering, Department of Artificial Intelligence,
Kyungpook National University, Daegu 41566, Republic of Korea

**Abstract.** Given a set of potential sites for locating facilities, the disruptive anti-covering location problem (DACLP) seeks to find the minimum number of facilities that can be located on these sites in such a way that each pair of facilities are separated by a distance which is more than $R$ from one another and no more facilities can be added. DACLP is closely related with anti-covering location problem (ACLP), which is concerned with finding the maximum number of facilities that can be located such that all the facilities are separated by a distance which is more than $R$ from each other. The disruptive anti-covering location problem is so named because it prevents the "best or maximal" packing solution of the anti-covering location problem from occurring. DACLP is an $\mathcal{NP}$-hard problem and plays an important role in solving many real world problems including but not limited to forest management, locating bank branches, nuclear power plants, franchise stores and military defence units. In contrast to ACLP, DACLP is introduced only recently and is a relatively under-studied problem. In this paper, two intelligent optimization approaches namely genetic algorithm (GA) and discrete differential evolution (DDE) are proposed to solve the DACLP. These approaches are the first heuristic approaches for this problem. We have tested the proposed approaches on a total of 80 DACLP instances containing a maximum of 1577 potential sites. The effectiveness of the proposed approaches can be observed from the results on these instances.

**Keywords:** Disruptive anti-covering location problem · Facility location · Genetic algorithm · Discrete differential evolution · Intelligent optimization algorithm

## 1 Introduction

Considering a set of potential sites where facilities can be located and individuals that interact with these facilities, the most common type of facility location

A. R. Molla et al. (Eds.): ICDCIT 2023, LNCS 13776, pp. 165–180, 2023.
https://doi.org/10.1007/978-3-031-24848-1_12

problems are the ones based on the interaction between facilities and individuals. However, depending on the problem under consideration, interaction among the facilities also plays a key role in deciding the sites where facilities can be located. The disruptive anti-covering location problem (DACLP) comes under the category of location problems involving facility-facility interactions where no two facilities can be located within a distance of $R$ from one another. In the DACLP jargon [18], a proper solution is defined as the one in which all non-facility sites are within the separating distance $R$ from one or more of the selected facilities. Obviously, no more facilities can be added to a proper solution. So, DACLP is concerned with finding the minimum number of facilities that can be located on a subset of sites while giving a proper solution. DACLP is derived from the more commonly known anti-covering location problem (ACLP) [16], which is concerned with finding a subset of facilities of maximum cardinality which forms a proper solution. The disruptive anti-covering location problem is so named as it prevents the "best or maximal" packing solution of the anti-covering location problem from occurring. Node and site have been used synonymously throughout this paper.

Niblett and Church [18] introduced DACLP for the first time in 2015 and proposed a model based on integer linear programming (ILP) for solving this problem. DACLP is an $\mathcal{NP}$-hard problem ([18]). There are many real world applications where DACLP can be used for finding the minimum number of facilities that can be located with the minimum separating distance requirement between each pair of facilities. Some of the applications of DACLP include but not limited to forest management [1], locating bank branches [18], nuclear power plants ([6,7]), franchise stores ([8,12]) and military defence units [4]. Further, in competitive environments where there are minimum separation requirements among facilities, DACLP can be used at the minimum expanse to prevent competitors from opening more facilities in an area. For example, if there is a minimum separation requirement between any two liquor stores in a city then opening of liquor stores by a company as per DACLP solution for this city at minimum cost will forbid competitors from opening any more stores in that city [13,18].

Over the past many years there have been several methods devised for solving ACLP ([2,3,11,14,16,17]). However, no method exists in the literature to solve DACLP other than the ILP proposed by Niblett and Church [18] while introducing DACLP. This served as the motivation to develop the intelligent optimization approaches described in this paper which are based on genetic algorithm (GA) and discrete differential evolution (DDE). In fact, these approaches are the first heuristic approaches for solving the DACLP. There exist numerous intelligent optimization techniques in the literature [15,21,25], and among them evolutionary techniques have a significant share. Evolutionary techniques are particularly successful in solving difficult discrete optimization problems (e.g. [5,20,22–24,27]). This fact has inspired us to develop the two evolutionary approaches for DACLP which is a discrete optimization problem. Though differential evolution and genetic algorithm both are evolutionary algorithms and utilize crossover and mutation, the similarity between our two approaches ends here as different solution encodings are used and also the crossover and mutation operators used by the two proposed approaches are completely different. The proposed approaches are executed on a total of 80 DACLP instances containing a maximum of 1577 potential sites. The effectiveness of the proposed approaches can be clearly seen from the results obtained on these instances.

The remaining part of this paper is organized as follows: Sect. 2 gives the formal definition of the problem. Section 3 presents the proposed DDE approach, whereas Sect. 4 describes the proposed GA approach for the DACLP. The results of the conducted experiments along with their analysis are presented in Sect. 5. Finally, Sect. 6 wraps up the paper by listing the contributions made along with some potential avenues for further research.

## 2   Problem Definition

Considering a set $V$ of $n$ potential sites where facilities can be located, i.e., $V = \{1, 2, \ldots, n\}$ ($|V| = n$), and $R$ is the minimum separating distance such that no two facilities are permitted within distance $R$ from one another, the disruptive anti-covering location problem can be formally defined as follows: For each site $u \in V$, the shortest distance between site $u$ and site $v \in V$ is given by $d_{uv}$. $Q_u$ represents the forbidden set of site $u \in V$, which denotes the set of sites whose shortest distance from $u$ is not more than $R$, i.e., $Q_u = \{v | v \in V \wedge d_{uv} \leq R \wedge u \neq v\}$. If a facility is located at site $u$ then we can not chose any site in $Q_u$ for locating another facility. A solution with set $S \subseteq V$ of facilities is called proper in case facilities are located in such a manner that no two facilities are located within distance $R$ from one another and all non-facility sites are within a distance $R$ from one are more facilities, i.e., $Q_u \cap S = \emptyset \ \forall u \in S$ and $(\cup_{u \in S} Q_u) \cap \{v\} \neq \emptyset \ \forall v \in (V \setminus S)$. DACLP seeks a proper solution with minimum number of facilities. Considering binary variables $x_v \forall v \in V$ that have value 1 if a facility is located at site $v$ ($x_v = 1$) and value 0 when no facility is located at site $v$ ($x_v = 0$) and $Y$ as a large positive integer, Niblett and Church [18] formulated the following mathematical model of DACLP:

$$min \ Z = \sum_{u \in V} x_u \tag{1}$$

subject to:

$$Y x_u + \sum_{v \in Q_u} x_v \leq Y, \ \forall u \in V \tag{2}$$

$$x_u + \sum_{v \in Q_u} x_v \geq 1, \ \forall u \in V \tag{3}$$

$$x_u \in \{0, 1\}, \ \forall u \in V \tag{4}$$

Here, Eq. 1 minimizes the number of sites where facilities are located and gives the DACLP's objective function. According to Eq. 2 if a site $u$ is selected for

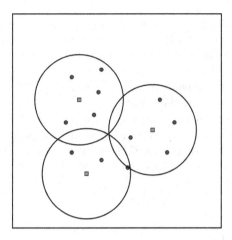

**Fig. 1.** DACLP illustration (Color figure online)

locating a facility (i.e. $x_u = 1$), then the $Yx_u = Y$, and, which causes the summation $\sum_{v \in Q_u} x_v$ to be 0, i.e., it enforces the constraint that no two facilities can be within distance $R$ from one another. Equation 3 enforces that either site $u$ is selected to locate a facility or a site $v \in Q_u$ which is within $R$ distance from $u$ is selected to locate a facility. Together Eqs. 2 and 3 make sure that the solution is proper. Constraint 4 restricts the variables $x_u \forall u \in V$ to binary values.

Consider the Fig. 1 which illustrates DACLP with an example. There are a total of 15 nodes having the following coordinates: A(10, 20), B(30, 15), C(20, 6), D(48, 10), E(30, 75), F(6, 40), G(50, 30), H(65, 35), I(80, 40), J(70, 55), K(28, 60), L(25, 45), M(10, 70), N(15, 55) and O(75, 20). Considering $R = 30$, where all the facilities are separated from each other by a distance of more than $R$, the subset {C, H, N} forms a feasible, proper solution satisfying the separating distance constraint. In the Fig. 1 we have depicted the nodes selected for facilities in green color and nodes not selected for facilities are marked in blue color.

## 3   DDE Approach for DACLP

A discrete differential evolution (DDE) based method for DACLP has been developed by us. The readers are requested to refer to [19] and [9] for an overview of differential evolution and discrete differential evolution. The salient features of our proposed DDE approach are presented in the subsequent subsections.

### 3.1   Solution Representation and Fitness

Given there are $n$ sites in total, each solution is considered as an $n$-bit vector where position $p$ has a value of 1 if the site $p$ is selected to locate a facility. Otherwise, position $p$ has a value of 0 if the site $p$ is not selected to locate a facility. For the fitness function, we have used the objective function of DACLP which is given in Eq. 1.

## 3.2    Generating Initial Population

To generate each member of the initial population, we have used a semi-greedy method, in which from a set of greedily ranked sites, we randomly select facilities and add them to the solution. Initially, all sites are unmarked. For each solution, starting with an empty set the afore mentioned step is repeated until the termination condition is satisfied. In each iteration, 3 unmarked sites with highest values of $|Q_u|$ are determined. Out of these 3 sites, we randomly select a site $u$ and set the corresponding bit to 1 making it part of the solution and the site $u$ is marked. Since all the sites in $Q_u$ are within distance of $R$ from $u$, all the sites in the set $Q_u$ are also marked, then an iteration completes. Once all of the $n$ sites are marked, this iterative process stops and one solution is generated. In this manner, a population of total $POP_{cnt}$ candidate solutions are generated.

## 3.3    DDE Framework

Given the initial population, our DDE approach uses an iterative process for solving the DACLP. In every iteration, the solutions in the population are considered one after the other. The solution currently under consideration is referred to as target solution. Mutation is performed on a single parent solution and that parent can be either the target solution or the global best solution found till that time or a candidate solution which is selected randomly from the population [26]. We have applied mutation on the global best solution with probability $p_m$ (i.e., with probability $1 - p_m$ no mutation is applied). We apply repair operation on the mutant, which is the resulting solution after mutation. Then the mutant solution and target solution are considered as the two parents for crossover operation, which generates a trial solution. The repair operation is applied on the trial solution as well. After that, according to the replacement policy it is decided whether the trial solution can replace the target solution or not. When each solution in the population is considered, we move on to the next iteration. These iterations continue as long as the termination condition is not satisfied. After all the iterations are over, the best solution found is considered as the solution generated by the DDE approach.

## 3.4    Mutation

In our approach, the global best solution is selected as the parent to perform mutation operation. As the solution is a bit vector, with probability $p_{mut}$ we flip each bit in the selected solution. We apply the repair operation (Sect. 3.6) on the mutant to make it feasible and proper.

## 3.5    Crossover

The mutant and the target solution are considered as two parents for the crossover. The resulting solution after crossover is referred to as trial solution. We have applied a simple uniform crossover operation. Let $f(S)$ represents the fitness of the solution $S$. With probability $p_{copy} = \frac{f(target_{sol})}{f(mutant)+f(target_{sol})}$, we have copied the bit values from the mutant to the trial solution or else bit values are copied from the target solution. As DACLP is a minimization problem, this policy results in copying of more bits to trial solution from better solution between mutant solution and target solution. If mutation has not been applied then crossover is mandatorily applied. Otherwise, we apply crossover with probability $p_c$, i.e., with probability $1-p_c$, mutant solution is copied to trial solution. Like mutant solution, repair operator is also applied on the trial solution.

### 3.6   Repair

In the repair operation, if the solution obtained through mutation/crossover is infeasible because the separating distance constraint among facility sites is not satisfied, we eliminate some facilities to make it a feasible solution. A randomized approach is followed for eliminating facilities from the given solution. We randomly select a site $u$ which is part of the current solution and mark all the sites which are in its forbidden set $Q_u$ as not being part of the solution. We repeat this step until the trial solution is made feasible.

After the trial solution becomes feasible, we check whether it is a proper solution or not. If it is not proper, then to add new facilities, we find the set $X_{rem}$ which contains the sites that can be part of the solution without making the solution infeasible. Then, a new set $Q'_v = Q_v \cap X_{rem} \ \forall v \in X_{rem}$ is computed. After that a site $u \in X_{rem}$ with the highest cardinality of $Q'_u$ is added to the solution. Then we update the sets $X_{rem}, Q'_u \forall u \in X_{rem}$ according to the latest changes in the solution. The repair procedure stops once $X_{rem}$ becomes empty.

### 3.7   Population Replacement Model

If the trial solution's fitness value after repair is less than the target solution's fitness value then the trial solution is added to the population in place of the target solution or else the trial solution is discarded. If trial solution replaces target solution, we also compare its fitness with the best solution fitness. And if the trial solution's fitness is less than that of the best solution we make this trial solution as the best solution.

The pseudo-code for our DDE approach is given in Algorithm 1. The mutation (Sect. 3.4), crossover (Sect. 3.5), repair (Sect. 3.6) and fitness computation (Sect. 3.1) operations are performed by the four functions *Mutation, Crossover, Repair* and *fitness* respectively. $r01$ is a uniform variate in $[0, 1]$.

## 4   GA Approach for DACLP

A steady-state genetic algorithm [10] is the another evolutionary approach that we have proposed for the DACLP. In the remainder of this paper we refer to this approach as GA. The following subsections present the important features of the proposed GA approach.

---

**Algorithm 1:** DDE algorithm for DACLP

---

Generate initial population;
$best_{sol} \leftarrow$ optimal solution from the initial population;
**while** *(termination condition remains unsatisfied)* **do**
    **foreach** *(target$_{sol} \in$ population)* **do**
        **if** *(r01 $\leq p_m$)* **then**
            $no\_mutation \leftarrow 0$; $mutant_{sol} \leftarrow$ Mutation($best_{sol}$);
        **else**
            $mutant_{sol} \leftarrow best_{sol}$;
            $no\_mutation \leftarrow 1$;
        **if** *((no_mutation $= 1$) or (r01 $\leq p_c$))* **then**
            $trial_{sol} \leftarrow$ Crossover($target_{sol}$, $mutant_{sol}$);
        **else**
            $trial_{sol} \leftarrow mutant_{sol}$;
        $trial_{sol} \leftarrow$ Repair($trial_{sol}$);
        **if** *fitness(trial$_{sol}$) $\leq$ fitness(target$_{sol}$)* **then**
            $target_{sol} \leftarrow trial_{sol}$;
            **if** *fitness(trial$_{sol}$) $\leq$ fitness(best$_{sol}$)* **then**
                $best_{sol} \leftarrow trial_{sol}$;

**return** $best_{sol}$;

---

## 4.1 Solution Representation and Fitness

A solution in our GA approach represents the sites selected for locating facilities as an ordered list. It is an efficient representation compared to bit-vector as it consumes less memory to store a solution and also requires less computation time in the overall operations. Even though ordered list causes sorting overhead, such an encoding allows the efficient implementation of variation operators like crossover and mutation as explained in corresponding subsections. The chromosome length in this representation is not fixed as in the bit-vector representation but the variation operators are designed accordingly.

For our GA approach also, we have taken the objective function as the fitness function in the same way as in our proposed DDE approach.

## 4.2 Initial Solution Generation

To generate the initial solutions, we have used a method which is a combination of a completely random method and a semi-greedy method. In this method, to begin with consider all the $n$ sites as unvisited and start with an empty set for the solution, then we follow an iterative procedure. In each iteration, with probability $\rho_{rd}$ we randomly chose an unvisited site $u$, and make it part of the solution. As part of the semi-greedy method, 5 unvisited sites with highest value of $|Q_v|$ are determined and one of these 5 sites is randomly selected and is added to the solution. Considering the newly added site as $u$, we mark the site $u$ and every site in its forbidden set as visited, and proceed to the next iteration. Till there are no unvisited sites remaining, this process is repeated. Based on the number of unvisited sites, following are some exceptions to the afore mentioned rules of selecting a site in an iteration. If there is only a single unvisited site, then it is added directly. If there are two sites which are unvisited, then the one having highest cardinality of the forbidden set is selected. On the other hand, if there are more than 2 and less than 6 unvisited sites, we randomly choose one site from among the unvisited sites. After generating a complete solution, we make it an ordered list by sorting.

### 4.3 Selection

The two parent solutions for crossover and a single parent for mutation are chosen using probabilistic binary tournament selection in which parameter $\rho_{pbt}$ gives the probability based on which the fitter of the two randomly chosen solutions from the population is selected to be a parent.

### 4.4 Crossover

As part of the crossover, we first determine the intersection set of sites present in two parents. As solutions are represented as ordered lists, time taken to find the intersection of the parent solutions $S_1$ and $S_2$ is only $\mathcal{O}(min(|S_1|, |S_2|))$ instead of $\mathcal{O}(|S_1|.|S_2|)$. We copy the sites from the intersection set to the child, as the sites occurring in both the parents have a higher chance of being part of several good solutions. After this, in a similar method followed in generating initial solutions, remaining sites are selected to be part of the child solution one at a time, but value of $\rho_{rd}$ can vary. We set the value of $\rho_{rd}$ to zero with probability $\rho_{add}$, otherwise the same $\rho_{rd}$ value as in initial solution generation is used.

### 4.5 Mutation

For every site present in the parent solution, we generated a uniform random number $u01 \in [0, 1]$. Only if $u01$ is less than $\rho_m$, the corresponding site is copied to the mutant otherwise it is not copied to the mutant. After repeating this for all the sites in the parent solution, we have followed the same method as in crossover to add other sites to the mutant.

In our GA approach, we have utilized crossover and mutation in a mutually exclusive manner. Crossover is utilized with probability $\rho_c$, and with the remaining probability of $1 - \rho_c$ mutation is utilized. The reason being, as part of crossover operator we retain the common sites which are in both the parent solutions so as to generate even better child solutions using these common sites. If mutation is applied after the crossover some of these sites which are common in both the parents will be deleted.

### 4.6 Population Replacement Model

We have used a steady-state population replacement model in our GA approach. In this model, every generation produces only a single child solution. The child solution is discarded if it is found to be the same as any of existing members of the current population. Otherwise it replaces the member with the worst fitness if its fitness is better than that of the worst fitness member.

---

**Algorithm 2:** GA for DACLP

---

Construct $ps$ initial solutions $X_1, X_2, \ldots, X_{ps}$;
$X_{best} \leftarrow$ Optimal solution among $ps$ initial solutions;
**while** *(termination condition remains unsatisfied)* **do**
    **if** *(u01 < $\rho_c$)* **then**
        $S_1 \leftarrow BTS(X_1, \ldots, X_{ps})$;
        **repeat**
            $S_2 \leftarrow BTS(X_1, \ldots, X_{ps})$;
        **until** *($S_1 \neq S_2$)*;
        $X_C \leftarrow Cross(S_1, S_2)$;
    **else**
        $S_1 \leftarrow BTS(X_1, \ldots, X_{ps})$;
        $X_C \leftarrow Mutate(S_1)$;
    $X_C \leftarrow$ Localsearch$(X_C)$;
    Include $X_C$ in the population as per replacement policy;
    **if** *($X_C$ is better than $X_{best}$)* **then**
        $X_{best} \leftarrow X_C$;
**return** $X_{best}$;

---

### 4.7 Local Search

After crossover/mutation, to further minimize the child solution fitness we performed a two-one exchange operation as part of the local search. In this local search, we replace a pair of sites in the solution with a single site only if the resulting solution is proper and feasible.

Algorithm 2 gives the pseudo-code for the proposed GA where the probabilistic binary tournament selection method (Sect. 4.3), crossover operator (Sect. 4.4), mutation operator (Sect. 4.5) and local search (Sect. 4.7) are carried out by four functions $BTS()$, $Cross()$, $Mutate()$ and $Localsearch()$ respectively. Further, $u01$ is a uniform random variate in $[0, 1]$ and $ps$ is the population size.

## 5 Experimental Results

Both of our proposed approaches, viz. DDE and GA have been implemented in C. Table 1 lists the different parameters involved and their corresponding values for DDE and GA based approaches both. The respective parameter values of both the proposed approaches DDE and GA are chosen empirically. We have run both our approaches on a Linux system with 8 GB RAM and 3.40 GHz Core-i5-7500 processor. For each test instance, we have performed 10 independent runs of DDE and GA. We have fixed the same maximum execution time for each run of both the proposed approaches DDE and GA. We have executed both DDE and GA for 10 s on those instances having number of nodes upto 100. On those instances having more than 100 and upto 500 nodes, we have run both the proposed approaches for 60 s, and on the remaining instances having number of nodes greater than 500, we have run the two approaches for 100 s.

**Table 1.** Parameters for DDE and GA

| DDE Parameters | | GA Parameters | |
|---|---|---|---|
| Parameter | Value | Parameter | Value |
| $POP_{cnt}$ | 250 | $ps$ | 250 |
| $p_m$ | 0.9 | $\rho_{rd}$ | 0.75 |
| $p_c$ | 0.9 | $\rho_{bts}$ | 0.8 |
| $p_{mut}$ | 0.02 | $\rho_c$ | 0.5 |
| | | $\rho_m$ | 0.75 |
| | | $\rho_a dd$ | 0.9 |

We have tested our approaches on 2 datasets namely Beasley's OR-Library[1] and the TSPLIB[2] which are first introduced in [2]. There are 40 instances in the dataset derived from OR library with the number of nodes in the range of 50 to 1000 and $R$ value in the range of 5 to 50. Similarly, there are 40 instances derived from TSPLIB with the number of nodes in the range of 51 to 1577 and $R$ values of TSPLIB instances are considered as mentioned in [2].

Table 2 presents the results obtained by our proposed approaches DDE and GA on OR-Library instances, while Table 3 presents the results obtained by our proposed approaches on TSPLIB instances. In both the Tables 2 and 3, the 1st column, *Instance*, is the dataset name. Column two, $R$, gives the distance within which no two facilities can be located. The least and average solution values of the DDE method over 10 independent runs are given in columns 3, 4 and the least and average solution values of the GA method over 10 independent runs are given in columns 5, 6 respectively. The least objective value across all techniques is highlighted in bold for easy identification. Table 4 provides the summary of results in terms of number of instances on which DDE obtained better solution ($<$), same solution ($=$) and worse solution ($>$) when compared with GA. This summary is provided for the least objective values and average objective values both.

On the OR-Library dataset, for the least objective value over 10 independent runs, out of the 40 instances DDE produced the smaller objective values on 9 instances and the same objective value as GA on 31 instances. For the average objective values of 10 independent runs on the OR library dataset, DDE produced the smaller average values as compared to GA on 16 instances and the same average values as the GA on 24 instances. On the TSPLIB dataset, for the least objective value out of 10 independent runs, out of the 40 instances DDE produced the smaller objective values on 10 instances and the same objective value as GA on 29 instances and only on one instance DDE has got a higher objective value than GA. On the same TSPLIB dataset, coming to the average objective values of 10 independent runs, DDE produced the smaller average values as compared to GA on 19 instances and the same average values as the GA on 21 instances.

---

[1] http://people.brunel.ac.uk/~mastjjb/jeb/orlib/esteininfo.html.
[2] http://elib.zib.de/pub/mp-testdata/tsp/tsplib/tsp/index.html.

**Table 2.** Results on OR-Library dataset for DDE and GA

| Instance | $R$ | DDE | | GA | |
|---|---|---|---|---|---|
| | | Least | Average | Least | Average |
| OR_50.1 | 5 | 42 | 42.00 | 42 | 42.00 |
| | 10 | 26 | 26.00 | 26 | 26.00 |
| | 25 | 8 | 8.00 | 8 | 8.00 |
| | 50 | 3 | 3.00 | 3 | 3.00 |
| OR_50.2 | 5 | 39 | 39.00 | 39 | 39.00 |
| | 10 | 26 | 26.00 | 26 | 26.00 |
| | 25 | 7 | 7.00 | 7 | 7.00 |
| | 50 | 3 | 3.00 | 3 | 3.00 |
| OR_100.1 | 5 | 60 | 60.00 | 60 | 60.00 |
| | 10 | 29 | 29.00 | 29 | 29.00 |
| | 25 | 7 | 7.00 | 7 | 7.00 |
| | 50 | 3 | 3.00 | 3 | 3.00 |
| OR_100.2 | 5 | 64 | 64.00 | 64 | 64.00 |
| | 10 | 31 | 31.00 | 31 | 31.00 |
| | 25 | 7 | 7.00 | 7 | 7.00 |
| | 50 | 3 | 3.00 | 3 | 3.00 |
| OR_250.1 | 5 | 104 | 104.00 | 104 | 104.00 |
| | 10 | 34 | 34.90 | 35 | 35.40 |
| | 25 | 8 | 8.00 | 8 | 8.00 |
| | 50 | 3 | 3.00 | 3 | 3.00 |
| OR_250.2 | 5 | 93 | 93.00 | 93 | 93.00 |
| | 10 | 33 | 33.00 | 33 | 33.20 |
| | 25 | 7 | 7.00 | 7 | 7.90 |
| | 50 | 3 | 3.00 | 3 | 3.00 |
| OR_500.1 | 5 | 114 | 114.30 | 117 | 117.90 |
| | 10 | 35 | 35.50 | 37 | 38.00 |
| | 25 | 8 | 8.00 | 8 | 8.40 |
| | 50 | 3 | 3.00 | 3 | 3.00 |
| OR_500.2 | 5 | 110 | 110.50 | 113 | 114.00 |
| | 10 | 35 | 35.20 | 37 | 37.40 |
| | 25 | 8 | 8.00 | 8 | 8.00 |
| | 50 | 3 | 3.00 | 3 | 3.00 |
| OR_1000.1 | 5 | 125 | 127.20 | 137 | 140.40 |
| | 10 | 38 | 39.70 | 41 | 42.40 |
| | 25 | 8 | 8.00 | 8 | 8.90 |
| | 50 | 3 | 3.00 | 3 | 3.00 |
| OR_1000.2 | 5 | 123 | 124.70 | 133 | 134.70 |
| | 10 | 38 | 39.60 | 42 | 43.20 |
| | 25 | 8 | 8.00 | 8 | 8.90 |
| | 50 | 3 | 3.00 | 3 | 3.00 |

To understand the difference between DACLP solution and ACLP solution, and how this difference varies with $R$, Fig. 2, provides the plots of DACLP and ACLP solutions for $R = 6$, $R = 15$ and $R = 30$ respectively on the *eil51* instance having 51 nodes. In these plots, the nodes selected for locating facilities are depicted in yellow color and nodes not chosen for locating facilities are depicted in brown color. We have not provided the plot for the case with $R = 3$, as both ACLP and DACLP solutions have the same number of facilities which

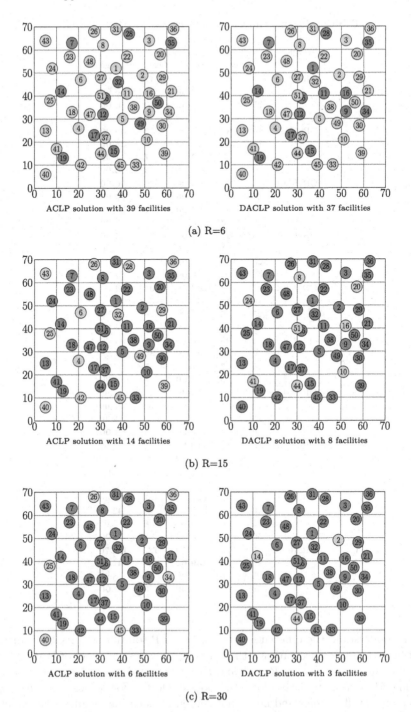

**Fig. 2.** Plots of ACLP and DACLP solutions on the *eil51* instance having 51 nodes for different values of R

**Table 3.** Results on TSPLIB dataset for DDE and GA

| Instance | $R$ | DDE | | GA | |
|---|---|---|---|---|---|
| | | Least | Average | Least | Average |
| eil51 | 3 | 50 | 50.00 | 50 | 50.00 |
| | 6 | 37 | 37.00 | 37 | 37.00 |
| | 15 | 8 | 8.00 | 8 | 8.00 |
| | 30 | 3 | 3.00 | 3 | 3.00 |
| rat99 | 8 | 82 | 82.00 | 82 | 82.00 |
| | 15 | 31 | 31.00 | 31 | 31.00 |
| | 38 | 7 | 7.00 | 7 | 7.00 |
| | 75 | 2 | 2.00 | 2 | 2.00 |
| rat195 | 11 | 106 | 106.00 | 106 | 106.00 |
| | 21 | 34 | 34.00 | 33 | 34.10 |
| | 52 | 7 | 7.00 | 7 | 7.00 |
| | 104 | 2 | 2.00 | 2 | 2.00 |
| pr299 | 222 | 84 | 84.00 | 84 | 84.60 |
| | 445 | 29 | 29.90 | 30 | 30.20 |
| | 1114 | 8 | 8.00 | 8 | 8.00 |
| | 2228 | 2 | 2.00 | 2 | 2.00 |
| d493 | 170 | 54 | 54.00 | 54 | 55.30 |
| | 340 | 18 | 18.00 | 18 | 18.40 |
| | 851 | 5 | 5.00 | 5 | 5.00 |
| | 1700 | 2 | 2.00 | 2 | 2.00 |
| u724 | 111 | 125 | 126.00 | 131 | 132.50 |
| | 222 | 37 | 37.10 | 37 | 38.60 |
| | 555 | 7 | 7.00 | 7 | 7.20 |
| | 1110 | 2 | 2.00 | 2 | 2.00 |
| pr1002 | 650 | 111 | 113.40 | 118 | 119.60 |
| | 1300 | 35 | 36.00 | 36 | 38.10 |
| | 3250 | 7 | 7.00 | 7 | 7.20 |
| | 6500 | 2 | 2.00 | 2 | 2.00 |
| pcb1173 | 120 | 149 | 151.30 | 158 | 161.30 |
| | 240 | 41 | 42.30 | 43 | 44.00 |
| | 600 | 7 | 7.00 | 7 | 7.80 |
| | 1200 | 2 | 2.00 | 2 | 2.00 |
| d1291 | 176 | 71 | 73.20 | 75 | 78.80 |
| | 352 | 23 | 24.00 | 25 | 25.30 |
| | 879 | 6 | 6.00 | 6 | 6.00 |
| | 1760 | 2 | 2.00 | 2 | 2.00 |
| fl1577 | 91 | 56 | 57.30 | 58 | 60.00 |
| | 182 | 27 | 27.10 | 28 | 28.40 |
| | 456 | 8 | 8.00 | 8 | 8.80 |
| | 910 | 2 | 2.00 | 2 | 2.00 |

**Table 4.** Summary table

| Dataset name | Least | | | Average | | |
|---|---|---|---|---|---|---|
| | < | = | > | < | = | > |
| OR-Library dataset | 9 | 31 | 0 | 14 | 26 | 0 |
| TSPLIB dataset | 10 | 29 | 1 | 19 | 21 | 0 |
| Overall | 19 | 60 | 1 | 33 | 47 | 0 |

is 50, and only node 46 is a non-facility node. DACLP solutions are obtained through approaches presented here. On the other hand, ACLP solutions were obtained by the approaches of [2]. As *eil51* instance is a small instance, all the approaches for ACLP/DACLP obtain the same ACLP/DACLP solutions. As the minimum separating distance $R$ increases, the difference in number of facilities being located is evident as DACLP gives the lower bound on the number of facilities and ACLP the upper bound. It can be observed that for the $R=30$ case on *eil51* instance, DACLP solution locates 3 facilities in comparison to 6 facilities located by using ACLP, which is 50% lesser number of facilities being located. It can also be observed that in case of ACLP solution, facility nodes tend to be located near the boundaries, whereas in case of DACLP solution, facility nodes tend to be more centrally located.

## 6   Conclusions and Future Work

DACLP is a recently introduced under-studied problem. As part of this work, we have devised two intelligent optimization approaches for DACLP, viz. DDE and GA approaches. We have tested the performance of our approaches on a total of 80 instances with upto 1577 sites. When the least objective value is considered, DDE produced smaller objective values than GA on 19 instances, same objective values on 60 instances, and a greater objective value on 1 instance. Similarly, for the average objective value, DDE produced smaller objective values on 33 instances, equal objective values on 47 instances. Overall, when the comparison is done with respect to least objective value DDE produced solutions of equal or better quality in comparison to GA on 79 out of the 80 instances and when the comparison is done with respect to average objective value on all the 80 instances DDE produced solutions of equal or better quality in comparison to GA. A dedicated repair operator applied in DDE possibly contributed to its better performance by removing additional facilities and that too in a diverse manner while making the solution proper thereby paving the way for a better exploration of the search space.

Our DDE and GA approaches are the first heuristic approaches for the DACLP. Hence, these approaches will be used in future as the baseline approaches for evaluating the performance of any new heuristic approach for DACLP. Similar intelligent optimization algorithms may be proposed for other related problems such as the minimum dominating set and minimum vertex cover problems, as well as their variations.

## References

1. Barahona, F., Weintraub, A., Epstein, R.: Habitat dispersion in forest planning and the stable set problem. Oper. Res. **40**(Supplement 1), S14–S21 (1992)
2. Chappidi, E., Singh, A.: Discrete differential evolution-based solution for anti-covering location problem. In: Tiwari, A., Ahuja, K., Yadav, A., Bansal, J.C., Deep, K., Nagar, A.K. (eds.) Soft Computing for Problem Solving. AISC, vol.

1392, pp. 607–620. Springer, Singapore (2021). https://doi.org/10.1007/978-981-16-2709-5_46

3. Chaudhry, S.S.: A genetic algorithm approach to solving the anti-covering location problem. Expert Syst. **23**(5), 251–257 (2006)

4. Chaudhry, S.S., McCormick, S.T., Moon, I.D.: Locating independent facilities with maximum weight: greedy heuristics. Omega **14**(5), 383–389 (1986)

5. Chaurasia, S.N., Singh, A.: A hybrid evolutionary algorithm with guided mutation for minimum weight dominating set. Appl. Intell. **43**(3), 512–529 (2015). https://doi.org/10.1007/s10489-015-0654-1

6. Church, R.L., Cohon, J.L.: Multiobjective location analysis of regional energy facility siting problems. Technical report, Brookhaven National Lab., Upton, NY (USA) (1976)

7. Church, R.L., Garfinkel, R.S.: Locating an obnoxious facility on a network. Transp. Sci. **12**(2), 107–118 (1978)

8. Current, J.R., Storbeck, J.E.: A multiobjective approach to design franchise outlet networks. J. Oper. Res. Soc. **45**(1), 71–81 (1994)

9. Das, S., Suganthan, P.N.: Differential evolution: a survey of the state-of-the-art. IEEE Trans. Evol. Comput. **15**(1), 4–31 (2011)

10. Davis, L.: Handbook of Genetic Algorithms. Van Nostrand Reinhold (1991)

11. Dimitrijević, B., Teodorović, D., Simić, V., Šelmić, M.: Bee colony optimization approach to solving the anticovering location problem. J. Comput. Civ. Eng. **26**(6), 759–768 (2011)

12. Erkut, E.: The discrete p-dispersion problem. Eur. J. Oper. Res. **46**(1), 48–60 (1990)

13. Grubesic, T.H., Murray, A.T., Pridemore, W.A., Tabb, L.P., Liu, Y., Wei, R.: Alcohol beverage distribution control, privatization and the geographic distribution of alcohol outlets. BMC Public Health **12**, 1015 (2012)

14. Khorjuvenkar, P.R., Singh, A.: A hybrid swarm intelligence approach for anti-covering location problem. In: Proceedings of the 2019 IEEE International Conference on Innovations in Power and Advanced Computing Technologies (i-PACT 2019), vol. 1, pp. 1–6. IEEE (2019)

15. Li, W., Wang, G.G., Gandomi, A.H.: A survey of learning-based intelligent optimization algorithms. Arch. Comput. Methods Eng. **28**, 3781–3799 (2021)

16. Moon, I.D., Chaudhry, S.S.: An analysis of network location problems with distance constraints. Manag. Sci. **30**(3), 290–307 (1984)

17. Murray, A.T., Church, R.L.: Solving the anti-covering location problem using lagrangian relaxation. Comput. Oper. Res. **24**(2), 127–140 (1997)

18. Niblett, M.R., Church, R.L.: The disruptive anti-covering location problem. Eur. J. Oper. Res. **247**(3), 764–773 (2015)

19. Pan, Q.K., Tasgetiren, M.F., Liang, Y.C.: A discrete differential evolution algorithm for the permutation flowshop scheduling problem. Comput. Ind. Eng. **55**(4), 795–816 (2008)

20. Pandiri, V., Singh, A., Rossi, A.: Two hybrid metaheuristic approaches for the covering salesman problem. Neural Comput. Appl. **32**(19), 15643–15663 (2020). https://doi.org/10.1007/s00521-020-04898-4

21. Pham, D.T., Karaboga, D.: Intelligent Optimisation Techniques. Springer, London (2000). https://doi.org/10.1007/978-1-4471-0721-7

22. Singh, A., Rossi, A., Sevaux, M.: Matheuristic approaches for Q-coverage problem versions in wireless sensor networks. Eng. Optim. **45**(5), 609–626 (2013)

23. Srivastava, G., Singh, A., Mallipeddi, R.: A hybrid discrete differential evolution approach for the single machine total stepwise tardiness problem with release dates. In: Proceedings of the 2021 IEEE Congress on Evolutionary Computation (CEC 2021), pp. 652–659. IEEE (2021)

24. Srivastava, G., Singh, A., Mallipeddi, R.: NSGA-II with objective-specific variation operators for multiobjective vehicle routing problem with time windows. Expert Syst. Appl. **176**, 114779 (2021)

25. Tao, F., Laili, Y., Zhang, L.: Brief history and overview of intelligent optimization algorithms. In: Configurable Intelligent Optimization Algorithm. SSAM, pp. 3–33. Springer, Cham (2015). https://doi.org/10.1007/978-3-319-08840-2_1

26. Tasgetiren, M.F., Pan, Q.K., Suganthan, P.N., Liang, Y.C.: A discrete differential evolution algorithm for the no-wait flowshop scheduling problem with total flow-time criterion. In: Proceedings of the 2007 IEEE Symposium on Computational Intelligence in Scheduling, pp. 251–258. IEEE (2007)

27. Valente, J., Moreira, M., Singh, A., Alves, R.: Genetic algorithms for single machine scheduling with quadratic earliness and tardiness costs. Int. J. Adv. Manuf. Technol. **54**, 251–265 (2011)

# Enhancing Robustness of Malware Detection Model Against White Box Adversarial Attacks

Riya Singhal[✉], Meet Soni, Shruti Bhatt, Manav Khorasiya,
and Devesh C. Jinwala[iD]

Sardar Vallabhbhai National Institute of Technology, Surat, Gujarat 395007, India
riyapsinghal@gmail.com

**Abstract.** Deep Neural Networks(DNNs) have made remarkable break-throughs in several fields such as computer vision, autonomous vehicles etc. Due to its adaptability to malware evolution, security analysts heavily utilise end-to-end DNNs in malware detection systems. Unfortunately, security threats such as adversarial samples cause these classifiers to output erroneous results. These adversarial samples pose major security and privacy risks since a malware detection model will mistakenly label a malware sample as benign. In this paper, we assess the resilience and reliability of our deep learning-based malware detection algorithm. We employed Malconv architecture for malware detection and classification, which was trained using the Microsoft Malware Dataset. We used the Fast Gradient Sign Method (FGSM), a white-box gradient-based attack, to generate adversarial samples for our malware detection model. Based on the performance of our model against this attack, we draw a comparative study between various mitigation techniques such as adversarial training, ensemble methodologies, and defensive distillation in order to analyse how capable they are at solving the problem at hand. Finally, we propose a novel approach - Iterative Distilled Adversarial Training - that combines two of these defence mechanisms, namely adversarial training and defensive distillation, in order to make our model more resilient to an adversarial attack in a white box setting. As a result, we drastically reduced the FGSM attack success rate by around 75% with only a small increase in training time. Additionally, unlike other multi-model defence strategies like ensemble learning, our technique uses one architecture while offering stronger defensive capabilities by relatively decreasing the success rate of attacks by 15%.

**Keywords:** Deep learning · Malware detection · Adversarial attack · Adversarial training · Ensemble learning · Defensive distillation

## 1 Introduction

The fast paced advancement in computer technology makes people worried about the security of the network systems. Malware threats have become ubiquitous

---

R. Singhal and M. Soni—These authors contributed equally to this work.

© The Author(s), under exclusive license to Springer Nature Switzerland AG 2023
A. R. Molla et al. (Eds.): ICDCIT 2023, LNCS 13776, pp. 181–196, 2023.
https://doi.org/10.1007/978-3-031-24848-1_13

and keep on growing in vertical (in terms of volumes and numbers) and horizontal manner (in terms of categorisations and capabilities). Moreover, malware developers and analysts are constantly competing with one another, with the developers building a malware detection system and the analysts creating malicious software that could bypass the detection systems [1]. Therefore, a robust system is required to automatically detect whether a given software is malicious or not in real-time without requiring domain expert knowledge in malware detection.

Deep Neural Networks(DNN's) have constantly proven their efficacy in dealing with complex and difficult machine learning problems. They have already outperformed traditional machine learning algorithms by precluding the need for feature extraction and selection, two crucial but time-consuming steps in the machine learning process and still giving equal or better accuracy [2]. As a result, domain expert knowledge and human intervention are no longer expected in order to build deep learning based malware detection systems.

However, researchers [3] have found empirical evidence demonstrating that these deep neural networks are indeed vulnerable to adversaries that can easily fool the model and cause them to misclassify, without actually being discovered by humans. This can be extremely hazardous for networks and computer systems using deep learning based malware detection systems, as adversaries can cause them to incorrectly classify a malicious sample as benign.

These adversarial examples, however, have defined a new direction of research that focuses primarily on studying the attack methods and how these adversarial samples are generated. There are various attack methods such as FGSM [4], L-BFGS [5], C&W(Carlini and Wagner) [3], Deepfool [6] to name a few, each having certain merits and demerits on how they generate these adversarial samples. The findings from this study can then be further employed to design defense mechanisms that will make the model resilient to such attacks.

As a result, we endeavored to design a robust and efficient deep learning based malware detection model that can detect various malware instances in real-time with a high level of accuracy while also withstanding adversarial attacks. We applied the white-box FGSM attack on our malware detection model and were able to successfully fool the model 98% of the time, demonstrating the severity of the adversarial attack. In order to mitigate these adversarial attacks, we evaluated existing defense mechanisms such as adversarial training [7,8], defensive distillation [9] and ensemble learning against FGSM attack. Our experimental analysis of the above mechanisms yielded two conclusions. Although the defensive distillation lowers the mis-classification rates of adversarial samples, the improvement observed is often insignificant. Moroever, adversarial training helps increase the generalization of the model provided that perturbations are carefully crafted to better understand how the adversarial samples bypass the malware detection system [10]. Also in existing literature not many authors have experimented with the combination of multiple defense mechanism as single system to provide defense against adversarial attacks, so in this paper we have tried to incorporate these combination of defense mechanisms in our study.

To incorporate the benefits of numerous defence mechanisms, we propose a novel approach - Iterative Distilled Adversarial training- that combines defensive distillation and adversarial training in a iterative fashion. This method significantly reduced the attack success rate (by 75%) as compared to previously mentioned single defense mechanisms.

This paper makes the following key contributions:

- We provide a comparative analysis of various mitigation techniques such as adversarial training, ensemble learning, and defensive distillation and evaluate their robustness against the FGSM attack in a white box scenario.
- We propose a novel mitigation technique called iterative distilled adversarial training that enhances the defensive capabilities of our malware detector against adversarial attacks by combing the advantages of the existing defensive mechanisms.

The remainder of paper is organized as follows. Section 2 discusses related work found in the existing literature. Section 3 provides background and preliminaries required for building a robust malware detection system. Section 4 describes the system model and our proposed defense approach against FGSM attack. Section 5 discusses the results obtained in our experimental evaluation. Finally conclusion and future work are given in Sect. 5.

## 2   Related Work

In this section, we summarise the existing literature discussing various adversarial attack mechanisms and defense strategies used for dealing with them.

[4] proposed Fast Gradient Sign Method (FGSM), a gradient-driven whitebox approach that generates adversarial samples using the classifier's loss function gradients. [11] proposed two whitebox methods to fool malware detectors. In Benign feature append method, perturbations are selected from benign files using GRAD-CAM method and are appended at the end of malware files to fool the detector. Enhanced BFA method is the combination of FGSM attack and BFA attack, in which instead of initializing perturbations as random-bytes they are taken from benign files to generate AEs more efficiently.

[12] utilized gradient-based FGSM attack and L-norm based C&W attack [13] method on image based malware detection system, and found that AEs generated for different Machine learning algorithms using these two attack methods could effectively bypass the image based malware detection system.

There are several known defenses that have been established previously in order to deal with these attacks. Few of them include adversarial training [7,8], ensemble methodologies, defensive distillation [9], GAN adversarial training.

[14] utilized the ensemble of 10 different models having the same architecture but with different weight initializations to provide defense against FGSM attack and C&W attack for images. [15] experimented with a GAN based adversarial training approach to provide defense against adversarial attacks. They utilized

LSGAN to produce adversarial samples for API based malware detection systems, which were later used for adversarial training to develop a robust malware detection model.

To decouple the generation of adversarial examples from trained model, [16] utilized ensemble adversarial training, although the error rates were similar for simple adversarial training and ensemble adversarial training for whitebox attacks (Iter-LL and R+Step-LL). Hence, this approach could not provide robustness against whitebox attacks. However, the adversarial training was improved by decoupling the AEs generation from the model that is being trained.

## 3    Background and Preliminaries

This section provides an overview of the types of Deep Learning Architectures, adversarial attack methods and pre-existing defense mechanisms used for combating adversarial attacks.

### 3.1    Deep Neural Network (DNN) Based Models

Following recent advancements in the disciplines of computer vision and NLP, the development of ML malware detection solutions has begun to move toward deep learning based solutions. These DL techniques have replaced the ML workflow's feature extraction and selection procedure with a completely trainable system. This section discusses the details about the DNN architectures for malware detection and classification that we have used in our paper.

**Malconv Architecture.** [2] presented a Malconv model which is a CNN based architecture, that is directly trained on PE files to eliminate the need of feature extraction and feature selection step. To better capture higher levels of location independent feature authors utilized convolutional activations in conjunction with global temporal max-pooling. Due to that, activations were generated regardless of the position of the feature. Also, Instead of training a network on raw bytes, a learnable embedding layer was employed to translate byte sequences to fixed length feature vectors, allowing shallow networks like this to generate activations for a larger range of input feature patterns (Fig. 1).

**Recurrent Convolutional Neural Network Architecture.** It is observed that sequential models are quite successful while dealing with textual data. They are often used in NLP problems. Recurrent Convolutional Neural Networks combine the benefits of both sequential models and convolutional models. The two types of sequential models we used in the RCNN mode are LSTM and GRU. This network architecture consists of three major layers:

- Convolutional layer to extract the local dependencies
- LSTM [17] or GRU [18] to extract the long-term or global dependencies from the binary files
- Classification layer to generate output probabilities

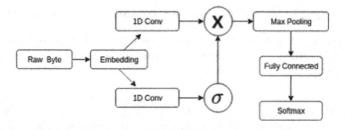

**Fig. 1.** Malconv model architecture [2]

## 3.2 Adversarial Attack

An adversary can try to tamper with the target model's output by manipulating the data set or the processing of data. Attack surface is generally the vulnerable points at which the attacker attacks the target model. Based on the knowledge about the model under attack, attack model generally falls under three categories: white-box attack, black-box attack and gray-box attack. Among the various adversarial attack methods available, this section focuses on the gradient based white box attack - Fast Gradient Sign Method(FGSM), that we have used in our experimentation.

[4] proposed FGSM, a gradient-driven whitebox approach that generates adversarial samples using the classifier's loss function gradients. The primary goal of FGSM is to generate AEs that can be used to breach the malware detection model by modifying an existing binary file. The method of creating AEs is iterative wherein minor perturbation bytes are added in gradient's direction to do FGSM attack. The main problem with this approach of adding perturbation works well for continuous input space, but they create problems for discrete input space such as byte (scalars from [0,255]), thus reconstruction from discrete to continuous input space is required.

Optimization problem to generate adversarial examples which are provided by [4]:

$$\bar{z} = \text{argmin}_{\bar{z}:||\bar{z}-z||_{p\leq\epsilon}} \bar{l}\left(\bar{z}, \bar{y}; \theta\right)$$

where $\bar{y} = $ desired target label,
$\epsilon = $ used to determine strength of the adversary
p = the norm value.

Assuming the loss function $\bar{l}$ is differentiable, then for max norm p = $\infty$ the solution is:

$$\bar{z} = z - \epsilon \cdot \text{sign}\left(\nabla_{\bar{z}}\bar{l}\left(\bar{z}, \bar{y}; \theta\right)\right)$$

utilizing $p = 2$ we get,

$$\bar{z} = z - \epsilon \cdot \left(\nabla_{\bar{z}}\bar{l}\left(\bar{z}, \bar{y}; \theta\right)\right)$$

### 3.3  Existing Mitigation Techniques Against White-Box Attack

In this section, we have discussed the existing and known defense methods against FGSM attack, namely adversarial training, ensemble learning and defensive distillation.

**Adversarial Training.** In adversarial training, adversarial samples are first generated for the model and then included into the original training set, so that model can learn from these adversarial samples about what type of perturbations are used to fool the model and increase the defensive capabilities against these adversarial samples.

**Defensive Distillation.** Defensive distillation is the technique of transferring the knowledge of one model to another model to provide the defense against adversarial attacks. In this paper we are using resilience approach for defensive distillation, in which knowledge of one network is transferred to other network with the same architecture to provide better generalization to model around training data points using soft labels obtained from first model. The training steps for this process are as follows:

- Train the first (original) model using input samples and their hard logits, in this temperature T is used for dividing outputs before passing it to softmax layer. Temperature controls the knowledge extracted by distillation process.
- Extract the probability distribution (soft labels) for each data point in training set using original model.
- Train second model (distilled) on same input data points but on soft labels obtained by first model, using the same temperature value which was used in first step.

At the end of these 3 steps a distilled model is obtained which has better generalization capabilities compared to original model on the training data points, and it provides resillience against perturbation.

**Ensemble Learning.** In supervised learning, ensemble methods are commonly used to enhance classifiers. The focus is generally on construction of a group of classifiers that may be used to classify a new data point by averaging their predictions, either weighted or unweighted. An ensemble must be both accurate and diverse in order to outperform a single classifier. A classifier is considered to be accurate if it performs better than random guessing, and a set of classifiers is said to be diverse if various classifiers produce different errors when presented with fresh data points.

# 4   Proposed Approach

## 4.1   System Model

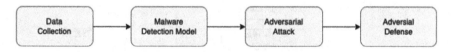

**Fig. 2.** Schematic diagram

Figure 2 provides our system overview for building a robust and resilient malware detection and classification system. The initial phase entails gathering data, in which we acquired samples of malicious and benign raw byte files of Windows executables. In step two, we constructed our baseline malware detection model that is trained on the dataset containing these raw byte files. If the system input is detected as a malware file, the classification model will be utilized to determine the malware family. In the third step, we evaluate the performance of our model against adversarial attacks. To delimit the scope of the study, our focus is only on whitebox attacks. We utilized FGSM based attacks to generate adversaries and cause our malware detector to misclassify. The fourth and the final step consists of implementing mitigation techniques developed as a countermeasure against the FGSM attack. Furthermore, we developed a novel technique in which known defense mechanisms are combined to produce a much more reliable solution. The components of the system are delineated in the Sects. 4.2, 3.2, 3.3 and 4.4.

## 4.2   Malware Detection and Classification Model

The Deep Learning based approach will be implemented as follows :

- Raw files will directly be fed as input to our model (for both training and testing purposes)
- The model will automatically extract the useful features
- The model will give the prediction whether the given file is malicious or benign (Fig. 3).

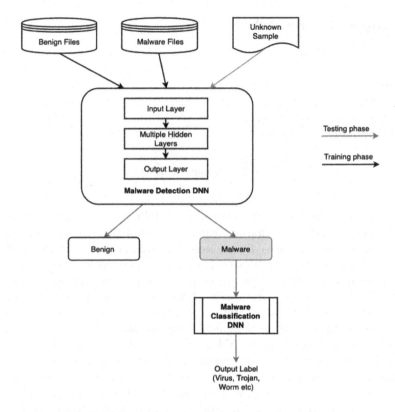

**Fig. 3.** Malware detection and classification system

This approach will make use of a Neural Network which can extract appropriate features from bytes data files and can also make the prediction. It eliminates the complex and time consuming task of feature extraction and selection.

Our aim was to perform two tasks, namely, malware detection and classification. Therefore, we trained two separate Deep Neural networks for the same. Initially, the input PE file would be passed through the malware detection model, which would identify the file as malicious or benign. If the file is predicted to be malicious, it will further be passed through the malware classification model to obtain the family of malware detected. We have used two deep learning architectures for the same which are Malconv and RCNN.

### 4.3   Attack Model

We have considered only evasion attacks as our attack surface, in which we only modify model inputs so that it can bypass the system. In our attack model we have chosen a white-box gradient based attack- FGSM which will modify the malware file such that it can bypass our malware detection/classification system.

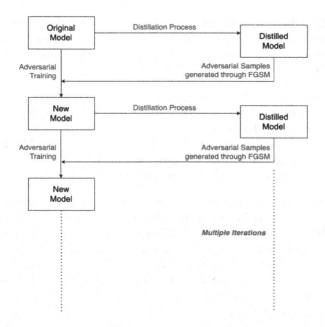

**Fig. 4.** Iterative Distilled Adversarial Training : Combination of Adversarial Training and Defensive Distillation

we have utilized max-norm FGSM and the perturbations have been added at the end of the file only, i.e. end-of-file injection (append attack) have been used. The size of perturbations was decided dynamically using the size of input binary file. These FGSM attack steps are run for multiple iterations (up to certain threshold) to successfully bypass the malware detection model.

In comparison to other attack methods, the FGSM attack has a low computing complexity and a very high transfer rate [19]. Furthermore, it is a non-targeted attack, in that it seeks to increase the likelihood of guessing the wrong label, but does not exactly specify which label should be outputted. As a result, it is well suited to models with small datasets and limited labels.

### 4.4    Adversarial Defense Model

In this paper, we have analysed and compared the performance of the existing defensive techniques such as adversarial training, ensemble methodologies and defensive distillation. Furthermore, an alternative defensive strategy is proposed, that combines advantages of the existing defense methodologies, namely, defensive distillation and adversarial training. In this method, multiple iterations are carried out, each consisting of model training, distillation model generation and adversarial sample generation.

**Iterative Distilled Adversarial Training.** We propose a novel approach that combines two known defensive mechanisms, namely, resilience defensive distillation with adversarial training. The implementation steps for this approach are described as below:

1. Train the original model on a training dataset.
2. Develop a distillation model using the soft labels obtained from the original model through distillation process.
3. Generate adversarial samples for the distilled model using FGSM attack
4. Apply adversarial training for the original model.
5. Repeat step-2 to step-4 for finite number of times

In Iterative distilled adversarial training, model improves during each iterations and it significantly increases the defensive capabilities of the model which is not possible in single defense methodology. Also it is noted in literature [10] that defensive distillation lowers the mis-classification rates of adversarial samples, but the improvement observed is often insignificant. Moreover, adversarial training helps increase the generalization of the model provided that perturbations are carefully crafted to better understand how the adversarial samples bypass the malware detection system. In order to overcome the drawbacks of both the defensive methods we have combined them in iterative fashion to propose this method (Fig. 4).

### 4.5 System Security Model

Out of all the potential adversarial attacks, we have delimited our scope of research to enhancing defensive capabilities of our malware detector against FGSM attack - a white-box, gradient-based, evasion attack. Additionally, for the experimental evaluation of existing and proposed mitigation measures against FGSM attack, we have primarily used the malconv architecture as a malware detection system.

## 5    Experiments and Results

In this section, we have presented the results obtained from our experimental evaluation.

### 5.1    Datasets

We have made use of 2 datasets for our experiment. The dataset used for malware classification is Microsoft malware Dataset released in 2015, for the Big Data Innovators Gathering Anti-Malware Prediction Challenge. It contains 10,868 labeled .asm and .bytes files of malwares belonging to 9 classes, namely: Ramnit, Lollipop, Kelihos_ver3, Vundo, Simda, Tracur, Kelihos_ver1, Obfuscator.ACY, Gatak. The dataset that we used for malware detection is a custom dataset consisting of 2 types of files - malicious and benign. For its preparation, we collected

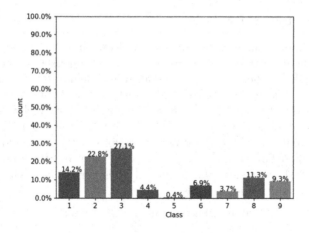

**Fig. 5.** Class distribution of microsoft malware dataset

dll, .exe and .net files from various softwares as benign files. We used files from the above mentioned Microsoft Malware Dataset for the malware files. A total of 8400 files were collected, consisting of 3900 benign files and 4500 malware files (Fig. 5).

## 5.2  Performance of Malware Detection and Classification Model

We have used Malconv architecture and GRU based RCNN for training our Malware Detection Model. The train-test split that we used throughout all our experiments was set as 80% data for training and 20% for validation. Also, the input type was byte files and the first $2^{20}$ bytes of each file were taken as input. The model classifies the input into one of the two classes- malware or benign.

We have used Malconv architecture for training our Malware Classification Model. The model classifies the input into one of the nine classes or families of malware present in our dataset. Table 1 shows the training and validation accuracy we obtained for our malware detection models as well as for our malware classification model, using the respective architectures.

**Table 1.** Detection and classification accuracy

| ModelAccuracy | Malware Detection | | Malware |
|---|---|---|---|
| | Malconv | RCNN(GRU) | Classification |
| Training | 99.5% | 99.3% | 99.5% |
| Validation | 99.1% | 98.7% | 97.15% |

## 5.3    Threat of FGSM Attack

Experimentation is done on two networks: Malconv and RCNN. The evaluation metrics used here is attack success rate(in%) which represents how many files, originally labelled malicious, are misclassified as benign, after the modification through the max norm ($p = \infty$) FGSM attack. We performed the attack on a total of hundred test files that were chosen randomly, all of which were originally malicious. We chose $\epsilon$ value to be 1, also we set the threshold for the FGSM attack to a maximum of ten iterations. Therefore, if the file could not bypass the detection model at the end of tenth iteration then the attack process will be halted. Table 2 shows results that obtained after performing FGSM attack on both our malware detection models.

**Table 2.** Success rate of FGSM attack

| DNN Architecture | Attack Success Rate% |
|---|---|
| Malconv Architecture | 97% |
| RCNN Architecture | 40% |

## 5.4    Defensive Capabilities of Various Mitigation Methods

We have performed the implementation related to mitigation techniques focusing on the Malconv based Malware Detection model. For all the defense methodologies that we have tested, we used a total of two hundred test files that were chosen randomly, all of which were originally malicious. The evaluation metrics used was attack success rate (in %), such that, the lower the attack success rate, the better the defense methodology is.

**Adversarial Training.** Firstly, we used the FGSM approach to create adversarial samples for 200 malware files. The Malconv model is then adversarially trained on a dataset consisting of both, entire training set and the previously generated adversarial samples. To verify the transferability property of adversarial samples, the above trained model is then tested against a separate collection of malware samples consisting of 200 files. Table 5 shows the outcomes for the same. As it is clearly evident, adversarial training performed inadequately and thus cannot provide resistance against AEs on its own.

**Ensemble Methodologies.** We trained 3 classifiers with the same network architecture (Malconv), but with different random initial input, creating a diverse set of 3 neural networks, with differentiating weights at each node. We used the following three different weight initialization methods:

(a) random initialization
(b) xavier uniform initialization and
(c) kaiming uniform initialization

Moreover, We trained 3 different classifiers with different architectures to create a diverse set of 3 neural networks. We used following 3 different architectures:

(a) Malconv
(b) RCNN
(c) RCNN with attention

In this defensive mechanism, Adversarial examples will be generated only for focused model (Malconv), and these Adversarial examples will be passed through above mentioned ensemble approaches to evaluate their defensive capabilities. The results of all the models used to create the ensemble are aggregated through simple averaging before passing them through the sigmoid function. Table 3 and Table 4 show the results that we obtained for the respective ensemble approaches. Based on the results, we could state that an ensemble of different architecture is a highly effective mitigation method. This is because of the reason that by utilizing different architectures, one model can encode information useful for malware detection that might not be extracted by other models. Thus, aggregation of information from different models through ensemble techniques proves to be an effective mitigation method against adversarial samples.

**Table 3.** Defensive capabilities of ensemble of same architectures

| Type of Initialization | Attack Success Rate% |
|---|---|
| Random Initialization | 97% |
| Xavier Uniform Initialization | 81% |
| Kaiming Uniform Initialization | 72% |
| *Ensemble of these 3 models* | *36%* |

**Table 4.** Defensive capabilities of ensemble of different architectures

| Architecture utilized | Attack Success Rate% |
|---|---|
| Malconv | 97% |
| RCNN | 40% |
| RCNN with attention | 43% |
| *Ensemble of these 3 models* | *16%* |

**Defensive Distillation.** In this approach, we trained the Malconv based Malware detection model, with temperature $T = 100$, on our training dataset with hard labels. Following that, we trained another Malconv based model on soft labels generated using the first model and using the same value of temperature. Here, for classification purposes, softmax activation was used instead of sigmoid activation in this case. Table 5 shows the results obtained by performing FGSM attack on the distilled Malconv based model. Based on the results, we could state that defensive distillation does not prove to be a strong defense against Adversarial samples. This may be due to the binary nature of the malware detection problem, as the soft labels generated were very close to the hard labels and hence, not much information could be encoded into the soft labels. However, it helps in generalizing the model.

**Iterative Distilled Adversarial Training.** For this novel approach, we merged the two known defensive mechanisms, namely adversarial training and defensive distillation, using an iterative process, thus incorporating the advantages of both the techniques. Malconv model obtained the strongest defensive capabilities against AEs generated by FGSM method after adopting iterative distilled adversarial training method. During our implementation, we were able to achieve good results after only two complete iterations of the entire training steps. Seperate malware files, not present in the dataset, were used for testing in this case as well. The outcomes are listed in the Table 5. It is evident that our proposed defensive method clearly surpasses the rest of the mitigation techniques implemented in this work, enabling our model with much better robustness against FGSM white box attack.

**Table 5.** Defensive capabilities of iterative distilled adversarial training

| Model | Attack Success Rate% |
| --- | --- |
| Malconv | 97% |
| Adversarial Training | 76% |
| Distilled Model | 75.5% |
| *Iterative Distilled Adversarial Training* | *1%* |

# 6   Conclusion

DL-based algorithms give higher accuracy than traditional and ML-based malware detection techniques while also reducing the time and computational power required for training the model, by eliminating the feature extraction and feature selection stages. However these DL-based models are sensitive to adversarial attacks, which makes real-world deployment of DNN systems difficult. Moreover, construction of powerful mitigation techniques that are resistant to adversarial attacks is still a work in progress.

Hence, in an attempt to solve this issue, we first checked the performance of our malware detection model against FGSM attack. This attack could successfully trick CNN based DL models (Malconv architecture) to classify the majority of the malicious files as benign. As a result, we examined different DL architectures to see how well they might withstand the attack while also including the benefits of sequential models such as RCNN. To further enhance the resistance of our DL models against FGSM attack, we applied known mitigation techniques namely adversarial training, ensemble learning and defensive distillation.

Adversarial training gave poor performance. Under ensemble methodologies, we created an ensemble of the same architecture with different weight initializations as well as an ensemble of three different architectures. The performance of the later method was discovered to be better. Furthermore, we devised a novel approach, Iterative Distilled Adversarial Training, in which we combined defense distillation and adversarial training to make our malware detector more robust to attacks. These defensive measures allowed us to significantly improve the DL-based model's sustainability and resilience.

## 7 Future Work

The presence of adversarial examples restricts the applications of deep learning. Constructing defenses that are resistant to adversarial examples is still a work in progress. As a future endeavor, we will test the defensive capabilities of the ensemble approach and our Iterative distilled adversarial training on other white box attacks, such as C&W attack. Moreover, we will investigate the mitigation techniques to make our model more impervious to attacks in black box setting.

## References

1. Gibert, D., Mateu, C., Planes, J.: The rise of machine learning for detection and classification of malware: research developments, trends and challenges. J. Netw. Comput. Appl. **153**, 102526 (2020). https://www.sciencedirect.com/science/article/pii/S1084804519303868
2. Raff, E., Barker, J., Sylvester, J., Brandon, R., Catanzaro, B., Nicholas, C.: Malware detection by eating a whole exe (2017)
3. Carlini, N., Wagner, D.: Towards evaluating the robustness of neural networks. In: 2017 IEEE Symposium on Security and Privacy (SP), pp. 39–57 (2017)
4. Kreuk, F., Barak, A., Aviv-Reuven, S., Baruch, M., Pinkas, B., Keshet, J.: Deceiving end-to-end deep learning malware detectors using adversarial examples, arXiv: Learning (2018)
5. Szegedy, C., et al.: Intriguing properties of neural networks, arXiv preprint arXiv:1312.6199 (2013)
6. Moosavi-Dezfooli, S., Fawzi, A., Frossard, P.: Deepfool: a simple and accurate method to fool deep neural networks, CoRR, vol. abs/1511.04599 (2015). http://arxiv.org/abs/1511.04599
7. Madry, A., Makelov, A., Schmidt, L., Tsipras, D., Vladu, A.: Towards deep learning models resistant to adversarial attacks (2017). https://arxiv.org/abs/1706.06083
8. Kurakin, A., Goodfellow, I., Bengio, S.: Adversarial machine learning at scale (2016). https://arxiv.org/abs/1611.01236

9. Papernot, N., McDaniel, P., Wu, X., Jha, S., Swami, A.: Distillation as a defense to adversarial perturbations against deep neural networks (2015). https://arxiv.org/abs/1511.04508
10. Grosse, K., Papernot, N., Manoharan, P., Backes, M., McDaniel, P.: Adversarial examples for malware detection. In: Foley, S.N., Gollmann, D., Snekkenes, E. (eds.) ESORICS 2017. LNCS, vol. 10493, pp. 62–79. Springer, Cham (2017). https://doi.org/10.1007/978-3-319-66399-9_4
11. Chen, B., Ren, Z., Yu, C., Hussain, I., Liu, J.: Adversarial examples for CNN-based malware detectors. IEEE Access **7**, 54 360–54 371 (2019)
12. Liu, X., Zhang, J., Lin, Y., Li, H.: ATMPA: attacking machine learning-based malware visualization detection methods via adversarial examples. In: 2019 IEEE/ACM 27th International Symposium on Quality of Service (IWQoS), pp. 1–10 (2019)
13. Carlini, N., Wagner, D.: Towards evaluating the robustness of neural networks (2016). https://arxiv.org/abs/1608.04644
14. Defending against adversarial examples. https://www.osti.gov/biblio/1569514. Accessed 27 May 2022
15. Wang, J., Chang, X., Wang, Y., Rodríguez, R., Zhang, J.: Lsgan-at: enhancing malware detector robustness against adversarial examples. Cybersecurity **4**, 38 (2021)
16. Tramèr, F., Kurakin, A., Papernot, N., Goodfellow, I., Boneh, D., McDaniel, P.: Ensemble adversarial training: attacks and defenses (2017). https://arxiv.org/abs/1705.07204
17. Hochreiter, S., Schmidhuber, J.: Long short-term memory. Neural Comput. **9**, 1735–80 (1997)
18. Cho, K., et al.: Learning phrase representations using RNN encoder-decoder for statistical machine translation (2014). https://arxiv.org/abs/1406.1078
19. Zhang, J., Li, C.: Adversarial examples: opportunities and challenges. IEEE Trans. Neural Netw. Learn. Syst. **31**(7), 2578–2593 (2020)

# Early Detection of Covid Using Spectral Analysis of Cough and Deep Convolutional Neural Network

Ramasamy Mariappan$^{(\boxtimes)}$ (iD)

V.R. Siddhartha Engineering College (Autonomous), Vijayawada 520007, India
`Prof.mariappan.r@gmail.com`

**Abstract.** Now-a-days, there are numerous techniques and ICT tools for the detection of Covid-19. But, these techniques are working with the help; of culminated or peak of symptoms. However, there is a demanding need for the early detection of Covid with self-reported symptoms or even without any symptoms, which makes it easier for further diagnosis or treatment. This research paper proposes a novel approach for the early detection of Covid with the spectral analysis of Cough sound using discrete wavelet transform (DWT), followed by deep convolution neural network (DCNN) based classification. The proposed method with the cough spectral analysis and Deep Learning based algorithm returns the covid infection probability. The empirical results show that the proposed method of covid detection using cough spectral analysis using DWT and deep learning achieves better accuracy, while compared to the conventional methods.

**Keywords:** Covid-19 · Cough sound · Deep learning · Discrete wavelet transform

## 1 Introduction

Currently, Covid-19 is detectable with Reverse Transcriptase Polymerase Chain Reaction (RT-PCR), which detects presence of genetic fragments of SARS-Cov-2 within secretions from nasal and pharyngeal epithelial mucus membrane. Employed techniques of RT-PCR and immunoglobulin presence detection methods have their own limitations of detection within a specific time period. Prior to detection through RT-PCR, no method is available to assess Covid-19 infection during incubation and after the onset of symptoms. Consequently, a high transmission rate has been reported and needs to be reduced for effective containment.

At present, Reverse Transcriptase Polymerase Chain Reaction (RT-PCR), which finds the presence of genetic fragments of SARS-Cov-2 within secretions from nasal and pharyngeal epithelial mucus membrane, is currently the only method for detecting Covid-19. The detection thresholds for the RT-PCR and immunoglobulin presence detection methods that are used each have their own restrictions. There is no technology available to evaluate Covid-19 infection during incubation and after the start of symptoms prior

A. R. Molla et al. (Eds.): ICDCIT 2023, LNCS 13776, pp. 197–207, 2023.
https://doi.org/10.1007/978-3-031-24848-1_14

to identification using RT-PCR. Since a high transmission rate has been noted, it must be decreased for containment to be effective.

Clinical manifestations of SARS-Cov-2 [1] appeared variable as compared to influenza. Symptoms of Covid-19 also vary slightly from region to region. Abdominal symptoms were more frequent in the USA than China. Asymptomatic, mild, and severe symptoms were observed in various studies. Asymptomatic or milder cases did not seek medical intervention; mild symptoms included a temperature >37.5 °C and dry cough initially and could develop to moderate symptomatic cases. Fever, cough, abdominal discomfort, and deranged blood biomarkers were recorded in moderate cases. Severe cases presented with shortness of breath, dyspnea, and tachypnea and required mechanical ventilation. Persistent cough, fever, and fatigue were associated symptoms of an underlying pathology or pre-existing pathology not restricted to cardiovascular issues, hypertension, liver compromise, and diabetes. Blood pO2 levels decreased. Blood biomarkers developed lymphopenia, thrombopenia, and elevated aminotransferases in moderate and severe cases. White blood cells deteriorated in severe cases and required mechanical ventilation. Persistent fever and characteristic consistent coughing—initially dry for several days followed by a productive cough—are the main features in patients with pre-existing respiratory infections; a few symptoms were variable with geographical regions.

In comparison to influenza, SARS-clinical Cov-2's symptoms seemed more varied. The signs and symptoms of Covid-19 also differ marginally by locale. The USA experienced more abdominal symptoms than China did. In numerous researches, asymptomatic, minor, and severe symptoms were noted. A temperature over 37.5 °C and a dry cough were minor symptoms that could progress to moderate symptoms in individuals that were asymptomatic or had milder symptoms. These cases did not seek medical attention. In moderate cases, symptoms such as fever, coughing, stomach pain, and abnormal blood biomarkers were noted. Tachypnea, dyspnea, and shortness of breath were symptoms of severe instances, and mechanical ventilation was necessary. The signs of an underlying pathology or pre-existing pathology, including but not limited to cardiovascular problems, hypertension, liver damage, and diabetes, included a persistent cough, fever, and exhaustion.

In the current study, it is proposed a novel method for the early detection of Covid-19 using Cough spectral analysis with discrete wavelet transform (DWT) and deep convolution neural network (DCNN). The rest of the article is organized as follows. Section 2 reviews the literature relevant to the current area of research in Covid detection. Section 3 defines the problem and describes the proposed system. Section 4 elaborates the implementation methodology, followed by results and discussion in Sect. 5. Finally, Sect. 6 concludes with the salient observations and challenges ahead for further work.

## 2    Literature Review

Yu-Ting Shen and Hui-Xiong Xu [2] proposed in ancient there are currently different technologies that may be utilised to improve the usage of the telemedicine system to improve and improve traditional public health techniques to cope with the COVID-19 outbreak. Because of that sophisticated technology, the COVID pandemic may create a compelling "case" for incorporating the potential advantages of telemedicine into real-world therapeutic activity. Telemedicine provides a fantastic chance to take use of new technologies while maintaining a constant focus on the object and efficiency. Ilana Harrus, Jessica [3] has proposed the power of Deep learning against the covid detection to be seen and ignored. AI applications have evolved from pre-epidemic use to drug diagnosis and development, to predict the spread of the disease and to monitor human movement. New applications are also designed to address new needs and major challenges in the epidemic, including the provision of health care.

Zhao, Jiang, and Qiu [4] created a model that detects covid instances with 92% accuracy. Our model, when compared to the architecture model, represents current performance in all of the criteria we've discussed. Tese guarantees that patients who do not have COVID-19 be treated as they are in most situations, reducing the possibilities of identifying illnesses that do not have COVID-19 and easing the strain on the health system. In addition, we tested the model's performance with little data and discovered that it was still working properly. Ting DSW, Lin H, Ruamviboonsuk P, Wong TY, Sim DA [5] have created a sophisticated and effective digital platform that can improve health communication by improving patient care and education, allowing for speedier decision, and reducing resource usage. Various machine learning domains, such as natural language processing, have been tested and employed in health care facilities to date. As a person-to-person chat agent between the user and the service provider, AI-based chatbots play a critical function. Our health-care system will be greatly influenced by this chatbot and response mechanism.

Qurat-ul-Ain Arshad, Wazir Zada Khan, Faisal Azam, Muhammad Khurram Khan [6] have proposed a standardized test called Polymerase Chain Reaction (PCR) tests to detect the COVID is expensive and it consumes more time and it is danger. However, to assist specialists and radiologists in diagnosing and diagnosing COVID-19, in-depth study plays an important role. Numerous research efforts have been made to develop deep learning strategies and techniques for diagnosing or classifying patients with COVID-19, and these strategies have been proven as great tool that can detect or diagnose COVID cases.

Wong ZSY, Zhou J, Zhang Q [1] proposed a review of the use of artificial intelligence, telehealth, which aligns with public health responses in the workplace of health care within COVID-19 disease. Systematic scoping reviews were conducted to identify potential symptoms. Other includes a more no of evidence on a variety of health and medical applications over the telephone. A large number of reports have investigated the use of artificial intelligence (AI) and analysis of big data, weaknesses in research design and translation intelligence, highlighting the need for continuous research of the real world.

Samer Ellahham [7] suggested that Deep learning introduced a limited support during the covid disaster. It can be used to teach patient, tests and early detection of symptoms such as the flu etc. with chatbot apps. AI can also be used to remotely control mild symptoms in patients with depressive conditions through the implementation of telemedicine in practice. For major roles such as treatment planning/diagnosis, medical supervision is highly recommended.

## 3　Proposed System

In the recent past, several researchers [8, 9] have been attempted to detect Covid from the cough sound, it is more challenging issue to find good and qualitative features of the cough signal, which are reflecting the Covid status. It remains to be open problem to find the salient features and signs to detect Covid characteristics, which can distinguish the covid related cough symptoms. The main objective of this paper is to foster a pre-screening technique that could prompt mechanized distinguishing proof of COVID-19 through the investigation of frequency domain analysis with comparative execution. As shown in Fig. 1, the proposed system consists of acquisition of cough sound, pre-processing of cough audio signal, frequency domain feature extraction relevant to covid characteristics and classification using machine learning algorithm. In the proposed work, we have the following modules.

- i.　Cough sound acquisition
- ii.　Pre-processing
- iii.　Feature extraction
- iv.　Prediction and classification using DCNN model

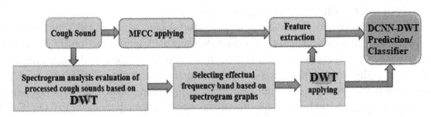

**Fig. 1.** Proposed system for Covid detection using Cough Sound

The main objectives of this proposed paper is follows.

- To develop a fast and accurate method for the programmed identification of Covid-19 using Cough spectral analysis.
- To obtain the joint time frequency domain features of the cough using DWT, prompting characteristic proof of Covid-19.
- To deploy an unique model of supervised method using deep convolutional neural network to analyze Covid-19 from the observed features.

### 3.1　Preprocessing

The process of reducing or suppressing noise from the cough sound signal is referred as pre-processing. The raw cough audio samples need to be pre-processed to enable to accept for further steps including feature extraction, classification, etc. The essential preprocessing steps followed are:

- Background noise removal
- Normalization
- Data augmentation

**Noise Removal**

The background noise removal is done by using noise filtering. The noise that associate in cough signals process are primarily Additive White Gaussian Noise (AWGN) with an even frequency distribution or the random noise. In our research work, we have considered the AWGN with the cough signals for noise filtering with the help of a first order Butterworth high pass filter (FOBHPF) in our method.

**Normalization**

It is the process of making the default values or data into standard scale. This is usually done when data attributes are at a different level. The standard formula for normalization is given by:

$$x_{std} = (x - x_{min})/(x_{max} - x_{min}) \tag{1}$$

**Data Augmentation**

This leads us to the following step in the data pre-processing data augmentation procedure. Many times, the amount of data we have is insufficient to adequately accomplish the classification task. In such circumstances, data augmentation is used. In deep learning problems, augmentation is frequently employed to improve the volume and variation of training data. Only the training set should be augmented; the validation set should never be augmented.

**3.2 Dynamic Feature Extraction**

The process of translating raw data into numerical features that can be processed while keeping the information in the original data set is known as feature extraction. It produces better outcomes than applying machine learning to raw data directly. The process of feature extraction consists of the following elements.

- Discrete wavelet transform (DWT)
- Spectral centroid
- Spectral analysis
- Zero Crossing Rate (ZCR)

**Chroma DWT**

A feature of a musical pitch class is the chroma feature. To include chroma characteristics, the Chroma DWT in the picture below required a short-term modification. Voice separation and signal structure are represented by the DWT. The highest values are displayed as a spike.

**Spectral Centroid**
The spectral centroid is a metric used to characterise a spectrum in digital signal processing. It shows where the spectrum's centre of mass is located. It has a strong perceptual link with the perception of a sound's brightness.

**Zero-Crossing Rate (ZCR)**
An audio frame's Zero-Crossing Rate (ZCR) is the rate at which the signal's sign changes during the frame. In other words, it's the number of times the signal's value changes from positive to negative and back, divided by the frame's length. ZCR is an indicator function and a key feature to classify percussive sounds. It is given by the formula,

$$zcr = \frac{1}{T-1} \sum_{t=1}^{T-1} 1\mathbb{R}_{<0}(S_t S_{t-1}) \tag{2}$$

where $s$ is a signal of length T and $1_{\mathbb{R}<0}$, is an indicator function.

**Spectral Analysis**
The difference between the upper and lower frequencies in a continuous range of frequencies is known as bandwidth. Because signals oscillate around a point, if the point is the signal's centroid, the sum of the signal's highest deviation on both sides of the point can be regarded the signal's bandwidth during that time frame. The mel spectrum is calculated by using DWT for the signal passing through the filter. The Mel frequency can be defined by,

$$f_{Mel} = 2595 \ \log_{10}\left(1 + \frac{f}{700}\right) \tag{3}$$

Finally, the feature extraction will be carried out by using Mel frequency cepstral coefficients (MFCC). The method of MFCC is applying the discrete signal on a window, with wavelet transform and computing logarithm of coefficients magnitude (cepstral coefficients), followed by warping frequencies to a mel scale as shown in Fig. 2. In our method, the Hanning window was utilised with a window size of 1024 and an overlap of 512. The size of the feature vector obtained from MFCC was 121 × 13.

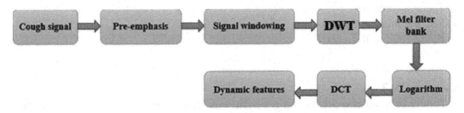

**Fig. 2.** Dynamic feature extraction using MFCC with DWT and DCT

### 3.3  Prediction and Classification

Deep learning [10, 11] is a developing branch of computation and prediction, which finds numerous applications in fields like medical, agriculture, forecasting the weather, the stock market, etc. The Deep learning based Convolutional neural networks (DCNN) are frequently utilised for both feature extraction and time series forecasting. We have used DCNN for feature extraction and long short term memory (LSTM) for time series multi-step forecasting of Covid based on Cough with multiple simultaneous inputs. In contrast to the majority of other forecasting algorithms, LSTMs may pick up on sequence nonlinearities and long-term relationships. Therefore, LSTMs are less concerned with stationary. The DCNN-LSTM model used for the covid-19 prediction is shown in Fig. 3.

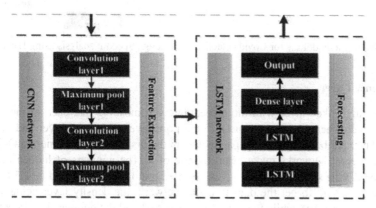

**Fig. 3.**  Feature extraction and Prediction with DCNN

## 4  Implementation Methodology

This section describes bout the methodology of implementation. It begins with capturing Cough audio signal capturing, preprocessing, feature extraction, training, validation, prediction and classification as shown in Fig. 4. For implementing this proposed system, we have used Kaggle and Github to collect audio and data sets as input. We recorded real-time cough sounds with Google Recorder and saved in.wav format after being converted from .mp3 format. The Audio recordings are preprocessed and noise filtered and transcribed in real time by the app.

### 4.1  Activation ReLU

In the neural network, the activation function is g which is responsible for converting the total input from node to local activation. The Relu function is active in the function of opening the fixed line unit. By default, this function returns the Relu unlock frequency (x, 0) while the smart 0-element element and the input tensor. It helps to prevent strong growth in computational needs in order to work on the network. Relu nets are well suited to represent convex activities.

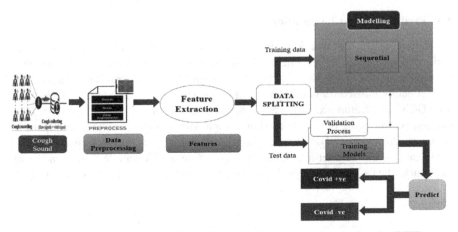

**Fig. 4.** Methodology for Covid detection with Cough Spectral analysis using DWT

## 4.2  Dense Layer

It is a highly linked layer with the preceding layer, meaning that each of the layer's neurons is connected to every neuron in the prior layer. The input layer and the first hidden layer are both supplied as parameters in the first component function Object(). Artificial neural networks make extensive use of it. The numbers of hidden neurons should be 2/3 the size of the input layer, and the input layer size should be 2/3 the size of the input layer. To get the density of the network, the density layer contains 576 input channel number, while the output number is 64 and the number of parameters is 36928. The main purpose of the dense layer is to separate the images based on the output from the convolution layers. Each layer in the neural network consists of neurons that comprise the rate at which they are implanted and the weighted mass transmitted by a non-linear function called activation function.

## 4.3  Sigmoid Activity

The estimated amount of input goes through the activation function and this output splits as the input is in the next layer. The sigmoid unit is a neural network and the activation function is a guarantee rather than the output of this unit which will remain between 0 and It is used to add a non-linearity to a machine learning model and performs tasks with great efficiency and to map out the real number in the possibilities. The scope of the sigmoid function is that the input value is between $-\infty$ and $+\infty$ while the output can only be between 0 and 1.

## 4.4  Dropout Layer

Dropout is also called Dilution/Drop-connect. Inputs not set to 0 are increased by 1/(1 - rate) which helps to prevent over-fitting. It is placed on top of the fully integrated layers only because they are the ones with the most parameters and over-aligning themselves causing over-fitting. It is a stochastic formulation. It is used during training to make it more dynamic in flexibility in data training after layers of convolution and after compilation of layers.

### 4.5 Convolution Layer

We can extract features from the our picture using convolution. Simply said, we break down our embedded picture into smaller tiles, make unique adjustments to these smaller tiles, and store the result as a new representation of our original tile. As a consequence, the most part interesting bits of the primary actual tile, i.e. the input picture components, are extracted for the each little output.

### 4.6 Max-Pool

Top integration is a blending function that chooses a substantial portion of the filter region covered by the feature map. As a result, the multi-component layer's output might be a f map that includes the features of the preceding feature. Kernel removes the most quantity of component possible throughout the Max Pooling conversion procedure.

### 4.7 Training and Validation

In order to attain a goal, ML algorithms [12] need training data. This training dataset will be analysed, the inputs and outputs will be classified, and the dataset will be analysed once again. An algorithm will effectively memorise all inputs and outputs in a training dataset if it is given enough time. The training dataset size and test sets are the process's key stop parameters. This is commonly stated as a percentage difference between training dataset 0 and 1 or test data sets. With A training set of 0.65 (65%) indicates that the remaining 0.35 (35%) is assigned to a test set.

A test set is a monitoring set that is used to evaluate on the performance of a model based on certain basic performance criteria. It is critical that the test set include no awareness from the training set. It will be difficult to determine if the algorithm has learnt to function normally from the training set or has just remembered it if the test set incorporates instances from the training set. Classification is a type of supervised machine learning and is a process of categorizing data into classes. Binary classification is about determine of a target variable is 0 or 1.

## 5 Results and Discussion

This section discusses the empirical results obtained from the collected datasets of Cough with the help of deep learning based models. The deep learning models tested are DenseNet-169, ResNet-50, InceptionV3, VGG-16 and VGG-19, applied with Adamax optimizer. The learning rates of different models and their obtained accuracies are tabulated in Table 1.

**Table 1.** Performance comparison of Deep learning models

| Model | Learning rate | Accuracy |
|---|---|---|
| VGG-19 | $10^{-4} *3$ | 93.6% |
| VGG-16 | 0.001 | 91.3% |
| Sequential (ANN) | $10^{-3} * 0.95$ | 89.7% |
| Densenet-169 | 0.14 | 91.7% |
| Resnet-50 | 0.002 | 90.1% |

## 6   Conclusion

In this research paper, we have devised a novel method for the early detection of the presence of corona virus using cough sound as an input with the help of deep learning model. The acquired cough signal was preprocessed and then applied with feature extraction with DCNN model followed by deep learning based prediction using LSTM model, which predicts the covid infection probability. The proposed DCNN-LSTM model was tested empirically and it identifies Covid-19 patients with cough sound to the extent of 93.6 % accuracy, which is better than the comparable similar methods. Also, it was observed that among the tested deep learning models with Adam-max optimizer, the VGG-19 model has performed better with the prediction accuracy of 93.6% .

## References

1. Wong, Z.S.Y., Zhou, J., Zhang, Q.: Artificial Intelligence for infectious disease big data analytics. Infect Dis Health. **24**, 44–48 (2019). https://doi.org/10.1016/j.idh.2018.10.002
2. Harrus, I., Wyndham, J.: Artificial intelligence and COVID-19: applications and impact assessment. Technical report prepared by under the auspices of the AAAS Scientific Responsibility, Human Rights and Law Program, 2021 (2021)
3. Zhao, W., Jiang, W., Qiu, X.: Fine-tuning convolutional neural networks for COVID-19 detection from chest X-ray images. Diagnostics **11**, 1887 (2021)
4. Ting, D.S.W., Lin, H., Ruamviboonsuk, P., Wong, T.Y., Sim, D.A.: Artificial intelligence, the internet of things, and virtual clinics: ophthalmology at the digital translation forefront. Lancet Digital Health 2(1), e8–e9 (2020)
5. Ting, D.S.W., Liu, Y., Burlina, P., Xu, X., Bressler, N.M., Wong, T.Y.: AI for medical imaging goes deep. Nat Med. **24**, 539–540 (2018). https://doi.org/10.1038/s41591-018-0029-3
6. Khan, W.Z., Azam, F., Khan, M.K.: Deep Learning Based COVID-19 Detection: Challenges and Future Directions. TechRxiv. Preprint. https://doi.org/10.36227/techrxiv.14625885 (2021)
7. Samer Ellahham, S., Ellahham, N.: Use of artificial intelligence for improving patient flow and healthcare delivery. J. Comput. Sci. Syst. Biol. **12**, 303 (2019)
8. Davenport, T., Kalakota, R.: The potential for artificial intelligence in healthcare. Future Healthc. J. **6**(2), 94–98 (2019)
9. Rousan, L.A., Elobeid, E., Karrar, M., Khader, Y.: Chest X-ray findings and temporal lung changes in patients with covid-19 pneumonia. BMC Pulm. Med. **20**(1), 1–9 (2020)

10. Cleverley, J., Piper, J., Jones, M.M.: The role of chest radiography in confirming covid-19 pneumonia. BMJ m2426 (2020)
11. Rahaman, M.M., Li, C., Yao, Y., Kulwa, F., Rahman, M.A., Wang, Q., et al.: Identification of COVID-19 samples from chest X-ray images using deep learning: a comparison of transfer learning approaches. J. Xray Sci. Technol. **28**(5), 821–839 (2020)
12. Wang, L., Lin, Z.Q., Wong, A.: Covid-net: A tailored deep convolutional neural network design for detection of COVID-19 cases from chest X-ray images. Sci. Rep. **10**(1), 1–12 (2020)

# Analysis of Tweets with Emoticons for Sentiment Detection Using Classification Techniques

Ravneet Kaur[(✉)], Ayush Majumdar, Priya Sharma, and Bhavana Tiple

School of Computer Science and Engineering, MIT World Peace University, Pune, India
ravneetkaur9978@gmail.com, ayushmajumdar6501@gmail.com,
priyasharma.3321@gmail.com, bhavana.tiple@mitwpu.edu.in

**Abstract.** Social networking services allow users to communicate with their friends and exchange ideas, photos, and videos that delineate their feelings. Sentiments are emotions that express a person's attitude, feelings, and worldview. This raises the possibility of analyzing individual moods and emotions in social network data in order to learn more about people's inclinations and perspectives when communicating online. Sentiment Analysis is the computational study of opinions, assessments, attitudes, subjectivity, and viewpoints represented in text. The emotive appraisal of a condition is a general evaluation of that condition that may be positive or negative depending on physical or mental reactions. In this paper, we attempt to evaluate tweets that contain both text and emoticons in order to determine whether they are positive or negative. This study looked at XGBoost, LinearSVC, Logistic Regression, and BernoulliNB algorithms, with XGBoost providing the highest accuracy of 87.841%. The paper's key contribution is the use of the XGBoost model for tweets that include emoticons, which also produced the greatest accuracy.

**Keywords:** Sentiment analysis · Social media · Tweet analysis · Classification techniques

## 1 Introduction

These days, communication relies not solely on text but also on the pictorial representations popularly known as emojis. Pictorial representations date back to the Egyptian scripts, where a sound was described by a symbol and a few symbols represented a word. In the advancing of times, drawing symbols that consumed a lot of time were replaced by drawing letters and words. These were quicker and to the point. But nowadays, with the introduction of social networking sites and emojis, their popularity is increasing, and a combination of text and emojis is rather popular to express thoughts and opinions.

SNS (Social Networking Sites) are an online community where people from all over the world, regardless of demographic or geographic disparities, may form networks with other individuals or organizations to share and express their ideas, opinions, and feelings. Twitter is a similar social networking platform that allows users to share news, information, and personal updates with other users in 140-character tweets or remarks, and it is utilized by a large number of people. Thoughts or tweets on any given topic can

A. R. Molla et al. (Eds.): ICDCIT 2023, LNCS 13776, pp. 208–223, 2023.
https://doi.org/10.1007/978-3-031-24848-1_15

be important in forming an opinion. This viewpoint can be of any type and be utilized for any purpose. Politics, brands, campaigns, and so on are some of the primary areas. Tweets about such themes may be interesting for analysis.

Sentiment analysis is a method that automatically extracts attitudes, opinions, viewpoints, and emotions from text, audio, tweets, and database sources using Natural Language Processing (NLP). In sentiment analysis, opinions in a text are categorized into "positive," "negative," and "neutral" categories. The terms subjectivity analysis, opinion mining, and appraisal extraction are also used to describe it. It is a technique for discovering the attitudes, viewpoints, and emotions expressed in a document by examining the emotional undertone of a string of words.

Emoticons are depictions of facial expressions such as a smile or frown that are generated by various keyboard character combinations and used to indicate the writer's feelings or intended tone. The capacity to extract sentiment and emotion insights from social data is a technique that enterprises all around the world are embracing. The objective of this research was to train a machine learning model to evaluate grammatical subtleties, cultural differences, extract emotions, and identify sentiment and meaning behind words using machine learning techniques. Various machine learning techniques were tried, tested, and implemented in this study. The novelty and contribution of this paper is thus to classify tweets containing emoticons for the first time, as emojis are usually neglected and removed as part of preprocessing the data, but emoticons, being an important factor in texting in these conventional years, have been made an integral part of this study to detect sentiments behind the tweets.

## 2  Related Work

The proposed aim of this study is to further increase the accuracy and robustness of Natural Language Processing models to classify sentiments based on textual data. The most popular application of sentiment analysis is to automatically determine if text data is good, negative, or neutral by detecting the polarity of the text data, such as a tweet, product review, or service request. It enables one to monitor what is being said on social media about one's product or services and can assist in identifying irate clients or unfavorable mentions before they worsen. In order to gather crucial data on how this field of study has rapidly expanded, particularly in recent years, the study has taken into account a number of previously published research projects that performed sentiment analysis using various machine learning techniques.

Alec Go et al., [1] presented a novel method for automatically classifying the emotions of Twitter messages. These messages were categorised either positive or negative in response to a query word. They discussed the outcomes of machine learning systems that classified the sentiment of tweets under remote supervision. As training data, they used Twitter messages with emoticons as noisy labels. Machine learning techniques (Naive Bayes, Maximum Entropy, and SVM) had an accuracy of more than 80% when trained with emoticon data.

A. U. Hassan et al., 2017 [2] have compared Naive Bayes, Support Vector Machine (SVM) and Maximum Entropy classifiers for phrase level sentiment analysis for measuring depression. They employed a feature selection method and a voting mechanism.

On Twitter and 20newsgroups datasets, they tested their suggested methodologies. They discovered that SVM outperformed Naive Bayes and Maximum Entropy classifier, with an accuracy of 91%.

Y. Chen et al., 2018 [3] used a deep learning LSTM network model to carry out sentiment analysis and prenatal depression detection tasks. It was proven that the deep learning LSTM network model could be used to screen for perinatal depression and yield results that were equivalent to those of the Edinburgh Postnatal Depression Scale Screening. Their approach improved the effectiveness of early detection of perinatal depression and was more objective than current manual screening techniques for perinatal users' actual conditions.

R. L. Rosa et al., 2019 [4] describes a Knowledge-Based Recommendation System (KBRS) that has a feature for tracking users' emotional health in order to identify those who might be experiencing psychological problems like stress or despair. Based on the monitoring outcomes, the KBRS was activated to deliver upbeat, serene, calming, or motivating messages to people with psychological difficulties. This was accomplished using ontologies and sentiment analysis. If the monitoring system notices a depressive interruption, the solution also includes a mechanism to send warning messages to authorised individuals. To identify words with depressing and stressful content, convolutional neural networks (CNN) and bi-directional long-short-term memory (BLSTM)- Recurrent Neural Networks(RNN) were employed. The accuracy of the suggested technique was 89% for depressed users and 90% for stressed users. According to the experimental findings, the suggested KBRS obtained a rating of 94% from customers who were extremely satisfied, as opposed to 69% for an RS that didn't use a sentiment metre or ontologies.

M. Deshpande et al., 2017 [5] objective was to analyze emotions in Twitter feeds using natural language processing, with a focus on depression. In order to identify depression, certain tweets were categorized as neutral or negative using a chosen word set. The class prediction technique used Naive-Bayes and Support Vector Machine classifiers. The major classification measures used to show the results were the F1-score, accuracy, and confusion matrix. The accuracy of the Mutinomial Naive Bayes was 83%, which was greater than SVM.

P. Arora et al., 2019 [6] proposed a novel way to identify health tweets for sadness and anxiety from all mixed tweets using Support Vector Regression (SVR) and Multinomial Naive Bayes algorithm as a classifier, which can help us determine health status in real-world situations. The accuracy of these analyzers was determined by validating tweets based on positive, neutral, and negative sentence scores. Support Vector Regression (SVR) classifiers outperformed the Multinomial Naive Bayes with an accuracy of 79.7%.

M. R. H. Khan et al., 2020 [7] used automatic sentiment analysis algorithms to identify happiness and sadness in people's sentiments. To prepare their dataset, they gathered Bengali postings from Facebook. Because collecting significant volumes of data in the aforementioned areas was problematic, they trained their model with a limited dataset and discovered greater accuracy and prediction. To create automated techniques, various classification methods such as Multinomial Naive Bayes, Support Vector Machine, KNearest Neighbors, Random Forest, Decision Tree, and XGBoost were used. Amongst these, the Multinomial Naive Bayes gave the maximum accuracy, which is 86.67%.

Abhilash Biradar et al., 2019 [8] provided a methodology for determining whether a person has depression using information from social media in a way that might use Twitter as a trustworthy source. They used a hybrid model that combined a backpropagation neural network (BPNN) model for classification and sentiment analysis methods like SentiStrength to provide train data. A single, narrowly targeted predictive algorithm analyzed patient Twitter activity to determine whether or not they were depressed.

Ayvaz et al., 2017 [9] examined the use of Emoji characters on social networks as well as their effects on text mining and sentiment analysis. In the analysis, they used Twitter as their information source. They gathered text information for a number of international positive and negative events in order to research the function of emoji characters in sentiment analysis. They discovered that using Emoji characters in sentiment analysis led to higher sentiment scores. Additionally, they found that the overall sentiments of positive opinions were more strongly influenced by the use of Emoji characters in sentiment analysis than those of negative opinions.

Chen et al., 2018 [10] proposed a new strategy for Twitter sentiment analysis with a focus on emojis. After learning bi-sense emoji embeddings under both positive and negative sentimTweets typically contain slang and lingoental tweets separately, they trained a sentiment classifier by attending to these embeddings with an attention-based LSTM network. Their findings demonstrated that the bi-sense embedding outperforms existing techniques and is effective for extracting sentiment-aware emoji embedding. Additionally, they used attention visualization to show that the bi-sense emoji embedding offered deeper insight into the attentional process, giving them a stronger understanding of the semantics and emotions.

A. P. Jain et al., 2016 [11] goal was to provide step-by-step instructions to use machine learning for performing sentiment analysis on Twitter data. Their study also provided details on the sentiment analysis method that was suggested. Their paper suggested a text analysis framework for Twitter data that was more flexible, quick, and scalable because it was built with Apache Spark. In the suggested system, the machine learning techniques Nave Bayes and Decision trees were employed for sentiment analysis. The results showed that the decision tree performs admirably, with 100 percent accuracy, precision, recall, and F1-Score.

M. S. Neethu et al., 2013 [12] attempted to analyse Twitter messages on electronic devices such as mobile phones and laptop computers using a machine learning technique. They showed off a brand-new feature vector for classifying tweets as positive or negative and extracting reviews of products. Twitter-specific properties were gathered in the initial step and added to the feature vector. Then, these characteristics were eliminated from tweets, and feature extraction was carried out once more as though it were on regular text. The feature vector also had these attributes. Several classifiers, including SVM, Maximum Entropy, Nave Bayes, and Ensemble classifiers, were used to evaluate how accurately the feature vector was classified. For the new feature vector, each of these classifiers delivered accuracy that was essentially identical.

G. Gautam et al., 2014 [13] on the basis of Twitter data, a number of machine learning techniques with semantic analysis were proposed for identifying sentences and product reviews. The main objective was to use a labelled Twitter dataset to assess a huge number of reviews. They preprocessed the dataset initially for this purpose, and then they extracted the adjective from the dataset that had some meaning—this was referred to as a feature vector. They then selected the feature vector collection and used Semantic Orientation-based WordNet in addition to machine learning-based classification techniques such as Naive Bayes, Maximum Entropy, and SVM. They discovered that SVM exposed to an unigram model outperformed SVM on its own, and that the naive bayes technique outperformed maximal entropy.

Y. Chandra et al., 2020 [14] collected the data from tweets and then processed them via machine learning classifiers. After the individual classifiers classified the "tweet", a voted classification procedure was utilized to determine the class of the "tweet" and the percentage confidence in it. The proportion of positive and negative tweets was calculated using the polarity classification method. Finally, Deep Learning Models for Tweet Classification were suggested. RNN, LSTM, and CNN RNN models were employed to categorize the tweets. Machine learning methods were outperformed by deep learning models like CNN-RNN, LSTM, and their many combinations.

Naresh, A. et al., 2021 [15] suggested an optimization-based machine learning technique to categorize Twitter data. Three stages were taken to complete the process. In the first stage, the data was gathered and preprocessed; in the second, it was optimized by removing pertinent features; and in the third, the updated training set was sorted into different classes using different machine learning techniques. The findings from each algorithm were distinct. When compared to other machine learning techniques, the proposed method, sequential minimum optimization with decision tree, produced a high accuracy of 89.47%.

## 3  Methodology

This study examines algorithms for evaluating tweets and deciding whether they should be positive or negative. As a result, several classification methods, including XGBoost, LinearSVC, Logistic Regression, and BernoulliNB, have been used to classify a single tweet's sentiment with the maximum level of accuracy.

Following is the outline of the research proposed in this study:

1. Data Gathering
2. Dataset Preprocessing
3. Model Training
4. Validation of results
5. Results & Inference

### 3.1  Data Gathering

The dataset being used is the sentiment140 dataset [16]. It contains 1,600,000 tweets extracted using the Twitter API. A significant number of tweets containing emoticons (Fig. 1). were extracted using the Twitter API. The tweets have been annotated (0 = Negative, 4 = Positive). There are both positive and negative tweets in the dataset. The sentiment has been recognised for the inclusion of both text and emojis; therefore, working with only text or only emojis will not provide the best possible accuracy. As a result, we decided to use a combination of text and emojis from the tweets, even though working with only one would have been a faster option. The dataset contains the following 6 fields:

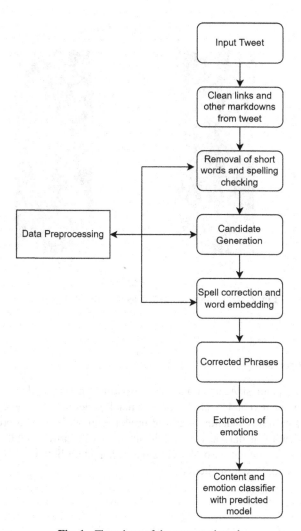

**Fig. 1.** Flowchart of the proposed work

- sentiment: the contrariety of the tweet (0 = negative, 4 = positive)
- ids: The tweet id (1467810672)
- date: The tweet date(Mon Apr 06 22:19:49 PDT 2009)
- flag: The query (lyx). If absent, then this value is NO_QUERY.
- user: The twitter user (scotthamilton)
- text: The tweet text (is upset that he can't update his Facebook by…)

As only the sentiment and text of the tweets are required, the rest of the attributes of the dataset are dropped. Furthermore, the sentiment field is being tweaked so that it has new values to reflect the sentiment. (0 = Negative, 1 = Positive) (Fig. 2).

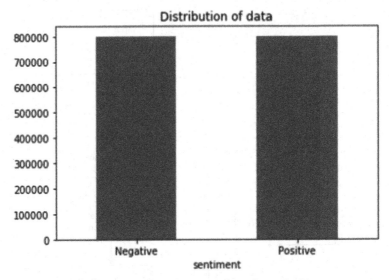

**Fig. 2.** Distribution of tweets as Negative or Positive

### 3.2  Data Preprocessing

Tweets typically contain slang and lingo, thus providing unstructured data. It needs to be cleaned of its unique text and symbols before a machine learning model can understand it. Building any complex machine learning model requires both data preparation and cleaning. The quality of data has a significant impact on how reliable the model is. Before preprocessing the data, world clouds were created for both negative and positive tweets to understand the dataset better.

The following measures were taken to preprocess the dataset acquired from the sources mentioned above (Figs. 3 and 4):

**Fig. 3.** Word cloud for negative tweets

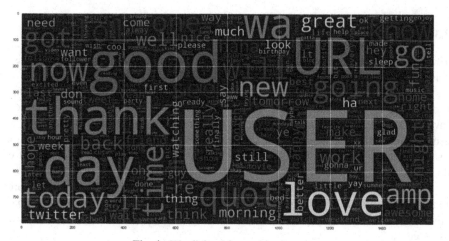

**Fig. 4.** Word cloud for positive tweets

- Removal of hyperlinks: All links, hyperlinks, and user-ids of both the author and any tagged accounts have been removed as they are assumed to not contribute much to the sentiment.
- Replacing emoticons with their meaning: Commonly used emoticons such as :) ;) :( and < 3 have been replaced with their meaning. In total, 11 most frequently occurring emoticons have been replaced with their meanings. In this study only 11 emoticons were replaced as they were the highest occurring and most widely repeated. More

than 11 emoticons exists in the dataset but due to their very low occurrence they were treated as outliers and discarded.
- Removal of punctuations: ',.*!' and other punctuation marks that are obsolete to the cause and working of the model were removed.
- Reducing text to lowercase: To avoid any disputes with character encoding in vectorizing, all text characters were converted to lowercase regardless of their origin.
- Tokenizing text: long phrases of text were split into arrays of tokens for further vectorization and processing.
- Lemmatization: Adjectives and tenses were reduced to their root word, called "lemma", considering their meanings. As all variations of the root word convey the same meaning, we do not require each of these to be converted into different vectors. For example, "caring" could be converted into "car" when used with stemming because it removes the last few characters from a word but with lemmatization it is converted into its meaningful root form "care" so its meaning is not changed and such an error does not occur.
- Removing stop words: Superannuated words like "this", "an", "a", "the" etc. that do not modify or contribute to the sentiment of the tweet were dropped from the tokenized arrays. Words such as "wa", "brb", "lol" and many more are used in a variety of contexts. As a result, these aforementioned colloquial terms were not removed with the other stop words as they were an integral part to the meaning of the sentence.

### 3.3 Model Training

The dataset was split into partitions for testing and training of the model. The test dataset was roughly 10% the size of the data used for training.

Since some of the Models used have no support for strings and data in textual formats, the TF-IDF Vectorizer is used. TF-IDF is an abbreviation for Term Frequency-Inverse Document Frequency Features. It transforms text into feature vectors that an estimator can utilize as input. Count Vectorizer provides frequency numbers in relation to vocabulary indexes, whereas TF-IDF takes into account the entire word weight of documents.

Four different classifiers were trained using the preprocessed dataset. These are: BernoulliNB, LinearSVC, Logistic Regression, and XGBoost. To arrive at the comparative results, the accuracy of every one of these, as well as the confusion matrix, was assessed.

### 3.4 Result Validation

The validation of the result is done using the F1 score and the accuracy score. Another crucial indicator in the evaluation of machine learning is the F1-score. By integrating two seemingly disparate criteria, accuracy and memory, it elegantly sums up a model's performance in terms of prediction. [17].

$$F1 - score = 2 \times \frac{(P \times R)}{(P + R)} \qquad (1)$$

Sometimes used as an alternative to F-score, Accuracy score is a binary classification scoring method that determines whether a response or returned information is correct or not.

$$Accuracy = \frac{TP + TN}{TP + TN + FP + FN} \qquad (2)$$

### 3.5  Results and Inferences

After the model has been trained, we assess its performance. The model that performs best or has the highest test accuracy on our dataset will be utilised to predict further tweets. Future predictions are generated using the XGBoost Classifier as it was determined to have the highest accuracy of 87.841%, beating all other algorithms.

## 4  Algorithms Used

### 4.1  Bernoulli Naive Bayes

The core principle of Naive Bayes is that each characteristic contributes independently and equally to the results. It is assumed that the feature pair is independent of it and that each feature is assigned equal weight (or importance). The Bernoulli Naive Bayes Classifier algorithm uses features, which are independent binary variables that describe whether or not a term appears in the material under evaluation. This technique is also a well-liked method for text classification tasks because it is comparable to the multinomial model in the classification process. However, the Bernoulli approach is solely concerned with determining whether a term is present or missing in the document under review, whereas the multinomial approach takes into account term frequencies. Eq. (1) [12].

$$P(x|C_k) = \prod_{i=1}^{n} p^{x_i} ki(1 - pki)^{(1-x_i)} \qquad (3)$$

where $p_{ki}$ is the likelihood that class Ck would produce the term $x_i$.

### 4.2  Linear Support Vector

The SVM model is a point-in-space representation of an example, with each cate- gory example projected to be separated by the greatest possible gap. SVMs may do non-linear classifications as well as linear classifications by implicitly mapping inputs to higher dimensional feature spaces. The main objective of the support vector ma- chine(SVM) algorithm is to find a hyperplane in an N-dimensional space( where N is the number of features) that distinctly classifies the data points. Hyperplanes serve as a dividing line or as judgement lines to help categorise the data points. Different classi- fications can be given to data points that lie on different sides of the hyperplane [18]. It is usually complex with multiple features, but with just two features, it is a line and with three features, it tends to be a two-dimensional plane. Either the feature vector for emotion analysis or the feature vector for text classification can be used to train the L1 Norm Soft Margin model shown below.

$$min\frac{1}{2}\|w\|_2^2 + C\sum_{i=1}^{m}\varepsilon_i$$
$$s.t.y^{(i)}(w^Tx^{(i)} + b) \geq 1 - \varepsilon_i \tag{4}$$
$$\varepsilon_i \geq 0$$

SVM's benefits make it effective for text classification. One of these benefits is its abil- ity to handle large features, which, in contrast to some of the algorithms mentioned above, helps in understanding larger features collectively. Another benefit is that SVM is resilient in the presence of sparse examples and that the majority of problems are linearly separable. [19].

## 4.3  Logistic Regression

One popular machine learning methodology that belongs to the supervised learning approach is logistic regression. From a collection of independent factors, it is used to forecast the categorical dependent variable. Except for how they are employed, Logistic Regression is very similar to Linear Regression. Linear regression is used to solve regression problems, whereas logistic regression is used to solve classification difficulties. Logistic regression is a key machine learning technique because it can generate probabilities and categorise data using both discrete and continuous datasets. The logistic model is a statistical model. It ends up modeling the probability of one event taking place by analyzing a linear combination of one or more independent variables together in an equation. The logistic regression learning algorithm can be derived by maximizing the following likelihood function:

$$L(\theta) = \prod_{i=1}^{m}\left(h_\theta\left(x^{(i)}\right)\right)y^{(i)}\left(1 - h_\theta\left(x^{(i)}\right)\right)^{1-y^{(i)}} \tag{5}$$

where $h_\theta$ is the sigmoid function. It calculates the likelihood of an event occurring by fitting data to a log function. [20]

$$\log(\frac{p}{1-p}) = \beta_0 + \beta(num) \tag{6}$$

In this case, the success ratio always looks to be more than 50% if $\log\left(\frac{p}{1-p}\right)$ is greater than zero

## 4.4  XGBoost

Boosting is a type of ensemble modeling. This technique aims to create a strong classifier out of a large number of weak ones. First, develop a model using the training data. Following that, a second model is built to attempt to remedy any faults in the first model.

This method is used repeatedly until either the maximum number of models are added or the full training dataset can be accurately predicted.

With gradient boosting, each prediction corrects the preceding predictor's inaccuracy. The residual error of each predictor is used to train it. This algorithm builds decision trees in a sequential fashion. All independent variables are weighted and incorporated into the prediction decision tree. The tree increases the weight of the variables that were predicted wrongly, and these variables are passed as labels to the second decision tree predecessor.

The boosting algorithm is defined as Eq. (7) [21]:

$$F_0(x) = argmin_\gamma \sum_{i=1}^{n} L(y_i, \gamma) \tag{7}$$

Iteratively computing the loss function's gradient:

$$r_{im=} -\alpha \left[ \frac{\delta(L(y_i, F(x_i)))}{\delta F(x_i)} \right]_{F(x)=F_{m-1}(x)} \tag{8}$$

where $\alpha$ is the learning rate. Each terminal node's multiplicative factor $\gamma_m$ is determined, and the augmented model Fm(x) is defined as:

$$F_m(x) = F_{m-1}(x) + \gamma_m h_m(x) \tag{9}$$

For this study, the XGBoost Classifier has been implemented with 4000 trees, with the 'gbtree' booster, max depth of 10 trees and a learning rate of 0.01.

## 5 Discussion

The dataset was trained on by the four aforementioned machine learning algorithms. The dataset was divided into training and testing sets to better evaluate the model as our objective was to estimate the performance of the model on unseen data points. To achieve the highest level of accuracy without overfitting the models, all algorithms were hyperparameter tweaked. The results of all the various experiments conducted are shown in Table 1.

**Table 1.** Validation of algorithms using precision, recall & f1 score.

| Sr.No. | Algorithm | Precision | Recall | F1-score | Accuracy |
|--------|-----------|-----------|--------|----------|----------|
| 1. | Bernoulli Naive Bayes | 0.805 | 0.800 | 0.800 | 0.80 |
| 2. | Linear SVC | 0.815 | 0.815 | 0.820 | 0.82 |
| 3. | Logistic Regression | 0.825 | 0.825 | 0.830 | 0.83 |
| 4. | XGBoost | 0.864 | 0.898 | 0.880 | 0.87 |

From the accuracy measures mentioned in Table 1, one can clearly conclude that the tree-boosting algorithm XGBoost returns the highest accuracy as well as a superior F1 score as compared to the other algorithms. The gradient boosted trees in XGBoost are non-parametric and results in a better overall fit on the dataset as compared to the other three machine learning models which are parametric in nature. XGBoost with its 87.84% accuracy is 4.84% greater than Logistic Regression which is the second highest accurate model and 5.64% greater than the unigram featured SVM model found in a previous study [1] (Figs. 5, 6, 7 and 8).

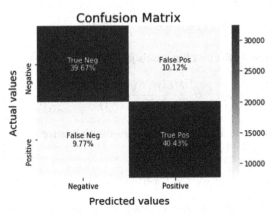

**Fig. 5.** Confusion Matrix for Bernoulli NB

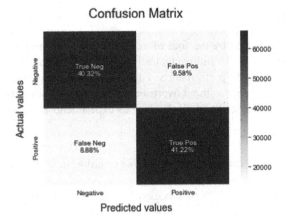

**Fig. 6.** Confusion Matrix for Linear SVC

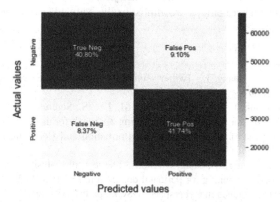

**Fig. 7.** Confusion Matrix for Logistic Regression

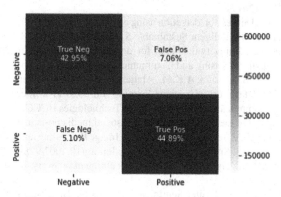

**Fig. 8.** Confusion Matrix for XGBoost Classifier

## 6  Conclusion and Future Work

The capacity to use Twitter as a tool for sentiment analysis, which enables us to determine the sentiment of a tweet written by a user, has been demonstrated in this paper. Several research difficulties were outlined at the beginning of this publication to help the reader understand the nature of our study. This paper compares four machine learning methods, XGBoost, LinearSVC, Logistic Regression, and BernoulliNB, to determine which algorithm predicts the sentiment of a tweet with the highest accuracy. XGBoost outperformed all other methods because it employs a gradient descent algorithm, which is why it is known as Gradient Boosting. The entire concept of XGBoost is to rectify the model's prior error, learn from it, and increase performance in the next step. Previous results [1] on the same dataset have been improved; the natural language processing model has been made more robust; and performance overall has been improved. In the future, if a specific user consistently tweets negative messages, the model can be extended to determine whether that user is depressed and in need of support. Also, the study could be further expanded to include emojis which would enable it to be more robust in these times, where emojis are almost exclusively and excessively used.

# References

1. Go, A., Bhayani, R., Huang, L.: Twitter sentiment classification using distant supervision. Processing **150** (2009)
2. Hassan, A.U., Hussain, J., Hussain, M., Sadiq, M., Lee, S.: Sentiment analysis of social networking sites (SNS) data using machine learning approach for the measurement of depression. In: 2017 International Conference on Information and Communication Technology Convergence ICTC, pp. 138–140 (2017)
3. Chen, Y., Zhou, B., Zhang, W., Gong, W., Sun, G.: Sentiment analysis based on deep learning and its application in screening for perinatal depression. In: 2018 IEEE Third International Conference on Data Science in Cyberspace (DSC), pp. 451–456 (2018)
4. Rosa, R.L., Schwartz, G.M., Ruggiero, W.V., Rodríguez, D.Z.: A knowledge-based recommendation system that includes sentiment analysis and deep learning. IEEE Trans. Industr. Inf. **15**, 2124–2135 (2019)
5. Deshpande, V., Rao: Depression detection using emotion artificial intelligence. In: 2017 International Conference on Intelligent Sustainable Systems ICISS, pp. 858–862 (2017)
6. Arora, P., Arora, P.: Mining twitter data for depression detection. In: 2019 International Conference on Signal Processing and Communication (ICSC), pp. 186–189 (2019)
7. Khan, R.H., Afroz, U.S., Masum, A.K.M., Abujar, S., Hossain, S.A.: Sentiment analysis from bengali depression dataset using machine learning. In: 2020 11th International Conference on Computing, Communication and Networking Technologies (ICCCNT), pp. 1–5 (2020)
8. Biradar, A., Totad, S.G.: Detecting depression in social media posts using machine learning. In: Santosh, K., Hegadi, R. (eds.) Recent Trends in Image Processing and Pattern Recognition. RTIP2R 2018, vol. 1037. Springer (2019). https://doi.org/10.1007/978-981-13-9187-3_64
9. Shiha, M., & Ayvaz, S.: The effects of emoji in sentiment analysis. Int. J. Comput. Electr. Eng. **9** (2017)
10. Chen, Y., Yuan, J., Luo, J.: Twitter Sentiment Analysis via Bi-sense Emoji Embedding and Attention-based LSTM (2018)
11. Jain, A.P., Dandannavar, P.: Application of machine learning techniques to sentiment analysis. In: 2nd International Conference on Applied and Theoretical Computing and Communication Technology (iCATccT), pp. 628–632 (2016)
12. Neethu, S., Rajasree, R.: Sentiment analysis in twitter using machine learning techniques. In: 2013 Fourth International Conference on Computing, Communications and Networking Technologies (ICCCNT), pp. 1–5 (2013)
13. Gautam, G., Yadav, D.: Sentiment analysis of twitter data using machine learning approaches and semantic analysis. In: Seventh International Conference on Contemporary Computing (IC3), pp. 437–442 (2014)
14. Chandra, Y., Jana, A.: Sentiment analysis using machine learning and deep learning. In: 2020 7th International Conference on Computing for Sustainable Global Development, pp. 1–4 (2020)
15. Naresh, A., Krishna, P.V.: An efficient approach for sentiment analysis using machine learning algorithm. Evol. Intel **14**, 725–731 (2021)
16. Sentiment140 Dataset. https://www.kaggle.com/datasets/kazanova/sentiment140
17. Sokolova, M., Japkowicz, N., Szpakowicz, S.: Beyond accuracy, f-score and ROC: a family of discriminant measures for performance evaluation. In: Sattar, A., Kang, B.-h (eds.) AI 2006. LNCS (LNAI), vol. 4304, pp. 1015–1021. Springer, Heidelberg (2006). https://doi.org/10.1007/11941439_114

18. Gandhi, R.: Support Vector Machine — Introduction to Machine Learning Algorithms (2018). https://towardsdatascience.com/support-vector-machine-introduction-to-machine-learning-algorithms-934a444fca47

19. Zainuddin, N., Selamat, A.: Sentiment analysis using support vector machine. In: I4CT 2014 - 1st International Conference on Computer, Communications, and Control Technology, Proceedings, pp. 333–337 (2014). https://doi.org/10.1109/I4CT.2014.6914200

20. Colianni, S., Rosales, S., Signorotti, M.: Algorithmic Trading of Cryptocurrency Based on Twitter Sentiment Analysis. https://cs229.stanford.edu/proj2015/029_report.pdf

21. Chen, T. et al.: XGBoost. In: Proceedings of the 22nd ACM SIGKDD International Conference on Knowledge Discovery and Data Mining. ACM (2016)

# Fine-Tuning of Multilingual Models for Sentiment Classification in Code-Mixed Indian Language Texts

Diya Sanghvi[1,2], Laureen Maria Fernandes[1,2], Siona D'Souza[1,2], Naxatra Vasaani[1,2], and K. M. Kavitha[1,2(✉)] (iD)

[1] St Joseph Engineering College, Mangaluru, India
{18cs038.diya,18cs111.siona,kavitham}@sjec.ac.in
[2] Visvesvaraya Technological University, Belagavi, India

**Abstract.** We use XLM (Cross-lingual Language Model), a transformer-based model, to perform sentiment analysis on Kannada-English code-mixed texts. The model was fine-tuned for sentiment analysis using the KanCMD dataset. We assessed the model's performance on English-only and Kannada-only scripts. Also, Malayalam and Tamil datasets were used to evaluate the model. Our work shows that transformer-based architectures for sequential classification tasks, at least for sentiment analysis, perform better than traditional machine learning solutions for code-mixed data.

**Keywords:** DICT-MLM · Task adaptive pre-training · Domain adaptive pre-training · Transfer learning · Transductive transfer · LSTM · Pseudo labelling

## 1 Introduction

Sentiment analysis is essential in many real-world applications such as stance detection, review analysis, recommendation system, and so on. In today's environment of data overload, companies have huge amount of customer feedback collected. Sentiment analysis helps these companies better understand customer emotions with minimal human intervention. In the most basic form, it involves taking a piece of text, whether it is a sentence, a comment, or an entire document, and returning a score that measures how positive or negative the text is. Since the COVID-19 lockdown, the number of social media users has increased extensively. Spurred by that growth, companies and media organizations are increasingly seeking ways to mine social media platforms for information on what people think and feel about their products and services. It is highly critical in macro-scale socio-economic phenomena, as for example, in predicting the stock market rate of a particular firm. Among other emerging uses of sentiment analysis is predicting the outcomes of popular political elections and surveys.

The main challenge in sentiment analysis of comments in microblogging sites is the incredible breadth of topics that is covered. It is observed that people comment about anything and everything. Therefore, to be able to build systems that identify the underlying sentiments, we need methods to quickly identify data that can be used for training. Another challenge is the prevalence of user-friendly interfaces that allow users to present content in their own native language or as code-mixed texts. Code-mixing is a way of writing texts using more than one language. Traditional methods for sentiment analysis focus majorly on monolingual texts. While machine learning methods like logistic regression, support vector machines, etc. perform very well on high-resource languages, they fail on code-mixed data. Sentiment analysis becomes more difficult when the data is noisy, code-mixed, and collected from social media. As an example consider the code-mixed sentence: "Yes bro nanu Chinese appsgalna delete madidhini." Translated data: "Yes bro, I have deleted Chinese apps." In the example,"appsgalna", "delete" are English words mixed with Kannada words in composing the sentence. It is observed that, in India, people prefer and heavily use code-mixed language while communicating on social media. Jose et al. discuss the relevance of code-switching in multilingual communities and social media engagement [1]. Some of the challenges faced when analyzing code-mixed data include no word order, spelling variations, creative spellings, abbreviations, no capitalization, incredible breadth of the topic covered and so forth. Scarcity of annotated code-mixed data required for sentiment analysis further limits the advances in the field.

In this paper we present our experiences on using XLM (Cross-lingual Language Model), a transformer-based model, to perform sentiment analysis of code-mixed texts in Kannada-English. We have compared and evaluated these models using evaluation metrics (F1-score), as well as standards such as dataset size, model parameters, etc. The model was fine-tuned for sentiment analysis using the KanCMD dataset, which consisted of comments by YouTube viewers in code-mixed Kannada-English. YouTube video comments offered a more realistic picture of public sentiment than conventional online articles and web blogs as YouTube has a much greater quantity of informative content than typical blogging sites. The model's performance was also assessed on English-only and Kannada-only scripts. Additionally, Malayalam and Tamil datasets were used to evaluate the model.

## 2    Related Literature

Although code-switching research started years ago, the non-availability of data was a major hurdle in solving problems. In a survey of current datasets for code-switching research [1], Jose et al. discuss a set of quality measures, viz. number of words, vocabulary size, the number of sentences, average sentence length etc. for evaluating and categorizing the dataset.

Hande et al. introduced Kannada CodeMixed Dataset (KanCMD), a multi-task learning dataset for sentiment analysis and offensive language identification and identify six combinations among code-mixed sentences, such as "no-code-mixing only Kannada written in Kannada", "Kannada written in the Latin script", "inter-sentential code-mixing", "code-switching at morphological level", "intra-sentential code-mixing", and "inter-sentential and intra-sentential mix". The dataset comprised of comments collected from YouTube using the YouTube Comment Scraper with keywords from 18 different videos ranging from movie trailers to current trends, such as India's prohibition of mobile applications, the India-China border dispute, Mahabharata, and transgenders. Annotations were collected from annotators using Google forms. The dataset consisted of 64,997 tokens, with a vocabulary size of 20,667 [2]. While extracting right sentiments from code-mixed data is a challenging task with the current methodologies, Shekhar et al. report on the artificial immune systems-based LSTM model to classify code-mixed data [3].

Chakravarthi et al. report on using various Machine Learning Algorithms such as Logistic Regression (LR), Support Vector Machine (SVM), Multinomial Naive Bayes (MNB), K-Nearest Neighbors (KNN), Decision Trees (DT) and Random Forest (RF) separately for sentiment analysis and offensive language detection. In classifying the code-mixed texts into one of the five groups: 'Positive', 'Negative', 'neutral', 'mixed feelings', and 'other languages', the authors report better accuracy, recall and F1-score for positive class than the negative class. Further, they report that the classification algorithms fared better when applied to sentiment analysis task than in detecting offensive words [4].

Devlin et al. [5] recognise pre-training of language model as an effective way of improving NLP tasks, both at the sentence and paragraph level, the latter aiming to predict relationships between sentences by analyzing them holistically, and at the token level by using tasks such as NER [6] and question-answering. They report feature-based [7] and fine-tuning [8] as two primary methodologies for applying pre-trained language representations to downstream tasks. Feature-based approaches use task-specific designs such as those proposed by ELMo [7] and pre-trained representations as additional features. Fine-tuning approaches such as the Generative Pre-trained Transformer (OpenAI GPT) [8] introduces minimal task-specific parameters, and is trained simply by fine-tuning all pre-trained parameters. Objectively, the two approaches coincide as both use unidirectional language models to learn language representations.

Using a Cloze-task inspired Masked Language Model (MLM) pre-training target [9], BERT bypasses the previously noted unidirectionality restriction. The purpose of the masked language model is to identify the original vocabulary id of the masked words purely based on their context by masking some tokens from the input at random. Unlike pre-training a left-to-right language model, the MLM goal allows combining the left and right contexts into the representation, allowing in pre-training a deep bidirectional Transformer. BERT [5] pre-trains deep bidirectional representations using masked language models, in contrast to using unidirectional language models [8]. In comparison, Peters et al. [7] employ a shallow concatenation of independently trained left-to-right and right-to-left LMs. They also show that pre-trained representations reduce the requirement for numerous task-specific architectures that are substantially developed. BERT is the first fine-tuning-based representation model to outperform numerous task-specific architectures on a wide range of sentence-level and token-level tasks. It advances the state-of-the-art for 11 NLP tasks[1]. BERT's model architecture is a multi-layer bidirectional Transformer encoder based on the original implementation described in Vaswani et al. [10] and released in the tensor2tensor[2] library.

---

[1] https://github.com/google-research/bert.

[2] https://github.com/tensorflow/tensor2tensor.

Studies have also proven the benefits of continuing pretraining on domain-specific unlabeled data. Nevertheless, these studies only look at one domain at a time and utilize a language model that is pre-trained on a smaller and less diversified corpus than the most recent language models. In one of the related works, Liu at al. focus on domains such as biomedical and computer science papers, news and reviews and experimented on eight different categorization tasks [11]. They attempted to study how domain adaptive pre-training compares to task-adaptive pretraining (TAPT) on a smaller but directly task relevant corpus. Although Task-adaptive pre-training has been proven to be beneficial [10], it is not commonly used with modern models. With or without domain-adaptive pretraining, they found that TAPT improves the performance of RoBERTa [11].

Peter et al. [12] went on to explore schemes to best adapt the pre-trained representations to diverse tasks. Sequential inductive transfer learning has two stages: pre-training (where the model learns a general-purpose representation of inputs) and adaptation (the representation is transferred to a new task) [13]. Adaptation has two main paradigms: feature extraction (the model's weights are 'frozen') and fine tuning. Both have benefits: they enable the use of task-specific model architectures. It may be computationally cheaper as features only need to be computed once. On the other hand, it is convenient as it may adapt a general-purpose representation to many different tasks. Two state-of-the-art pre-trained models have been compared, ELMo [7] and BERT [5], across seven diverse tasks, including entity recognition, natural language inference (NLI), and paraphrase detection (PD). These werere evaluated on five different tasks, utilizing several standard datasets: NER, Sentiment analysis (SA), NLI, PD, and Semantic textual similarity (STS). Both ELMo and BERT outperform the sentence embedding method significantly, except on the semantic textual similarity tasks (STS). In conclusion, feature extraction and fine tuning with BERT models have similar performances.

In a survey focused on pre-trained models, Min et al. [14] discuss the use of large pre-trained language models in solving NLP tasks via pre-training then fine-tuning, prompting or text generation approaches. They also report on the approaches that use pre-trained models to generate data for training augmentation or other purposes. They organize the works that leverage PLMs for NLP into the following three paradigms- (i) Pre-train then fine-tune : perform general purpose pre-training with a large unlabeled corpus, and then perform a small amount of task-specific fine-tuning for the task of interest, (ii) Prompt-based learning : prompt a PLM such that solving an NLP task is reduced to a task similar to the PLM's pre-training task (e.g. predicting a missing word), or a simpler proxy task (e.g. textual entailment). Prompting can usually more effectively leverage the knowledge encoded in the PLMs, leading to few-shot approaches, and (iii) NLP as text generation : Reformulate NLP tasks as text generation, to fully leverage knowledge encoded in a generative language model such as GPT-2 [8] and T5 [15].

Neto et al. discuss on using Ensemble of language models for Sentiment Analysis of Code-Mixed English-Hindi Tweets [16]. The approach applied consisted of training and then using the predictions of four models: MultiFiT [17], BERT [5], ALBERT [18], and XLNet [19]. After retrieving prediction values, the ensemble calculates an average of all softmax values from them. For MultiFiT, the method used is based on Universal Language Model Fine-tuning (ULMFiT) [10] whose goal is to increase efficiency when modeling languages other than English. The implementation of BERT used two steps: pre-training and fine-tuning. ALBERT, A Lite BERT for Self-supervised Learning of Language Representations, provides an option for parameter reduction, reducing the number of hours of training required, which consequently increases the operational costs. XLNet is a model that uses a bidirectional learning mechanism, doing that as an alternative to word corruption via masks implemented by BERT. Results were reported for each model and an ensemble using a combination of results of the four models, XLNet, BERT, ALBERT, and MultiFiT. Ensemble produces the best F1 and the XLNet model represents the best result among the other models. Using only MultiFiT and BERT architectures, the results were only 66.5%.

Since hate speech detection [20] is closely related to sentiment classification, in the sense that they are both sequence classification problems, we look at the recent advances in offense detection task. In one of the related literature, Multilingual BERT [5] models with pseudo labeling and ensemble techniques have been used. For the purpose, the authors use two transformer-based models, DistilmBERT (multilingual) and Indic-BERT, as well as a non-transformer based model, ULMFiT [10]. Chinnappa et al. [21] proposed a unified framework to predict hope speech in the English, Tamil, and Malayalam datasets. The experimental results showed detecting hope speech is difficult regardless of the language, and that code-mixing and transliterations in Tamil and Malayalam increases the complexity of the problem. This paper was insightful since it showed us approaches concerning language identification for code mixed data. With a training dataset of roughly 6500 samples for Sinhala-English code-mixed data, the authors empirically demonstrated that a newly developed Capsule+biGRU classifier [22] outperformed the classifier based on the English-BERT [5] and XLM-R [23]. They demonstrated that the efficacy of contextual embedding models on code-mixed text classification is dependent on a variety of circumstances, and that they might be inferior when compared to text classification using neural models other than transformers. This finding implies that classical deep learning techniques are still viable, at least for text classification on code-mixed data including extremely low-resource languages that are under-represented in big multilingual embedding models.

## 3   Methodology

We chose the Cross-lingual Language Model (XLM); particularly, the one pre-trained using MLM (Masked Language Modeling) called xlm-mlm-100-1280 (trained using 100 languages including Kannada and English) provided by HuggingFace transformers. Unlike BERT which uses pairs of sentences, XLM uses streams of an arbitrary number of sentences and truncates once the length is 256. This is where "languages with the same script or similar words provide better mapping" comes into the picture

### 3.1   Baseline

Existing methods implemented in previously examined papers were replicated by us to get a better understanding and also to establish a baseline for comparative analysis. The classifiers used were Gaussian Naive Bayes, KNN, Logistic Regression, Decision Tree and Random Forest. These traditional machine learning algorithms have been used in making predictions [2] and follow the workflow as depicted in Fig. 1 (left).

## 3.2 Transformer Architecture

After reviewing the available literature and based on the experimental results obtained using traditional machine learning models mentioned in the preceding subsection, we conclude that transformer based models perform better than traditional machine learning architectures for the task considered in this paper. Considering tasks on code-mixed data, the available literature have been suggestive of approaches analogous to sentiment analysis task (specifically w.r.t the target class), those being sequence classifiers (e.g., hope detection, offensive speech classification, etc.). Hence, for our chosen dataset, we carry out the task of sentiment analysis on Kannada-English code-mixed data using pretrained XLM and other models, fine tune them for our chosen task, and analyze their performance on the same.

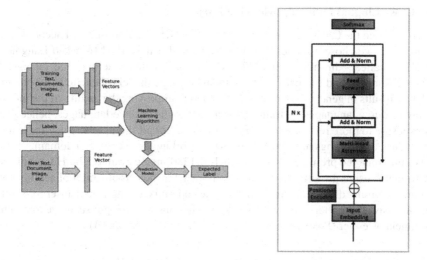

**Fig. 1.** Traditional machine learning workflow for classification (left) and Proposed Transformer based Architecture [24] (right)

The diagram on the right in Fig. 1 shows the architecture of a transformer in general, an architecture that completely shuns recurrence and resorts to attention mechanisms. Adapting to an architecture with attention mechanisms have proven to be much more efficient than recurrent architectures. The transformer block follows a stacked encoder-decoder architecture with multi-headed attention and feed forward layers. Self-attention is computed several times in transformer's architecture, thus referred to as multi-head attention. This approach collectively attends to information from different representations at different positions. For instance, consider the phrase "Ask Powerful Questions". To calculate self-attention of the first word 'Ask', the scores for all words in the phrase with respect to 'Ask' is to be computed, which then determines the importance of other words when certain words are being encoded into the input sequence. The scores are divided by the square root of the dimension of the key vector. The score of the first word is calculated using dot-product attention, as the dot product of the query vector $q_1$ with keys $k_1$, $k_2$, and $k_3$ of all words in the input sentence. The scores are then normalised using the softmax activation. The normalised scores are then multiplied by vectors $v_1$, $v_2$, and $v_3$ and summed up to obtain the self-attention vector $z_1$. It is then passed to feed-forward network as input. The vectors for the other words are calculated in a similar way in dot-product attention. Softmax is an activation function that transforms the vector of numbers into a vector of probabilities. We use Softmax function when we want a discrete variable representation of the probability distribution over $n$ possible values. Softmax activation function over $K$ classes is represented as follows:

$$softmax(z) = \frac{e^{z_i}}{\sum_{j=1}^{K} e^{z_j}} \tag{1}$$

### 3.3   Cross-Lingual Language Model (XLM)

We have chosen the Cross-lingual Language Model (XLM), proposed by FacebookAI[3]. In particular, we chose a model that was pre-trained using MLM (Masked Language Modeling) called 'xlm-mlm-100-1280' (Masked language modeling, 100 languages including Kannada and English). Cross-lingual language model is beneficial for obtaining better results in generic downstream tasks and in improving the quality of the model for low-resource languages by training on similar high-resource languages, hence getting exposure to more relevant data. MLM is a powerful pre-training strategy for learning sentence embeddings and specifically while working on specialized domain.

We fine-tune the pre-trained 'xlm-mlm-100-1280' made available by HuggingFace transformers[4] using the simpletransformers library for the classification of sentiments. First, the pre-trained model is fine-tuned on the code-mixed language dataset for sentiment analysis. Thereafter, it is used to make predictions on the input sentence to output a sentiment as either 'Positive', 'Negative' or 'Mixed' (Figs. 2 and 3).

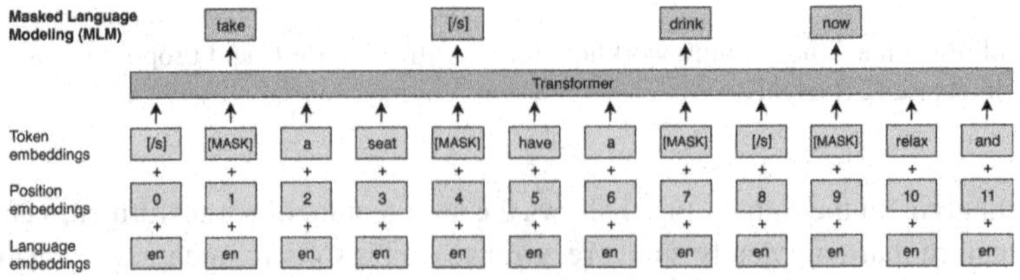

**Fig. 2.** Pre-training XLM model for English language using MLM technique

---

[3] https://towardsdatascience.com/xlm-enhancing-bert-for-cross-lingual-language-model-5aeed9e6f14b.

[4] https://huggingface.co/docs/transformers/model_doc/xlm.

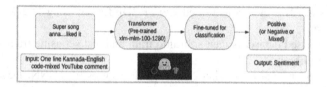

**Fig. 3.** Proposed model workflow

# 4 Experimental Setup

## 4.1 Dataset

We used the Kannada CodeMixed Dataset (KanCMD), a multi-task learning dataset for sentiment analysis and offensive language identification [2]. It contains comments made by YouTube viewers in code mixed text rather than monolingual text. It was originally intended for two tasks: sentiment analysis and offensive phrase detection for Kannada, a language with limited resources. A minimum of three annotators annotated each sentence as 'Positive', 'Negative', 'Mixed', 'Neutral' or 'Not in intended language'. For our experiments, we used data classified as 'Positive', 'Negative', and 'Mixed'. Class-wise distribution of the comments in Kannada-English dataset is: Positive-3291, Negative-1481 and Mixed-678; in Malayalam-English is Positive-5565, Negative-1394, Mixed-794 and in Tamil-English dataset is: Positive-24501, Negative-5190 and Mixed-4852 respectively. The dataset was cleaned and classified under three labels 'Positive', 'Negative', and 'Mixed'. The class 'Not in intended language' proves redundant here as our intention was to improve the performance for sentiment analysis using Kannada-English code mixed data. We also tested our model on similar datasets of Malayalam-English and Tamil-English code-mixed text [4].

## 4.2 Comparision on Existing Methods

To establish a baseline for comparative analysis, existing methods implemented in previously mentioned papers were experimented. The Python IDE available on Kaggle was used for this purpose. The classifiers used for this experiment were Gaussian Naive Bayes, KNN, Logistic Regression, Decision Tree, and Random Forest. They were made available through the scikit-learn library. We also used numpy and pandas libraries to manipulate the DataFrames. The following steps were carried out:

1. Firstly, we loaded the dataset using the read_csv() method available in the pandas library. After this, we dropped the rows with 'unknown' and 'not-language' labels.
2. Using the value_counts() method with normalization enabled, we obtained the percentage of each label in our dataset.
3. Using LabelEncoder() from sklearn.preprocessing, we encoded the sentiment labels via its fit_transform() method.
4. Then, we converted the 'comment' column data to numerical vectors using CountVectorizer() from sklearn library. Using its fit_transform() method, we got the bag-of-words form of these vectors.

5. The bag of words is then transformed into TF-IDF format using TfidfTransformer() from sklearn.
6. The preprocessed dataset is then split into train and test set using train_test_split() from sklearn library.
7. The classifiers namely, Gaussian NB, Logistic Regression, KNN, Decision Tree, Random Forest are applied on the preprocessed train set.
8. Lastly, the test set is used for predictions. F1-score and log loss of each of the classifier's performance on the test set is displayed.

### 4.3  XLM Model

We used the pre-trained version of the XLM model, made available by the simple-transformers[5] library. This was implemented using the Python IDE available on Kaggle notebooks. This library is based on the Transformers library by HuggingFace and allows to quickly train and evaluate transformer models. As with the existing methods, we used numpy and pandas to manipulate the DataFrames and carried out these experiments for Kannada, Malayalam, and Tamil code-mixed text. Also, we ran the model on these languages individually and analyzed the results for all. The steps performed are as follows:

1. The input data was formatted to make it fit our chosen model's input structure. First, the comments with 'Not Kannada' and 'Unknown' sentiments were removed. Then, the duplicates were dropped and missing values were replaced. Then, the DataFrame was reduced to only have the comment text and the corresponding sentiment as its columns.
2. The sentiment column of the DataFrame was reformatted to have only numeric labels where 0 corresponds to 'Mixed', 1 to 'Negative', and 2 to 'Positive'. Using this, a labels column was created. This column consisted of a 3-tuple for each comment whose values indicated whether the comment was positive, negative, or mixed in sentiment or not.
3. The model chosen for sentiment analysis was the XLM. A transformer-based multi-class text classification model usually starts with a transformer and then adds a classification layer on top of it. Each class is represented by $n$ output neurons in the classification layer. The pre-trained model was fine-tuned by choosing the parameters and hyperparameters. This was done after several iterations of observing what fits the model best and gives us the optimal accuracy. The various parameters available are:
   - model_type: Specifies one of the model types from the supported models (e.g. bert, electra, xlnet).
   - model_name: Defines the specific architecture to be used, as well as the training weights to be used. This might be a HuggingFace Transformers compatible pre-trained model, a community model, or the link to a directory containing HuggingFace Transformers compatible pre-trained models.
   - num_labels (int, optional): Represents number of labels or classes in the dataset.

---

[5] https://simpletransformers.ai/.

- weight (list, optional): A list of length num_labels containing the weights to assign to each label for loss calculation.
- args (dict, optional): Default args values is used if this parameter is not specified. If specified, it should be a dict containing the args values that should be changed in the default args.
- use_cuda (bool, optional): It allows using the GPU, if available. Setting to False forces the model to use CPU only.
- cuda_device (int, optional): Specifies the GPU that should be used. By default, uses the first available GPU.
- kwargs (optional): Provides proxies, force download, resume download, cache dir, and other options related to the 'from pretrained' implementation.
4. The model was evaluated on the test sets and its classification reports were studied.

### 4.4   XLM Optimisation by Fine Tuning Parameters on Kannada-English Dataset

**Train Batch Size.** The train batch size defines the number of samples that will propagate through the network. An advantage of not propagating all samples at once is that the training procedure will require lesser memory and time. The other parameters remained constant and the model was trained on different values of train_batch_size after choosing an appropriate range for the same.

**Maximum Sequence Length.** This parameter specifies the maximum number of tokens in the input. The number of tokens is greater than or equal to the number of words in the input. The transformer does not accept tokens beyond this length. We kept the other parameters constant and trained the model on different values of max_seq_length after choosing an appropriate range for the same.

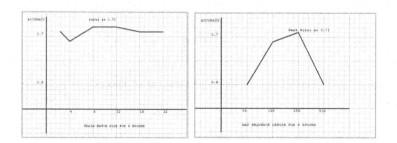

**Fig. 4.** Graph of accuracy vs train_batch_ size, max_seq_length

Figure 4 depicts the classification reports for every 2 epochs for each value of the train batch size (left) and maximum sequence length (right) chosen in these runs.

**Learning Rate.** Learning rate is a hyper-parameter that controls how much we are adjusting the weights of our network with respect to the loss gradient. The lower the value, the slower we travel along the downward slope. The model was trained on different values of gradient_accumulation_steps after choosing an appropriate range for the same and keeping the other parameters constant.

**Gradient Accumulation Steps.** Gradient accumulation steps means running a configured number of steps without updating the model variables while accumulating the gradients of those steps and then using the accumulated gradients to compute the variable updates. The model was trained on different values of gradient_accumulation_steps after choosing an appropriate range for the same and keeping the other parameters constant.

Figure 5 depicts the classification reports for every 2 epochs for each value of the learning rate (left) and gradient accumulation steps (right) chosen in these runs.

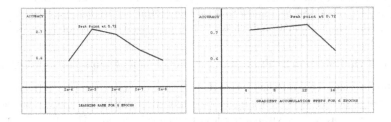

**Fig. 5.** Graph of accuracy vs learning_rate and gradient_accumulation_steps

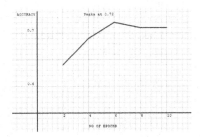

**Fig. 6.** Graph of accuracy vs number_of_epochs

**Number of Epochs.** One epoch means that each sample in the training dataset has had an opportunity to contribute to the model parameter values. An epoch consists of one or more batches. The right number of epochs depends on the inherent perplexity (or complexity) of the dataset. A good rule of thumb is to start with a value that is 3 times the number of columns in the data. Hence, we started with 2 epochs. The model was trained on different values of number_of_epochs after choosing an appropriate range for the same and keeping the other parameters constant. Figure 6 depicts the classification reports every 2 epochs.

# 5  Results and Discussion

## 5.1  Performance of Various ML Methods

Given below in Tables 1, 2 and 3 are the accuracies, F1-scores and log loss of the chosen classifiers for Kannada, Malayalam and Tamil code-mixed texts respectively. These are calculated from precision and recall, where precision is the ratio of true positives to the total number of positives including those not identified correctly, and recall is the number of true positive results divided by the number of all samples that should have been identified as positive. We observe that Logistic Regression performs the best but with a heavy log loss, which indicates that it predicts with higher probabilities of values staying away from the actual prediction. Random forest also has a competitive accuracy which can be attributed to the fact that it is suitable for dealing with the high dimensional noisy data in text classification.

**Table 1.** Comparison of existing machine learning methods for Kannada

| Algorithm | Accuracy (F1-score) (Existing) | Accuracy (F1-score) (Current) | Log Loss (Current) |
|---|---|---|---|
| Gaussian NB | – | 0.45 | 0.8331 |
| Logistic Regression | 0.57 | 0.69 | 18.7875 |
| K Nearest Neighbors | 0.43 | 0.58 | 0.7073 |
| Decision Tree | 0.52 | 0.62 | 9.5676 |
| Random Forest | 0.55 | 0.68 | 10.0179 |

**Table 2.** Comparison of existing machine learning methods for Tamil

| Algorithm | Accuracy (F1-score) (Existing) | Accuracy (F1-score) (Current) | Log Loss (Current) |
|---|---|---|---|
| Gaussian NB | – | 0.30 | 23.9865 |
| Logistic Regression | 0.57 | 0.71 | 0.7361 |
| K Nearest Neighbors | 0.43 | 0.32 | 10.1498 |
| Decision Tree | 0.52 | 0.63 | 12.1811 |
| Random Forest | 0.55 | 0.71 | 0.9466 |

We look at the F1-scores for comparison because F1-score is a better metric when there are imbalanced classes, as in our case. The Fig. 7 shows the expected and predicted sentiments for the corresponding Kannada-English code-mixed text. From the results, we can see that it classifies Positive' sentiment more accurately than the other two. This can be proved from the classification reports above by looking at the class-wise scores as well. The reason behind this is the fact that the number of rows for the 'Positive' label are more than that of 'Negative' and 'Mixed'. Moreover, 'Mixed' sentiment has the least number of rows which gives rise to its performance being worse than the rest.

**Table 3.** Comparison of existing machine learning methods for Malayalam

| Algorithm | Accuracy (F1-score) (Existing) | Accuracy (F1-score) (Current) | Log Loss (Current) |
|---|---|---|---|
| Gaussian NB | – | 0.57 | 14.7864 |
| Logistic Regression | 0.71 | 0.78 | 0.5292 |
| K Nearest Neighbors | 0.43 | 0.24 | 1.3550 |
| Decision Tree | 0.52 | 0.71 | 8.4450 |
| Random Forest | 0.55 | 0.79 | 0.6128 |

The fine-tuned XLM model works better than traditional machine learning methods mainly due to the following:

- They can understand the relationship between sequential elements that are far from each other.
- The language model was pre-trained in English, Kannada, Malayalam and Tamil (and 96 other languages). Having been trained on similarly structured languages the model grasping and understanding ability increases. It does this using Byte-Pair Encoding (BPE) that splits the input into the most common sub-words across all languages, thereby increasing the shared vocabulary between languages. This is a common pre-processing algorithm.
- They process sentences as a whole and learn relationships between words thanks to multi-head attention mechanisms and positional embeddings.

The model was also trained on Malayalam and Tamil code-mixed text using the same parameter values. Figure 8 shows the comparative analysis of various methods for the 3 languages. We observe that XLM outperforms all the other methods for every language.

## 5.2 Evaluation of Model's Performance on Different Types of Code-Switching

Our model which has been fine-tuned for code-mixed sentiment analysis is then evaluated on the three main possible types of code switching text, i.e., mixed script, English-only script, and Language-only script. We filtered and compiled the required subsets for the experiments from the test set. Figure 9 summarizes these results and shows the trend in performance for English-only, Language-only and mixed script inputs for all the 3 languages. Following are our observations:

- Tamil and Malayalam have good performances owing this to their larger dataset size.
- Intuitively, language-only sentences are easier for the model to classify as they have been individually trained for each language.
- The models perform the lowest on English sentences and most of them are sentences in their respective languages but written in English script which is more challenging to the model.
- It gives an average performance for mixed scripts.

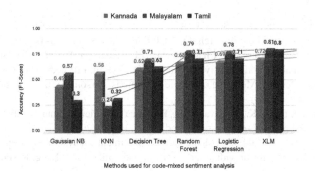

|  | Sentence | Expected | Predicted |
|---|---|---|---|
| 5317 | Yes bro nanu Chinese apps gaina delete madidhini . E thara innu jasti video madi bro . E thara good message viedlo madthairi | Positive | Positive |
| 547 | @Prithvi Vishwanath nie D boos bagi matadaku ugyote bekale.... dam bekale.... Kuri shetra ascar nominasahan filam gota? takat idre ond icy haatika filam tegiro nodana? jol d boos | Negative | Negative |
| 677 | @Troll Stupid Fans dagar yaru nin Amma na | Negative | Positive |
| 440 | @Nandi Parthasarathi ಸಂದೀಸಿ ಆಪ್ಟ್ ನೆಮ್ಮ ತುಲುಸಿ ಗೆಬ್ಬ ಪಾಸ್ಸಿ | Negative | Negative |
| 6702 | ಪತ್ತಾ, ಈ ಮೂತಿ ಪ್ರಸ್ಟ್ ಕೆಂಯುನ್ ನ ಜೆಗಂತ್ ಮಾಯಾತ್ತ ಅಸ್ಸೆ,ಕೆರು ಟ್ಯಂಣ್ ಮಾಣ | Positive | Positive |
| 587 | @Saathwik H R aadru avrige loss illa biron production house avru cleanaagi agreement maadkobeku alvaaa brooooo..... | Mixed | Negative |
| 1056 | Allige Ella beda just ondhu rally attend madfi | Negative | Positive |
| 4932 | Trailer ultimate ... ಸಿಂಬಿಮಾನ್ಯೇ ಚೆಸ್ಸ್ರ್ಯಾ?ರಲ್ಯುಹುಂಬು ಹೊಂಟ್ಟಂತ್ ಮ್ಯಾನ್ ಚೆತ್ತತಂಡಂತ್ತ ಒ.ಳ್ಸ್,ಂಬಾಗಲೇ... ಕ್ಯ್ಪಿ,ನ್ಯ,ಂತ್ ಡೇಪಂಣಂತ್. Look super ⬛⬛ | Positive | Positive |
| 7143 | ಸೂಪತ್ ಎಂಡಿಂಟ್,ಂ ಗುರು ಇದ ಲೀತ ಎಂಡಿಂಯ,ಂತ್ ಮಾಣ ಚೆಗಾ ಉರ್ಡೂತ್ಕೆಂಬೇಗ | Positive | Positive |
| 4393 | Super guru ninu.. | Positive | Positive |

**Fig. 7.** Predictions for Kannada-Engish (top), Tamil-Engish (bottom left) and Malayalam-Engish (bottom right) code-mixed dataset

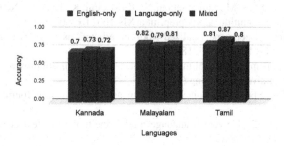

**Fig. 8.** Comparison of performance of various methods for the 3 languages

**Fig. 9.** Evaluation of model's performance on different types of code-switching for 3 languages

# 6 Conclusion and Future Work

Our work shows that transformer-based architectures for sequential classification tasks, at least for sentiment analysis, perform better than traditional machine learning solutions for code-mixed data. Performance could be improved using larger standardized datasets. Furthermore, attempts could be made to carry out tasks like NER or POS tagging for code-mixed data using similar architectures.

Other emerging transformer-based models could be used for the same task having the fine-tuned XLM model as a baseline. Methods like feature extraction could be tried instead of fine-tuning. Similar sequence classification tasks such as, hope speech detection, hate speech detection, etc. could be carried out with confidence for code-mixed data using an implementation similar to the one done in this work.

**Acknowledgements.** We acknowledge Mr. Adeep Hande, Mr. Ruba Priyadharshini and Mr. Bharathi Raja Chakravarthi for providing us the KanCMD dataset.

# References

1. Jose, N., Chakravarthi, B.R., Suryawanshi, S., Sherly, E., McCrae, J.P.: A survey of current datasets for code-switching research. In: 2020 6th ICACCS, pp. 136–141. IEEE (2020)
2. Hande, A., Priyadharshini, R., Chakravarthi, B.R.: Kancmd: Kannada codemixed dataset for sentiment analysis and offensive language detection. In: Proceedings of the Third Workshop on Computational Modeling of People's Opinions, Personality, and Emotion's in Social Media, pp. 54–63 (2020)
3. Shekhar, S., Sharma, D.K., Agarwal, D.K., Pathak, Y.: Artificial immune systems-based classification model for code-mixed social media data. IRBM (2020)
4. Chakravarthi, B.R., et al.: Dravidiancodemix: sentiment analysis and offensive language identification dataset for dravidian languages in code-mixed text. Lang. Resour. Eval. 1–42 (2022)
5. Devlin, J., Chang, M.W., Lee, K., Toutanova, K.: Bert: pre-training of deep bidirectional transformers for language understanding. arXiv preprint arXiv:1810.04805 (2018)
6. Lample, G., Ballesteros, M., Subramanian, S., Kawakami, K., Dyer, C.: Neural architectures for named entity recognition. arXiv preprint arXiv:1603.01360 (2016)
7. Peters, M.E., et al.: Deep contextualized word representations. corr abs/1802.05365 (2018). arXiv preprint arXiv:1802.05365, 1802
8. Radford, A., Narasimhan, K., Salimans, T., Sutskever, I., et al.: Improving language understanding by generative pre-training (2018)
9. Taylor, W.L.: Cloze procedure: a new tool for measuring readability. Journal. q. **30**(4), 415–433 (1953)
10. Howard, J., Ruder, S.: Universal language model fine-tuning for text classification. arXiv preprint arXiv:1801.06146 (2018)
11. Liu, Y., et al. Roberta: a robustly optimized bert pretraining approach. arXiv preprint arXiv:1907.11692 (2019)
12. Peters, M.E., Ruder, S., Smith, N.A.: To tune or not to tune? adapting pretrained representations to diverse tasks. arXiv preprint arXiv:1903.05987 (2019)
13. Pan, S.J., Yang, Q.: A survey on transfer learning. IEEE Trans. Knowl. Discov. Data Eng. **22** (10) (2010)

14. Min, B., et al.: Recent advances in natural language processing via large pre-trained language models: a survey. arXiv preprint arXiv:2111.01243 (2021)
15. Raffel, C., et al.: Exploring the limits of transfer learning with a unified text-to-text transformer. J. Mach. Learn. Res. **21**(140), 1–67 (2020)
16. dos Santos Neto, M.V., Amaral, A., Silva, N. and da Silva Soares, A.: Deep learning Brasil-NLP at semeval-2020 task 9: sentiment analysis of code-mixed tweets using ensemble of language models. In: Proceedings of the Fourteenth Workshop on Semantic Evaluation, pp. 1233–1238 (2020)
17. Eisenschlos, J. M., Ruder, S., Czapla, P., Kardas, M., Gugger, S., Howard, J.: Multifit: Efficient multi-lingual language model fine-tuning. arXiv preprint arXiv:1909.04761 (2019)
18. Lan, Z., Chen, M., Goodman, S., Gimpel, K., Sharma, P., Soricut, R.: Albert: A lite bert for self-supervised learning of language representations. arXiv preprint arXiv:1909.11942 (2019)
19. Yang, Z., Dai, Z., Yang, Y., Carbonell, J., Salakhutdinov, R.R., Le, Q.V.: Xlnet: generalized autoregressive pretraining for language understanding. Adv. Neural Inf. Process. Syst. **32** (2019)
20. Tula, D., et al.: Bitions DravidianLangTech-EACL2021: ensemble of multilingual language models with pseudo labeling for offence detection in dravidian languages. In: Proceedings of the First Workshop on Speech and Language Technologies for Dravidian Languages, pp. 291–299 (2021)
21. Chinnappa, D.: dhivya-hope-detection@ lt-edi-eacl2021: multilingual hope speech detection for code-mixed and transliterated texts. In: Proceedings of the First Workshop on Language Technology for Equality, Diversity and Inclusion, pp. 73–78 (2021)
22. Chathuranga, S., Ranathunga, S.: Classification of code-mixed text using capsule networks. In: Proceedings of the International Conference on RANLP 2021, pp. 256–263 (2021)
23. Conneau, A., et al.: Unsupervised cross-lingual representation learning at scale. arXiv preprint arXiv:1911.02116 (2019)
24. Vaswani, A., et al.: Attention is all you need. Adv. Neural Inf. Process. Syst. **30** (2017)

# Landslide Classification Using Deep Convolutional Neural Network with Synthetic Minority Oversampling Technique

S. Sreelakshmi$^{(\boxtimes)}$ and S. S. Vinod Chandra

Department of Computer Science, University of Kerala, Thiruvananthapuram, India
{sreelakshmis,vinod}@keralauniversity.ac.in

**Abstract.** Landslides are one of the world's most devastating and catastrophic natural disasters affecting human life and the economy. Many machine learning-based studies are reported on analyzing, classifying, and predicting Landslides, but there are countless avenues where these techniques must be developed to their full potential. This work proposes a deep convolutional neural network for classifying landslide data. The synthetic minority over-sampling method is employed on the dataset to address the class imbalance issue. A total of six shallow-learning algorithms and one deep-learning algorithm were used for baseline comparison. The proposed DCNN approach outperformed all the baselines chosen with an improvement of 2.1% in the f1-score. This study shows that deep learning would be better for building models capable of classifying landslides on real-world datasets.

**Keywords:** Landslide classification · Machine learning · Deep convolutional neural network · Synthetic minority oversampling · Landslide prediction

## 1 Introduction

Landslides are one of the most impactful natural catastrophes that occur globally, causing widespread human and socio-economic loss that exceeds the capability of the victims or area to cope using its own resources [20]. According to World Bank, it is estimated that over 300 million people live in a 3.7 million square kilometer stretch of inland terrain that is susceptible to landslides [2]. It is estimated that over 56,000 people worldwide died in landslides between 2004 and 2016 [8]; these numbers have been raised in recent years and are expected to grow exponentially in the future [13]. These factors highlight how crucial it is to have sophisticated techniques for landslide analysis. Over the last few years, there are considerable advancements happened in information and communication technologies that enabled machine learning techniques to be implemented for analyzing landslides that result in managing disasters effectively [23]. Machine learning algorithms are being widely implemented by researchers in dealing with different types of analysis such as landslide prediction [18], risk analysis [11], and inventory mapping [3].

Recent advancements in machine learning and intense learning techniques resulted in dealing with large quantities of data to train models for tasks such as classification and regression. The disaster management domain, specifically landslide studies, has witnessed widespread adoption of deep learning techniques for better analysis [19]. In many cases, the deep learning algorithms are found to be outperforming standard machine learning algorithms in terms of accuracy measures [5]. The deep learning algorithms such as convolutional neural networks (CNN) [9] and multi-layer perceptron (MLP) have a greater ability to handle large datasets and to unearth latent themes or patterns concealed in the data. Numerous studies have been reported in the machine learning and landslide literature on temporal forecasting of landslides, mapping landslide risk, and detecting landslides, which are powered by deep learning techniques [23]. This work proposes a deep convolutional neural network (DCNN) approach for classifying landslides using a publicly available dataset. The class imbalance issue and outlier identification on the dataset is performed, and then a DCNN will be trained for classifying landslides. The major contributions of this paper can be summarized as follows:

- Discusses the applications of machine learning approaches for landslide classification
- Critically reviews some of the state-of-the-art approaches for machine learning-powered landslide classification
- Proposes a deep convolutional neural network (DCNN) approach for the landslide classification tasks, and
- Compares and discusses the results with some benchmark algorithms followed by a detailed discussion.

## 2  Related Studies

**Table 1.** Summary of landslide studies using machine learning approaches

| Reference | Model | Objective |
|---|---|---|
| (Tang et al., 2022) [21] | Random Forest | Landslide susceptibility mapping |
| (Hussain et al., 2022) [10] | CNN | Landslide detection |
| (Al-Najjar et al., 2019) [1] | Logistic Regression | Landslide susceptibility mapping |
| (Kim et al., 2018) [12] | Decision Tree | Landslide susceptibility mapping |
| (Nguyen et al., 2017) [15] | KNN | Landslide modeling |
| Pham et al., 2017) [17] | SVM | Landslide susceptibility mapping |
| (Tarantino et al., 2007) [22] | MLP | Landslide change detection |

During the past few years, significant advances in machine learning and geographic information systems have facilitated the detection of landslide-like disasters. Various models and algorithms for landslide prevention, detection, and classification have been proposed. These are very useful for decision-makers to mitigate social and economic losses [16]. Field surveys, which are a part of the geomorphological investigation of the research region and are traditionally used for landslide detection and mapping, are often difficult to access and inefficient. Numerous computer interpretation techniques are employed for geoscience information extraction jobs to increase the effectiveness and precision of landslide detection. To realize the intelligent acquisition of geoscience topic information, computer interpretation is the comprehensive use of geoscience analysis, remote sensing, geographic information systems, pattern recognition, and artificial intelligence technology. Statistics-based and heuristic methods are two categories of computer interpretation techniques used for landslides. The weight of evidence, logistic regression, and the analytic hierarchy process is statistical procedures; when new methods are utilized, they can be used as a benchmark. Advanced techniques are frequently used by heuristic or machine learning approaches to construct relationships by comparing landslide and non-landslide feature relationships. An automated modelling technique for data analysis is machine learning. To create an analysis model, it can learn the fundamental relationships that already exist in the data. Through an iterative learning process, they can deliver precise and predictable results [5].

For the geographical and temporal prediction of geohazards, several machine learning-based models have been proposed, including individual and ensemble techniques. Artificial neural networks (ANNs), random forests (RFs), support vector machines (SVMs), logistic regression (LR), decision trees (DTs), K-nearest neighbours (KNN), and Naïve Bayes (NB) models are typical examples of individual techniques. Geohazards' spatial and temporal prediction has recently been applied using deep learning techniques, such as autoencoders, convolutional, and recurrent neural networks, and the accuracy of predictions for such events has increased. In particular, traditional machine learning algorithms and deep learning have been widely used to reduce and predict landslide hazards, carry out landslide susceptibility evaluations, and create warning systems. The most common uses of machine learning for landslide mitigation include landslide susceptibility evaluations utilizing all kinds of machine learning approaches. Recent studies have also concentrated on using different machine learning models to create landslide alerts, for instance, by assessing rainfall thresholds and forecasting displacement [14]. A summary of some of the most notable and recent works reported in the literature on landslide analysis using machine learning techniques is given in Table 1.

## 3    Materials and Methods

### 3.1    Dataset

This study uses a publicly available labelled dataset [4] for implementing the proposed approach for landslide classification. A snapshot of the dataset with labels (refer to column "Landslide") is shown in Table 2. A "0" in the label column denotes "No Landslide", and "1" in the label column denotes "Landslide".

There are a total of 17 features such as *Aspect, Curvature, Earthquake, Elevation, Flow, Lithology, NDVI (Normalized Difference Vegetation Index), NDWI (Normalized Difference Water Index), Plan, Precipitation, Profile, Slope, Temperature, Humidity, Rain, Moisture,* and *Pressure* in the dataset. There are 181236 data points that correspond to the label "0" (Not Landslide) and 9654 data points that correspond to the label "1" (Landslide). The dataset is imbalanced, and we have used the *imbalanced − learn* library available at https://pypi. org/project/imbalanced-learn/ for implementing the SMOTE [7] algorithm. The parameter *random_state* is set as 2, and all the other hyperparameters were set as *default*. After splitting the dataset into training and testing samples, there are 6723 data points with label "1" and 126900 with label "0", and after over-sampling with SMOTE, both label "1" and label "2" has got 126900 data points which are balanced. A heatmap showing the correlation between different landslide features used in this study is given in Fig. 1.

**Table 2.** A snapshot of the dataset used for the experiment (10 samples)

| Landslide | Aspect | Curvature | Earthquake | Elevation | Flow | Lithology | NDVI | NDWI | Plan | Precipitation | Profile | Slope | Temperature | Humidity | Rain | Moisture | Pressure |
|---|---|---|---|---|---|---|---|---|---|---|---|---|---|---|---|---|---|
| 0 | 2.00 | 3.33 | 1.67 | 4.00 | 2.67 | 2.33 | 3.00 | 2.67 | 3.00 | 2.67 | 2.67 | 2.33 | 18.21 | 84.33 | 26668.92 | 31.25 | 1017.90 |
| 0 | 4.00 | 2.67 | 2.33 | 2.00 | 2.33 | 1.67 | 2.33 | 2.67 | 2.67 | 4.33 | 3.33 | 2.00 | 19.17 | 73.91 | 27103.17 | 75.94 | 1013.19 |
| 0 | 3.00 | 2.67 | 3.00 | 2.00 | 2.00 | 2.00 | 2.67 | 3.00 | 3.00 | 5.00 | 3.33 | 2.00 | 22.79 | 78.17 | 26689.17 | 51.23 | 1016.09 |
| 0 | 3.00 | 2.67 | 2.67 | 2.67 | 3.00 | 1.33 | 3.33 | 2.67 | 2.67 | 5.00 | 3.00 | 2.67 | 23.30 | 81.47 | 18678.75 | 72.09 | 1017.48 |
| 0 | 2.67 | 3.67 | 2.33 | 3.67 | 1.67 | 3.67 | 2.67 | 3.00 | 3.00 | 3.67 | 2.00 | 3.33 | 22.45 | 80.01 | 26713.92 | 53.48 | 1013.23 |
| 1 | 2.33 | 1.00 | 2.00 | 1.33 | 3.00 | 2.00 | 2.00 | 4.00 | 1.33 | 3.33 | 4.33 | 1.00 | 19.94 | 64.19 | 51195.42 | 47.00 | 1019.37 |
| 1 | 2.33 | 2.67 | 1.67 | 3.00 | 1.33 | 3.33 | 3.00 | 2.67 | 2.67 | 4.00 | 3.67 | 3.33 | 17.96 | 60.16 | 53333.33 | 54.50 | 1020.60 |
| 1 | 2.00 | 3.00 | 2.00 | 3.00 | 1.33 | 2.33 | 2.00 | 3.67 | 3.33 | 3.67 | 3.67 | 3.00 | 17.99 | 70.47 | 53506.92 | 47.75 | 1019.84 |
| 1 | 4.00 | 3.67 | 2.00 | 1.33 | 2.33 | 1.67 | 3.33 | 3.00 | 4.00 | 3.33 | 3.33 | 2.67 | 21.29 | 69.02 | 51537.42 | 52.46 | 1017.45 |
| 1 | 3.00 | 3.33 | 2.00 | 2.33 | 2.00 | 2.33 | 4.00 | 2.00 | 3.33 | 3.67 | 3.00 | 2.33 | 23.38 | 49.43 | 40401.75 | 35.85 | 1021.46 |

### 3.2   Normalization

Normalization of data is considered one of the important stages of any machine learning task, specifically when dealing with multi-dimensional datasets. This step enhances the numerical stability of the model, often reducing the time taken for training and, in many cases, reducing the error. Since the true distributions of the features considered are not known in advance, the feature normalization technique is applied before the training stage. We used the Standard Scalar method for feature normalization. This technique eliminates the mean and then scales its variance to one as the normalized value depends on the mean and variance. The scaling of attributes from zero to one has several advantages, such as being linear, reversible, and scalable.

### 3.3   Oversampling

One of the critical challenges in machine learning, especially in the classification task, is to deal with imbalanced classes. Random oversampling and random undersampling models are used to avoid the curse of imbalance. But they have disadvantages such as increasing the likelihood of overfitting. Also, duplicating

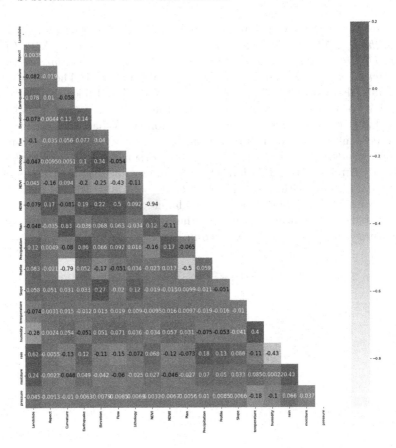

**Fig. 1.** A heatmap showing the correlation between different landslide features in the dataset

data samples from the majority or minority classes does not add new information to the model. Synthetic Minority Oversampling Technique (SMOTE) is a technique for dealing with class imbalance by adding recent examples that are synthesized from the existing data [7]. SMOTE randomly chooses a minority class instance and then finds its k-nearest minority class instances. Then a synthetic example is created at a randomly selected point between the models in feature space. Since the dataset chosen for this proposed approach is found to be highly imbalanced, SMOTE technique is used to address the class imbalance issue.

### 3.4    Baselines

In order to compare the classification performance of the DCNN-based landslide approach, several standard machine learning algorithms such as *Logistic Regression, Decision Tree, Support Vector Machine, Naive Bayes, K-Nearest Neighbor,* and *Random Forest,* and *Multi-Layer Perceptrons* from deep learning are chosen as baselines [6]. A very brief description of each of these baselines is given below:

- Logistic Regression (LR): is used to describe data and to explain the relationship between one dependent binary variable and one or more independent variables.
- Decision Tree (DT): creates the classification model by building a tree data structure called a decision tree. Each node in this tree specifies a test on an attribute, and each branch descending from that node corresponds to one of the possible values for that attribute.
- Support Vector Machine (SVM): The goal of the SVM algorithm is to create the best line or decision boundary that can segregate n-dimensional space into classes so that we can easily put the new data point in the correct category in the future. This best decision boundary is called a hyperplane.
- Naive Bayes (NB): is a popular classification algorithm used for the analysis of categorical text data. The algorithm is based on the Bayes theorem and predicts the tag of a text by computing the probability of each tag for a given sample and then gives the tag with the highest probability as output.
- Random Forest (RF): creates multiple decision trees using bootstrapped datasets of the original data and randomly selecting a subset of variables at each step of the decision tree. The model then selects the mode of all of the predictions of each decision tree.
- K-Nearest Neighbor (KNN): is a non-parametric machine learning algorithm that fits under the category of lazy learner algorithms. This algorithm takes into consideration the similarity between the new data point and available data points and put the new point into the bucket that is most similar to the available categories.
- Multi-Layer Perceptron (MLP): consists of fully connected dense layers that transform any input dimension to the desired dimension and every node uses a sigmoid activation function. The sigmoid activation function takes real values as input and converts them to numbers between 0 and 1 using the sigmoid formula.

## 4   Proposed Approach

This section details the proposed deep convolutional neural network (DCNN) approach for landslide classification. The overall workflow of the proposed approach is shown in Fig. 2. The first phase deals with the data collection and this work uses publicly available data containing different features of landslides such as *Aspect, Curvature, Earthquake, Elevation, Flow, Lithology, NDVI (Normalized Difference Vegetation Index), NDWI (Normalized Difference Water Index), Plan, Precipitation, Profile, Slope, Temperature, Humidity, Rain, Moisture,* and *Pressure,* to classify the landslide. The second phase of the proposed approach deals with data pre-processing to normalize the data and also to eliminate the class imbalance issues. Convolutional neural networks (CNN) or deep convolutional neural networks (DCNN) is a type of neural network that exhibits its robust

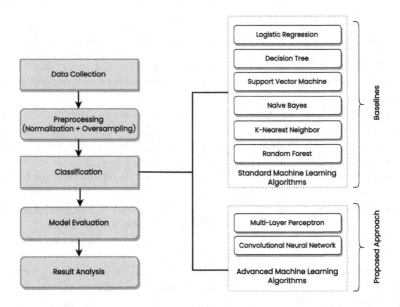

**Fig. 2.** Overall workflow of the proposed approach

capability in feature extraction that mimics the human brain. The structure of a deep convolutional neural network is shown in Fig. 3. The input layer of a DCNN comprises of several neurons and each neuron indicates a landslide feature such as *Aspect*, and *Curvature*. Notationally, let's assume that $l = l_1, l_2, ..., l_N$ refer to the input data where $N$ represents the number of landslide features present in the data.

$$C_j = \sum_{i=1}^{N} f(w_j * v_i + b_j), j = 1, 2, ..., k \tag{1}$$

$$f(x) = tanh(x) = \frac{e^x - e^{-x}}{e^x + e^{-x}} \tag{2}$$

Consider Eq. 1 and Eq. 2: if $f$ is a non-linear activation function, then $*$ is treated as the convolution operator, and $k$ is the number of convolutionary kernels. In the equation, $w_j$ represents the weight and $b_j$ denotes the choice. The size of the feature vector can be reduced by combining the outputs of an $N = 1$ patch from the previous layer. The proposed DCNN architecture uses max pooling which is the most widely used pooling method and the fully connected layers reorganize the local representations resulting from the convolution and pooling operations. The reverse spread algorithm employed in all DCNN layer variables to reduce the loss and the loss function is defined as given in Eq. 3.

$$Loss = -\frac{1}{m} \sum_{i=1}^{m} [y_i log(\hat{y}i) + (1 - y_i)log(1 - \hat{y}i)] \tag{3}$$

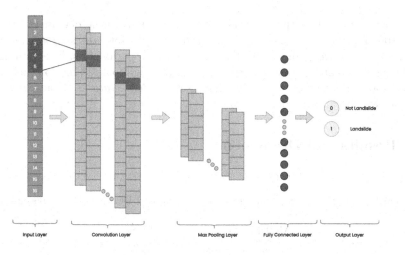

**Fig. 3.** Structure of a deep convolutional neural network for landslide classification

where $m$ denotes the amount of input landslide data, and for the $ith$ input sample, the variables $yi$ and $\hat{y}i$ signify the true and predicted label, correspondingly. Until the convergence is reached for the loss value, the factors are updated iteratively in the DCNN.

## 5    Experiment

All the machine learning algorithms discussed in this paper have been implemented using Lenovo Ideapad S340-14IIL with a Processor Intel(R) Core(TM) i5-1035G1 CPU @ 1.00 GHz, 1190 MHz, 4 Cores, 8 Logical Processors, and 8 GB memory. JupyterLab environment is used for scripting the experiments, and $scikit-learn$ library available at https://scikit-learn.org/stable/ is used for implementing all the baseline algorithms along with MLP, and $tensorflow$ available at https://www.tensorflow.org/ is used for implementing the deep convolutional neural network model. This work uses 6 shallow-learning baselines (Logistic Regression, Decision Tree, Support Vector Machine, Naive Bayes, Random Forest, and K-Nearest Neighbor) and one deep learning baseline (Multi-Layer Perceptron) for comparing the performance of the proposed DCNN-based landslide classification model. The dataset is divided into 70-30 splits where 70% of the data will go for training and the remaining 30% for testing.

The Logistic Regression used the default values for all the hyperparameters, and no parameter tuning was done. Still, for the Decision Tree classifier, we have used 'gini' for the $criterion$, 'best' for the $splitter$, '2' for $min\_samples\_split$, and '1' for $min\_samples\_leaf$ hyperparameters. We have used the default values provided by the $scikit-learn$ library for all the other parameters. To train the Support Vector Machine (SVM), we have set the value of $class\_weight$ parameter as 'balanced', and for the Naive Bayes classifier model, we have used the default hyperparameters. For the random forest classifiers we have set '100' for *n_estimators*, 'gini' for *criterion*, '2' for *min_samples_split*, '1' for *min_samples_leaf*, and 'auto' for *max_features* parameters. For training the K-Nearest Neighbors classifier, the number of neighbours is set to '3'. For the Multi-Layer Perceptron model, we have set *random_state* as '1' and *max_iter* as 300, which will run 300 epochs.

# 6    Results and Discussion

**Table 3.** Performance comparison of the proposed DCNN with the baselines

| Model name | F1-Score |
| --- | --- |
| Logistic Regression | 91.56 |
| Decision Tree | 95.03 |
| Support Vector Machine | 89.30 |
| Naive Bayes | 70.28 |
| Random Forest | 96.36 |
| K-Nearest Neighbors | 93.65 |
| Multi-Layer Perceptron | 93.53 |
| Proposed DCNN | 98.46 |

This section details the results obtained for the experiment given in Sect. 5, followed by a detailed discussion. Table 3 shows the f1-score comparison of the chosen baselines and the proposed DCNN approach. The Logistic Regression algorithm scored 91.56%, and Decision Tree obtained 95.03% for the f1-scores. The Support Vector Machines and Naive Bayes scores were 89.30% and 70.28%, respectively, and the Naive Bayes algorithm scored the lowest f1-score out of all the baselines chosen, followed by SVM. For the Random Forest and K-Nearest Neighbor classifiers, the f1-score was 96.36% and 93.65%, respectively. It is observed that among all the baselines chosen, the Random Forest classifier performed better in terms of f1-score followed by the decision tree. The deep learning baseline model Multi-Layer Perceptron (MLP) has scored 93.53% f1-score, and the proposed deep convolutional neural network model has attained

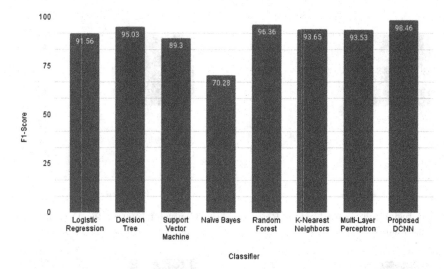

**Fig. 4.** Comparison of F1-score for the proposed DCNN with the baselines

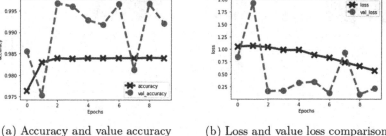

(a) Accuracy and value accuracy comparison of the proposed DCNN model with the baselines

(b) Loss and value loss comparison of the proposed DCNN model with the baselines

**Fig. 5.** Accuracy, value accuracy, loss, and value loss comparison of the proposed DCNN model with the baslines

an f1-score of 98.46% which shows that the deep learning classifier outperforms all the other baselines in this context for landslide classification. A comparison graph showing f1-score for the baselines and the proposed DCNN model is shown in Fig. 4, and the accuracy, value accuracy, loss, and value loss comparison for the proposed DCNN model is given in Fig. 5. The confusion matrices for all the baselines chosen are shown in Fig. 6.

250     S. Sreelakshmi and S. S. Vinod Chandra

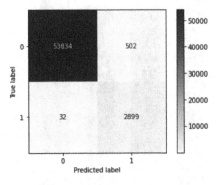

(a) Confusion matrix for Logistic Regression classifier

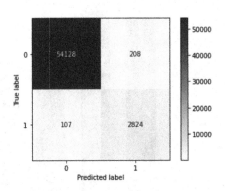

(b) Confusion matrix for Decision Tree classifier

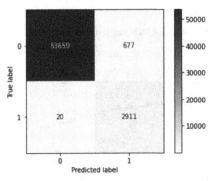

(c) Confusion matrix for Support Vector Machine classifier

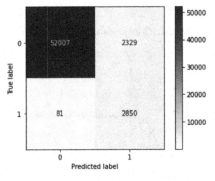

(d) Confusion matrix for Naive Bayes classifier

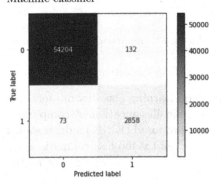

(e) Confusion matrix for Random Forest classifier

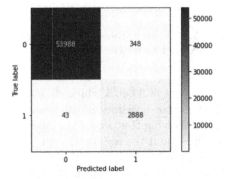

(f) Confusion matrix for K-Nearest Neighbor classifier

**Fig. 6.** The confusion matrices for Logistic Regression, Decision Tree, Support Vector Machine, Naive Bayes, Random Forest, and K-Nearest Neighbor classifier models

# 7    Conclusions

Landslides, one of the most devastating natural calamities, cause significant losses to lives and the economy. Early analysis and classification of landslides are crucial to avoid the impact of this catastrophic phenomenon. This work proposed a deep convolutional neural network for landslide classification using publicly available landslide datasets containing landslide features. Compared with six shallow-learning and one deep-learning baseline, the proposed DCNN approach significantly outperforms the baselines in the f1-score. The experiment concludes that deep learning techniques such as DCNN would be a better choice for building better models for landslide classification that would enable early predictions. The results are promising, and our work can be extended to construct early prediction models with more future data collected from landslide-prone areas.

**Acknowledgement.** The authors would like to thank the researchers and the staff members of Machine Intelligence Research Lab at the Department of Computer Science, University of Kerala, for providing all the facilities and support for carrying out a research of this scale.

# References

1. Al-Najjar, H.A., Kalantar, B., Pradhan, B., Saeidi, V.: Conditioning factor determination for mapping and prediction of landslide susceptibility using machine learning algorithms. In: Earth Resources and Environmental Remote Sensing/GIS Applications X, vol. 11156, pp. 97–107. SPIE (2019)
2. Anderson, M.G., Holcombe, E.: Community-Based Landslide Risk Reduction: Managing Disasters in Small Steps. World Bank Publications (2013)
3. Anshori, R.M., Samodra, G., Mardiatno, D., Sartohadi, J.: Volunteered geographic information mobile application for participatory landslide inventory mapping. Comput. Geosci. **161**, 105073 (2022)
4. Aslam, B., Zafar, A., Khalil, U.: Development of integrated deep learning and machine learning algorithm for the assessment of landslide hazard potential. Soft Comput. **25**(21), 13493–13512 (2021)
5. Cai, H., Chen, T., Niu, R., Plaza, A.: Landslide detection using densely connected convolutional networks and environmental conditions. IEEE J. Sel. Top. Appl. Earth Observ. Remote Sens. **14**, 5235–5247 (2021)
6. Chandra, S., Hareendran, S., et al.: Machine Learning: A Practitioner's Approach. PHI Learning Pvt. Ltd. (2021)
7. Chawla, N.V., Bowyer, K.W., Hall, L.O., Kegelmeyer, W.P.: SMOTE: synthetic minority over-sampling technique. J. Artif. Intell. Res. **16**, 321–357 (2002)
8. Froude, M.J., Petley, D.N.: Global fatal landslide occurrence from 2004 to 2016. Nat. Hazard. **18**(8), 2161–2181 (2018)
9. Gu, J., et al.: Recent advances in convolutional neural networks. Pattern Recogn. **77**, 354–377 (2018)
10. Hussain, M.A., et al.: Landslide susceptibility mapping using machine learning algorithm validated by persistent scatterer In-SAR technique. Sensors **22**(9), 3119 (2022)

11. Ibrahim, M.B., Mustaffa, Z., Balogun, A.L., Sati, H.I.: Landslide risk analysis using machine learning principles: A case study of bukit antrabangsa landslide incidence. J. Hunan Univ. Nat. Sci. **49**(5) (2022)

12. Kim, J.C., Lee, S., Jung, H.S., Lee, S.: Landslide susceptibility mapping using random forest and boosted tree models in pyeong-chang, korea. Geocarto Int. **33**(9), 1000–1015 (2018)

13. Kirschbaum, D., Stanley, T., Zhou, Y.: Spatial and temporal analysis of a global landslide catalog. Geomorphology **249**, 4–15 (2015)

14. Ma, J., et al.: A comprehensive comparison among metaheuristics (MHS) for geohazard modeling using machine learning: insights from a case study of landslide displacement prediction. Eng. Appl. Artif. Intell. **114**, 105150 (2022)

15. Nguyen, Q.K., Tien Bui, D., Hoang, N.D., Trinh, P.T., Nguyen, V.H., Yilmaz, I.: A novel hybrid approach based on instance based learning classifier and rotation forest ensemble for spatial prediction of rainfall-induced shallow landslides using gis. Sustainability **9**(5), 813 (2017)

16. Nhu, V.H., et al.: Landslide susceptibility mapping using machine learning algorithms and remote sensing data in a tropical environment. Int. J. Environ. Res. Public Health **17**(14), 4933 (2020)

17. Pham, B.T., Tien Bui, D., Prakash, I., Nguyen, L.H., Dholakia, M.: A comparative study of sequential minimal optimization-based support vector machines, vote feature intervals, and logistic regression in landslide susceptibility assessment using gis. Environ. Earth Sci. **76**(10), 1–15 (2017)

18. Rajabi, A.M., Khodaparast, M., Mohammadi, M.: Earthquake-induced landslide prediction using back-propagation type artificial neural network: case study in northern iran. Nat. Hazards **110**(1), 679–694 (2022)

19. Sreelakshmi, S., Chandra, S.V.: Machine learning for disaster management: insights from past research and future implications. In: 2022 International Conference on Computing, Communication, Security and Intelligent Systems (IC3SIS), pp. 1–7. IEEE (2022)

20. SS, V.C., Shaji, E.: Landslide identification using machine learning techniques: review, motivation, and future prospects. Earth Sci. Inform., 1–28 (2022)

21. Tang, X., Tu, Z., Wang, Y., Liu, M., Li, D., Fan, X.: Automatic detection of coseismic landslides using a new transformer method. Remote Sensing **14**(12), 2884 (2022)

22. Tarantino, C., Blonda, P., Pasquariello, G.: Remote sensed data for automatic detection of land-use changes due to human activity in support to landslide studies. Nat. Hazards **41**(1), 245–267 (2007)

23. Tehrani, F.S., Calvello, M., Liu, Z., Zhang, L., Lacasse, S.: Machine learning and landslide studies: recent advances and applications. Nat. Hazards, 1–49 (2022)

# ALPR - An Intelligent Approach Towards Detection and Recognition of License Plates in Uncontrolled Environments

Akshay Bakshi[1], Sudhanshu Gulhane[2], Tanish Sawant[1], Vijay Sambhe[1], and Sandeep S. Udmale[1(✉)] (iD)

[1] Department of Computer Engineering and Information Technology,
Veermata Jijabai Technological Institute (VJTI), Mumbai 400019, Maharashtra, India
{apbakshi_b19,tmsawant_b19}@ce.vjti.ac.in,
{vksambhe,ssudmale}@it.vjti.ac.in
[2] Department of Electronics and Telecommunications Engineering,
Pune Institute of Computer Technology (PICT), Pune 411043, Maharashtra, India
sudhanshugulhane072@gmail.com

**Abstract.** Most existing Automatic License Plate Recognition (ALPR) approaches focus on images containing approximately frontal views. The considerable variation of LP across complicated environments and perspectives remains a massive challenge for a robust ALPR. This work proposes a comprehensive ALPR paradigm emphasizing unrestricted express screenplays in which the LP may be significantly influenced by diverse shooting angles, illumination circumstances, and complicated surroundings. This system integrates a Spatial Transformer Network, which can catch and repair numerous distorted LPs in an image so that all the plates are consistently aligned. Then, a convolutional neural network is sketched to determine LP characters containing various font styles and sizes. We evaluated the system with a data set containing annotations for a challenging LP image set from multiple areas and acquisition states. The experimental outcomes reveal that our proposed ALPR paradigm attains adequate recognition accuracy compared to existing methods.

**Keywords:** Convolutional Neural Network (CNN) · License Plate (LP) · Spatial Transformer Network (STN) · YOLO

## 1 Introduction

Automatic License Plate Recognition (ALPR) routines offer various applications, including identifying stolen vehicles, monitoring traffic, smart toll collection, etc. [9,23]. The recent advancements in deep learning (DL) and parallel computing have contributed to achieving excellent performance in several digital image/video applications, such as optical character recognition and object detection and recognition, which have tremendously improved ALPR systems. Recently, convolutional neural network (CNN) has achieved exceptional performance and have been the primary machine learning approach for LP detection

A. R. Molla et al. (Eds.): ICDCIT 2023, LNCS 13776, pp. 253–269, 2023.
https://doi.org/10.1007/978-3-031-24848-1_18

and recognition [2,6,10–13,15,25,27]. Several ALPR commercial systems have also been employing DL methods. They are usually integrated with web services and large data centers to process millions of vehicle images daily and constantly improve the system. Some of the example systems to be mentioned are: OpenALPR[1], Sighthound[2], and Amazon Rekognition[3].

Moreover, the CNN-based object identification routines have become famous for LPR with the establishment of DL. Typically, faster regions with CNN (R-CNN) [21], Single Shot MultiBox Detector (SSD) [14], and You only look once (YOLO) [18] models are employed. Faster R-CNN [21], a modified version of R-CNN and fast R-CNN that forgoes time-consuming strategy, i.e., selective search, allows the architecture to understand the area manifestos. In this work, a NN has been utilized to forecast the region proposals rather than a particular search procedure to determine the area manifestos on the feature map (FM). Praveen Ravirathinam and Arihant Patawari in [16] demonstrated the effective handling of faster R-CNN in the detection of LP. The proposed model could also detect titled and non-rectangular plates. The mAP of their model went relatively low since it could not catch small-scale images. The study in [11] presented a robust object detection model using Fast Yolo and Yolov2 to detect LP in simple and realistic conditions.

Despite the advances in this field, most approaches focus on recognizing LP in controlled environments, assuming a frontal view of the vehicles and LP. The current challenges in ALPR include image distortion, image quality degradation, weather (snow, rain, etc.), variable illumination conditions, etc. A more permissive picture-gathering setting (e.g., a police car using a camera to track down an unlawful vehicle) could result in slanting vision. In such cases, the LP may be severely distorted and, thus, challenging to recognize, for which even existing standard commercial solutions struggle.

This paper proposes a comprehensive ALPR paradigm capable of performing well over various unrestricted capture screenplays and camera arrangements. We integrate a transformation module to estimate and rectify the distortion and improve the character recognition performance. An additional contribution is the collection of images from natural scenes, which cover various challenging scenarios and contain substantial LP distortions. The proposed system could also discover and identify LPs in independent test data sets using the same configuration. The data sets employed in this assignment are publicly available, and the samples can be obtained from SSIG-SegPlate database [4] and the application-oriented license plate (AOLP) data set [7].

## 2    Materials

This section provides background information about the various vital components employed in the proposed work (Fig. 1).

---

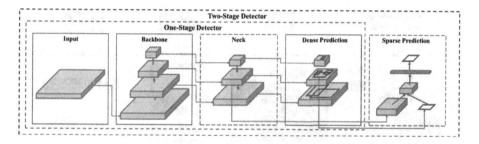

**Fig. 1.** Yolov4 architecture [1]

## 2.1 You Only Look Once (YOLO)

YOLO is one of the one-stage object detector approaches. YOLOv2 [19] model has been built upon YOLO with several incremental enhancements, such as batch normalization, excellent resolution, and anchor boxes. To perform better on smaller objects, YOLOv3 [20] improved upon earlier models by including the bounding box prediction with an objectness score. Also, it attaches links to the backbone network layers and performs predictions at three different degrees of granularity. YOLOv4 [1], a two-stage detector with multiple components, is currently an upgraded version of earlier generations. The higher versions of YOLO are volatile to use as a black box for our proposed methodology. The Yolov4 model consists of Backbone, Neck, and Head, as shown in Fig. 1.

**Backbone:** It consists of the CSPDarknet53 model, which detects objects with higher accuracy. Also, it includes the CSPDarknet53 model because it enhances through the MISH and other activation functions [1].

**Neck:** It consists of a spatial pyramid pooling layer (SPP) and Path Aggregation Network (PAN). SPP plays a crucial role when detecting objects of various scales for adequate context information and, thus, sits between CSPDarknet53 and PAN. It adds a spatial pyramid pooling layer in place of the last pooling layer, which comes after the final convolutional layer. A maximum pool is applied to a sliding kernel of various sizes. The result is then created by concatenating the FMs generated by different kernel sizes [1].

Further, the PAN network's capacity to reliably maintain spatial information, which aids in the proper localization of pixels for mask generation, was chosen, for example, segmentation in YOLOv4. The properties which make PAN so accurate are Bottom-up Path Augmentation, Adaptive Feature Pooling, and Fully-Connected Fusion Network [1].

**Head:** Bounding box location and categorization has performed using the head (Dense prediction). The procedure is the same as that described for Yolo v3; hence, it detects the score and the bounding box coordinates (x, y, height, and

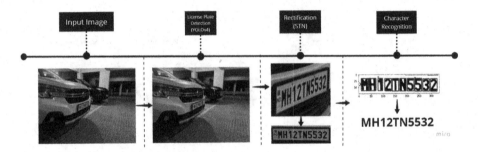

**Fig. 2.** Proposed system pipeline.

width). The algorithm splits the input image into several grid cells and uses anchor boxes to forecast the likelihood that each cell will contain an object. The result is a vector containing the bounding box coordinates and the class probabilities [1].

## 2.2 Spatial Transformer Network

Though CNN's defined as a powerful class of models, they are nonetheless constrained by their inability to be computationally and parameter-efficiently spatially invariant to the input data. The Spatial Transformer Network (STN) [8], a novel teachable module, explicitly permits the spatial modification of information within the architecture. Its differentiable module can be added to current convolutional architectures. It enables the NNs to actively modify FM spatial relationships based on the FM itself without changing the optimization procedure or adding additional training supervision.

## 3    Proposed Methodology

The proposed structure is demonstrated in Fig. 2 and comprises three main steps: LP Detection, LP Transformation and Rectification, and Character Recognition Network. Given an input image, the custom-trained YOLOv4 model detects LPs in the scene. The detections are cropped and forwarded to an STN to rectify LP images with diverse orientations and surroundings details. The corrected images have a uniform orientation and paltrier surrounding noise. These favorable and repaired detections are presented to a Character Recognition Network.

### 3.1    License Plate Detection

Detection of LPs is an essential phase in the ALPR process; hence we adopted a reliable model to carry it out. To select the best algorithm, we defined the criteria as 1) The algorithm must have an acceptable performance and recall rate because even a small amount of missed detection will cause the LP detection process to

**Fig. 3.** Examples of detected LP from testing data set

perform worse. 2) For real-time detection to be reliable, the method must have a high calculation speed. 3) Additionally, since their use in practical applications won't be hampered, the calculating costs should be reasonable. As a result, we carefully chose YOLOv4 as our network for LP detection. When comparing the cost and speed of calculations, the YOLOv4 algorithm is quite effective. Figure 3 reveals that we have refined the YOLOv4 model configurations according to our requirements to specialize it for LP detection. Since we need only one class, i.e., LP, for object detection, we altered the number of classes from 80 to 1 and, thus, the modified value of maximum batch size according to the below formula,

$$max\_batches = min(training\_images, min(classes * 2000, 6000)); \quad (1)$$

Secondly, we altered the number of filters in the convolutional layers using the formula below.

$$filters = (classes + 5) * 3; \quad (2)$$

Thus, we employ a reconfigured model for the detection of LPs.

### 3.2   Spatial Transformer Network (STN)

The STN suggested in [8] is a differentiable and self-contained module. Thus, it has been added to current convolutional architectures. It streamlines the

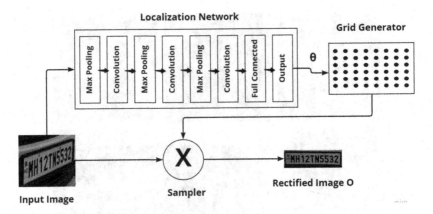

**Fig. 4.** Structure of STN [8].

subsequent classification work and improves classification results. It strengthens a model's spatial invariance against non-rigid deformations such as translations, scaling, rotations, and cropping. The suggested model is more resistant to various shooting angles and noises since the input LP photos are first rectified with the trained STN to those with a consistent orientation and reduced noise. Figure 4 shows that it is divided into three divisions. 1) The localization network (LN) derives the affine transformation parameter $\theta$ by extracting the key attributes from the input image $I$. 2) The initial grid is transformed into a new sampling grid by the grid generator based on the input $\theta$. 3) The sampler samples the $I$ by the new grid to create the rectified picture.

**Localization Network:** The LN accepts the input FM $U \epsilon \mathbb{R}^{H \times W \times C}$ with height ($H$), width ($W$), and channels ($C$) and outputs ($\theta$), the parameters of the transformation $\mathbb{T}_\theta$ operated to the FM: $\theta = f_{loc}(U)$. The proportion changes depending on the parameterized kind of transformation; for example, the size of an affine transformation is six dimensions. A final regression layer must be present in the LN function $f_{loc}()$ to obtain the transformation parameters, but it can be fully connected or convolutional.

$$\begin{bmatrix} x' \\ y' \end{bmatrix} = \begin{bmatrix} \theta_{11} & \theta_{12} \\ \theta_{21} & \theta_{22} \end{bmatrix} \begin{bmatrix} x \\ y \end{bmatrix} + \begin{bmatrix} \theta_{13} \\ \theta_{23} \end{bmatrix} \qquad (3)$$

The affine transformation matrix is represented by $\mathbb{A}_\theta$.

$$\mathbb{A}_\theta = \begin{bmatrix} \theta_{11} & \theta_{12} & \theta_{13} \\ \theta_{21} & \theta_{22} & \theta_{23} \end{bmatrix} \qquad (4)$$

**Table 1.** LN Configuration.

| Type | Configuration | | | |
|---|---|---|---|---|
| Input | Gray-scale distorted LP image | | | |
| Layer | Filters | Kernel size | Stride size | Padding |
| Max_pool_1 | – | $2 \times 2$ | $2 \times 2$ | $0 \times 0$ |
| Conv2D_1 | 20 | $5 \times 5$ | $1 \times 1$ | $0 \times 0$ |
| Max_pool_2 | – | $2 \times 2$ | $2 \times 2$ | $0 \times 0$ |
| Conv2D_2 | 20 | $5 \times 5$ | $1 \times 1$ | $0 \times 0$ |
| Max_pool_3 | – | $2 \times 2$ | $2 \times 2$ | $0 \times 0$ |
| Conv2D_3 | 20 | $5 \times 5$ | $1 \times 1$ | $0 \times 0$ |
| Fully connected | 100 hidden units, tanh activation | | | |
| Output | 6 hidden units, linear output activation | | | |

The LN structure summarized in the Table 1 consists of 3 sets of max-pooling and convolutional layers with a fully connected layer, and finally one output layer.

**Parameterised Sampling Grid:** Every pixel of the input LP image has a corresponding vector of coordinate, i.e., $K_i = (x_i, y_i)^T$ with the pixel index $i$. A multiplication operation is performed on *theta*, and $Ki$ to obtain the affine converted vector of coordinate, i.e., $K'_i = (x'_i, y'_i)^T$. It is expressed as

$$\begin{pmatrix} x'_i \\ y'_i \end{pmatrix} = A_\theta \begin{pmatrix} x_i \\ y_i \\ 1 \end{pmatrix} = \begin{bmatrix} \theta_{11} & \theta_{12} & \theta_{13} \\ \theta_{21} & \theta_{22} & \theta_{23} \end{bmatrix} \begin{pmatrix} x_i \\ y_i \\ 1 \end{pmatrix} \tag{5}$$

$K' = (K'_1, K'_2, ..., K'_i, ..., K'_{W \times H})$ are set up to obtain the grid generator's final output, where $W$ and $H$ in our experiments are 270 and 70, respectively.

**Differentiable Image Sampler:** In order to generate the rectified image O, the sampler samples the original image using the sampling grid $K'$. Bilinear interpolation, a differentiable module, is used in this sampling process. The STN, which may be trained end-to-end alongside other sections of the model, comprises the LN, the parameterized grid generator, and the image sampler. The STN is created by combining the LN, parameterized grid generator, and image sampler. It can be trained end-to-end with other model components. Please refer to [8] for further information.

### 3.3  Character Recognition

The recognition process consists of three parts: (1) Preprocessing the rectified image output of STN; (2) Character Segmentation; (3) Recognition of segmented characters.

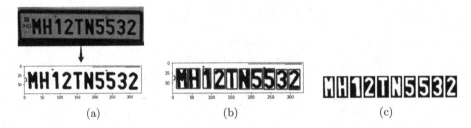

**Fig. 5.** (a) Binary conversion of the detected plate. (b) Bounding rectangles containing contours. (c) Binary images of segmented characters.

**Preprocessing Stage:** The rectified LP image is processed to make the character extraction easier. With a single 8-bit channel and values ranging from 0–255, where 0 and 255 indicate black and white, the input image is transformed into a grayscale image. This image is then further altered to become a binary image, where each pixel has a value of either 0 or 1, as shown in Fig. 5(a). Black is represented by the value 0, and white by the value 1. A threshold with a value between 0 and 255 is used to achieve it. We set the threshold value at 200 value. A pixel over 200 value in the grayscale image will be given a value of 1; otherwise, the value is 0.

The binary image is further processed for erosion. Erosion [5] is a technique applied to eliminate unwanted pixels from the object's boundary, i.e., pixels that have a value of 1 but should contain a value of 0. First, it considers each pixel in the image, then its neighbors (kernel size determines the number of neighbors). The pixel only receives a value of 1 if all of its neighbors also have values of 1, otherwise, it receives a value of 0.

The noise-free image is further processed for dilation. Dilation [5] fills up the absent pixels, i.e., pixels that should have a value of 1 but have a value of 0. Every pixel in the image is first taken into account, followed by its neighbors (kernel size determines the number of neighbors); a pixel is given a value of 1 if at least one of its neighbors is also a 1.

Discovering every contour in the input image is essential for extracting the individual characters from the LP. Curves with the same hue or intensity that connect all the continuous points (along the boundary) are called contours. After locating each contour, we examine it individually and determine the size of each bounding rectangle, as shown in Fig. 5(b). Once we have the dimensions of the bounding rectangles, we adjust the parameters and filter the necessary rectangles that contain the required text.

$$W = range\{0, \frac{input\_length}{character\_count}\} \tag{6}$$

$$L = range\{\frac{W}{2}, 4 * (\frac{W}{5})\} \tag{7}$$

Using the above equations, we perform a dimension comparison. The rectangles accepted have width and length in the range specified. To achieve this, we

**Table 2.** The layout of the designed CNN.

| Type | Configuration | | | |
|------|------|------|------|------|
| Input | 220 × 70 × 1 rectified image | | | |
| Layer | Filter size | Kernel size | Stride size | Padding |
| Conv2D_1 | 64 | 3 × 3 | 1 × 1 | 1 × 1 |
| Batch_norm_1 | – | – | – | – |
| ReLU_1 | – | – | – | – |
| Max_pool_1 | – | 2 × 2 | 2 × 2 | 0 × 0 |
| Conv2D_2 | 128 | 3 × 3 | 1 × 1 | 1 × 1 |
| Batch_norm_2 | – | – | – | – |
| ReLU_2 | – | – | – | – |
| Max_pool_2 | – | 2 × 2 | 2 × 2 | 0 × 0 |
| Conv2D_3 | 256 | 3 × 3 | 1 × 1 | 1 × 1 |
| ReLU_3 | – | – | – | – |
| Conv2D_4 | 256 | 3 × 3 | 1 × 1 | 1 × 1 |
| Batch_norm_4 | – | – | – | – |
| ReLU_4 | – | – | – | – |
| Max_pooling_3 | – | 2 × 2 | 2 × 2 | 0 × 0 |
| Conv2D_5 | 512 | 3 × 3 | 1 × 1 | 1 × 1 |
| ReLU_5 | – | – | – | – |
| Conv2D_6 | 512 | 3 × 3 | 1 × 1 | 1 × 1 |
| Batch_norm_5 | – | – | – | – |
| ReLU_6 | – | – | – | – |
| Max_pool_4 | – | 2 × 2 | 2 × 2 | 0 × 0 |
| Dropout | Rate: 0.4 | | | |
| Flatten | – | – | – | – |
| Dense | Units: 128, Activation: ReLU | | | |
| Dense | Units: 36, Activation: Softmax | | | |

perform dimension comparison by accepting only rectangles that have width in a range of 0, (length of input)/(number of characters) and length in the range of (width of the input)/2, 4* (width of the input)/5. This process results in segmenting all the characters as binary images, as shown in Fig. 5(c).

**Recognition of Segmented Characters:** CNN, a trainable feature extractor, has recently achieved significant success in computer vision problems. The success of CNNs results from advancements in two technical areas: developing methods to prevent overfitting and creating more robust models [3,17]. CNNs are formed of artificial neurons with self-optimizing properties, making them capable

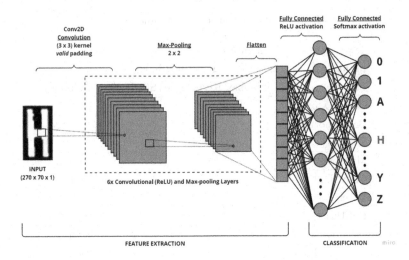

**Fig. 6.** Architecture of the proposed CNN

of extracting and classifying features from images more precisely than any other algorithm. Since the LP text consists of various font styles and sizes, we trained a more powerful deep network for this task. We want to give the model a more instinctive comprehension of the text or character. Among these fundamental characteristics lower-level text features like character labels and explicitly placed text pixels. We propose a Deep LPR CNN to accomplish this by training it on highly supervised text information at multiple levels, including segmentation of character regions, character labels, and text/non-text binary information. The additional supervised information provides the model with more specific textual features, enabling it to do tasks of high-level classification and low-level region segmentation. It allows our model to systematically recognize where and what the character is, which is crucial to make a reliable decision.

Table 2 and Fig. 6 display the detailed configuration and structure of the proposed CNN. The number of channels, stride, padding, and kernel sizes are similar to the VGGNET [26]. Other LP Recognition Tasks [12,13] have successfully applied these configurations.

## 4   Results and Discussion

The proposed ALPR paradigm is verified for effectiveness; thus, Tensorflow and Keras frameworks have been utilized to implement the model. Our system configuration for evaluation is as follows: Intel core 9th Gen i7 CPU, NVIDIA GeForce GTX 1650Ti with 4 GB memory, and RAM of 16 GB.

### 4.1   Data Sets Description

As per our work, a general data set for distorted LP images is unavailable. The use of robust DL algorithms in the smart recognition of distorted LP is hampered

**Fig. 7.** Data set samples of distorted LPs.

**Fig. 8.** Data set samples of characters of various fonts.

by the absence of enough images. To effectively train our custom YOLOv4, we created a data set of vehicles with deformed LP in different shooting angles and complex backgrounds, as shown in Fig. 7. The images were collected from google images and natural scenes. We collected 3000 images of vehicles with various LP styles and annotated them to train the model.

We have a data set of 37,623 images to train our CNN model. The data set includes letters (A–Z) and numbers (0–9) with 50+ unique fonts that are commonly found on various LP, as shown in Fig. 8. To make the model resistant to various oblique views, data augmentation methods, including random rotation and perspective transformations, were used. Therefore, each class of alphabet or digit contains 1045 images of size 28 × 28. We randomly select 33,861 character images for training and the remaining 3762 images for testing. Besides, Table 3 provides the comparative analysis of various data sets.

## 4.2   Result Analysis

The objective is to create a method that works well in several uncontrolled situations but simultaneously functions adequately in controlled ones (such as primarily frontal views). We have selected four online data sets: AOLP (RP), SSIG, and OpenALPR (EU and BR), which, as shown in Table 3, cover a wide range of scenarios. We have considered two variables: LP angles (frontal and oblique), as well as the separation between the vehicle and the camera (close

**Table 3.** Comparative analysis of various data sets.

| Data sets | LP angle | Images | Vehicle Dist |
|---|---|---|---|
| AOLP (Road Patrol) | Frontal + oblique | 611 | Close view |
| SSIG (test set) | Frontal | 804 | Medium, distant |
| OpenALPR (BR) | Frontal | 108 | Close view |
| OpenALPR (EU) | Frontal | 104 | Close view |
| Proposed data set | Oblique | 100 | All views |

**Table 4.** Performance analysis and comparison for multiple data sets.

| Methods | AOLP (RP) | SSIG test | OpenALPR | | Proposed data set |
|---|---|---|---|---|---|
| | | | EU | BR | |
| Proposed method (with no STN) | 83.11% | 82.01% | 92.88% | 89.71% | 70.67% |
| Proposed method (with STN) | **96.56%** | **89.55%** | 91.35% | 92.69% | **85.00%** |
| OpenALPR (See footnote 1) | 69.72% | 87.44% | **96.30%** | 85.96% | 75.32% |
| Sighthound (See footnote 2) | 83.47% | 81.46% | 83.33% | **94.73%** | 50.98% |
| Severo et al. [11] | – | 85.45% | – | – | – |
| Wang et al. [13] | 88.38% | – | – | – | – |
| Shen et al. [12] | 83.63% | – | – | – | – |
| G.S. Hsu et al. [6] | 85.70% | – | – | – | – |

view, intermediate view, distant view). Although these data sets cover various scenarios, a more general-purpose data set for challenging scenes is still a limitation. Thus as an additional contribution from our collected images, we have selected and manually annotated a set of 104 images that cover various challenging scenarios. The images contain substantial LP distortions but are still viewable to humans. A few images are shown in Fig. 7.

**Experimental Results:** This section expresses the experimental outcome analysis of the proposed ALPR mechanism and the comparison with other implemented methods. To testify to the overall performance of the presented model, we take the percentage of accurately identified LPs ($CL$) from the total number of testing LP images ($TL$). The recognition accuracy is given by

$$A = CL/TL \tag{8}$$

A point to note is that all the test data sets have been tested on the same network. No additional fine-tuning was performed to the network for a specific data set.

**Table 5.** mAP comparison of proposed YOLOv4.

| Models | mAP |
|---|---|
| Proposed YoloV4 | **90%** |
| YOLOv3 [22] | 89% |
| YOLOv2 [24] | 76.8% |

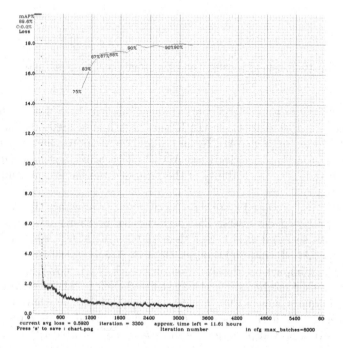

**Fig. 9.** Training performance of custom YOLOv4.

Table 4 indicates that the proposed method performs well with various data sets. Compared to other alternatives, it is superior on AOLP (RP) and SSIG Test data sets. The AOLP (RP) and SSIG Test data sets manifest the performance of 96.56% and 89.55% on the proposed method. The variation in performance on AOLP (RP) data set is approximately 27.0% for different approaches. Similarly, it is nearly 8.0% for the SSIG Test data set. Also, the error rate reduction due to the proposed method is 88.63% and 79.19%, respectively, compared to OpenALPR and Sighthound. Table 4 shows the comparison with other implemented systems. Our system has achieved recognition rates comparable to commercially available systems representing controlled scenes, where the LPs have frontal views and less complicated environments. Our system has achieved the best performance in AOLP RP and the proposed oblique LPs data sets.

Furthermore, the proposed ALPR method performance on the OpenALPR data set is inferior compared to other alternatives. The proposed ALPR app-

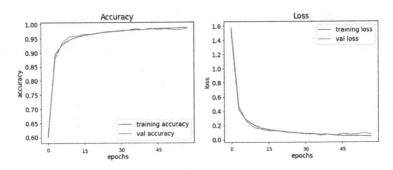

**Fig. 10.** Training accuracy and loss analysis of proposed model.

roach attains more than 90.0% performance but less than 4.95% and 2.04%, respectively, compared to OpenALPR and Sighthound methods. In addition, the proposed method, OpenALPR, and Singhhound approaches vary by 7.01%, 26.58%, and 13.27%, respectively, on AOLP (RP), SSIG Test, and OpenALPR data sets. It indicates the stability of the proposed mechanism in comparison to other alternatives.

Moreover, the proposed system presents superior outcomes than other mechanisms on proposed data sets. The performance of 85.0% is attained for the proposed data set, and it is better than 10.0% and 35.0%, respectively, compared to OpenALPR and Sighthound. Besides, it is essential to note that STN has a beneficial impact on identification outcomes. We remove the STN module from the proposed mechanism to demonstrate the effect. The recognition performance in oblique scenes of AOLP and the presented data sets have a significant gap, as seen in Table 4. This performance difference demonstrates how STN contributes to improved performance in identifying distorted LP.

Table 5 and Figure 9 illustrates the training performance of the custom Yolov4 model. The model achieved 90.0% mAP with 2800 iterations which outperformed the Yolov2 and Yolov3 used in [22,24]. Also, Fig. 10 indicates that the model is not overfitted on given input data. The continuous decrease in error trend is observed for the proposed model. Besides, character recognition performance is analyzed by a confusion matrix, and it is illustrated in Fig. 10. It is observed that the presented APLR method misclassified the 'O' and '0'.

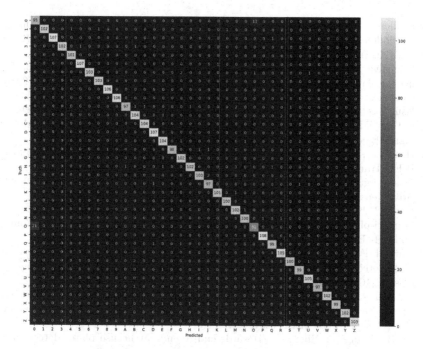

**Fig. 11.** Confusion matrix of the character recognition model per class.

# 5    Conclusion

This work demonstrated a comprehensive approach for ALPR in uncontrolled environments. Results indicate that the presented ALPR paradigm performs significantly better than the existing methods in challenging data sets with License Plates captured at severely oblique viewpoints. The use of the spatial transformer network, which aids in rectifying the distorted license plates, is the primary contribution of this work. This step helps the Recognition Network (Convolutional Neural Network) to understand the character patterns in a simplified way because it has to deal with far minimal distortion. Besides, we generated a complex data set by augmenting the images to detect license plates in skewed views. Currently, the system proposed can recognize the license plate number in English. For future work, we intend to enhance the current paradigm to recognize multilingual license plates written in the Devanagari language.

# References

1. Bochkovskiy, A., Wang, C.Y., Liao, H.Y.M.: YOLOv4: optimal speed and accuracy of object detection. arXiv preprint arXiv:2004.10934 (2020)
2. Bulan, O., Kozitsky, V., Ramesh, P., Shreve, M.: Segmentation-and annotation-free license plate recognition with deep localization and failure identification. IEEE Trans. Intell. Transp. Syst. **18**(9), 2351–2363 (2017)

3. Dhillon, A., Verma, G.K.: Convolutional neural network: a review of models, methodologies and applications to object detection. Progr. Artif. Intell. **9**(2), 85–112 (2020)
4. Gonçalves, G.R., da Silva, S.P.G., Menotti, D., Schwartz, W.R.: Benchmark for license plate character segmentation. J. Electron. Imaging **25**(5), 053034 (2016)
5. Gonzalez, R.C.: Digital Image Processing. Pearson Education India (2009)
6. Hsu, G.S., Ambikapathi, A., Chung, S.L., Su, C.P.: Robust license plate detection in the wild. In: 2017 14th IEEE International Conference on Advanced Video and Signal Based Surveillance (AVSS), pp. 1–6. IEEE (2017)
7. Hsu, G.S., Chen, J.C., Chung, Y.Z.: Application-oriented license plate recognition. IEEE Trans. Veh. Technol. **62**(2), 552–561 (2012)
8. Jaderberg, M., Simonyan, K., Zisserman, A., et al.: Spatial transformer networks. Adv. Neural Inf. Process. Syst. **28** (2015)
9. Kaur, P., Kumar, Y., Gupta, S.: Artificial intelligence techniques for the recognition of multi-plate multi-vehicle tracking systems: a systematic review. Arch. Comput. Methods Eng. **29**, 4897–4914 (2022)
10. Kurpiel, F.D., Minetto, R., Nassu, B.T.: Convolutional neural networks for license plate detection in images. In: 2017 IEEE International Conference on Image Processing (ICIP), pp. 3395–3399. IEEE (2017)
11. Laroca, R., et al.: A robust real-time automatic license plate recognition based on the yolo detector. In: 2018 International Joint Conference on Neural Networks (IJCNN), pp. 1–10. IEEE (2018)
12. Li, H., Wang, P., Shen, C.: Towards end-to-end car license plates detection and recognition with deep neural networks. CoRR abs/1709.08828 (2017)
13. Li, H., Shen, C.: Reading car license plates using deep convolutional neural networks and lstms. arXiv preprint arXiv:1601.05610 (2016)
14. Liu, W., et al.: SSD: single shot multibox detector. In: Leibe, B., Matas, J., Sebe, N., Welling, M. (eds.) ECCV 2016. LNCS, vol. 9905, pp. 21–37. Springer, Cham (2016). https://doi.org/10.1007/978-3-319-46448-0_2
15. Montazzolli, S., Jung, C.: Real-time Brazilian license plate detection and recognition using deep convolutional neural networks. In: 2017 30th SIBGRAPI Conference on Graphics, Patterns and Images (SIBGRAPI), pp. 55–62. IEEE (2017)
16. Ravirathinam, P., Patawari, A.: Automatic license plate recognition for Indian roads using faster-RCNN. In: 2019 11th International Conference on Advanced Computing (ICoAC), pp. 275–281. IEEE (2019)
17. Rawat, W., Wang, Z.: Deep convolutional neural networks for image classification: a comprehensive review. Neural Comput. **29**(9), 2352–2449 (2017)
18. Redmon, J., Divvala, S., Girshick, R., Farhadi, A.: You only look once: unified, real-time object detection. In: Proceedings of the IEEE Conference on Computer Vision and Pattern Recognition, pp. 779–788 (2016)
19. Redmon, J., Farhadi, A.: YOLO9000: better, faster, stronger. In: Proceedings of the IEEE Conference on Computer Vision and Pattern Recognition, pp. 7263–7271 (2017)
20. Redmon, J., Farhadi, A.: YOLOv3: an incremental improvement. arXiv preprint arXiv:1804.02767 (2018)
21. Ren, S., He, K., Girshick, R., Sun, J.: Faster R-CNN: towards real-time object detection with region proposal networks. Adv. Neural Inf. Process. Syst. **28** (2015)
22. Sahu, C.K., Pattnayak, S.B., Behera, S., Mohanty, M.R.: A comparative analysis of deep learning approach for automatic number plate recognition. In: 2020 Fourth International Conference on I-SMAC (IoT in Social, Mobile, Analytics and Cloud)(I-SMAC), pp. 932–937. IEEE (2020)

23. Saidani, T., Touati, Y.E.: A vehicle plate recognition system based on deep learning algorithms. Multimed. Tools Appl. **80**(30), 36237–36248 (2021). https://doi.org/10.1007/s11042-021-11233-z
24. Silva, S.M., Jung, C.R.: License plate detection and recognition in unconstrained scenarios. In: Proceedings of the European Conference on Computer Vision (ECCV), pp. 580–596 (2018)
25. Silva, S.M., Jung, C.R.: Real-time license plate detection and recognition using deep convolutional neural networks. J. Vis. Commun. Image Represent. **71**, 102773 (2020)
26. Simonyan, K., Zisserman, A.: Very deep convolutional networks for large-scale image recognition. arXiv preprint arXiv:1409.1556 (2014)
27. Xie, L., Ahmad, T., Jin, L., Liu, Y., Zhang, S.: A new CNN-based method for multi-directional car license plate detection. IEEE Trans. Intell. Transp. Syst. **19**(2), 507–517 (2018)

# Towards Railway Cable Infrastructure Protection: Turning Cross-Sectional Explorative Analytics to Answers

Moyahabo Rossett Mohlabeng[1] and Isaac O. Osunmakinde[2(✉)]

[1] School of Computing, College of Science, Engineering and Technology, University of South Africa, Pretoria 0003, South Africa
[2] Computer Science Department, College of Science, Engineering and Technology, Norfolk State University, Norfolk, VA 23504, USA
ioosunmakinde@nsu.edu

**Abstract.** The railway cable infrastructure has become a major concern worldwide due to attacks and failures causing substantial economic losses and destruction. This research seeks to rapidly detect cable threats and send timely alerts by integrating cross-sectional explorative analytics with a three-stage detection approach in a sensor network used on the railway cable infrastructure protection. Existing approaches have delayed gaps that are addressed in this paper. Experimental results using real-life sensors and publicly available data demonstrate that the proposed detection approach is sufficiently reliable, accurate and robust for detecting suspicious behavior on railway cable infrastructure. The results of this paper can be used as a reference guide to understand railway cable infrastructure protection gaps and enrich the literature on IoT (internet-of-things) in securing railway cable infrastructure.

**Keywords:** Railway · Cable · Detection · Sensor networks · Explorative · Data analytics · Infrastructure · Sagging · Theft · IoT

## 1 Introduction

Recent societies depend on the technologies and capabilities of critical infrastructure, such as transport systems, health care, power systems, telecommunication and others, to perform many functions. The attacks and vandalism happened in the form of worldwide. Hence, these attacks and vandalism caused enormous damage, financial loss and failure of railway cable infrastructure, resulting in the many deaths of people, accidents and excessive financial loss [1].

Critical railway infrastructure still lacks adequate protection control mechanisms to defend it against failure, destructions and security threats. Countries have suffered a huge rate of cable theft over several years, which had a negative impact on the countries' economy and globally. Thus, cable theft led to enormous financial loss and attacks on critical railway infrastructure [1]. According to [2, 3], the incidence of criminal actions and attacks threatening cable infrastructure increased in continent between 2013/2014

A. R. Molla et al. (Eds.): ICDCIT 2023, LNCS 13776, pp. 270–289, 2023.
https://doi.org/10.1007/978-3-031-24848-1_19

and 2014/2015, from 4 182 to 78 171 incidents, causing destruction and damage to the cable infrastructure. During 2016/2017 attacks started becoming more dangerous and started targeting all sizes of the railway infrastructure, varying from small to large stations. These attacks destroyed large amounts of critical cable infrastructure, causing the death of people and triggering chaos. Figure 1 shows the statistics of cable crime cases in several countries in 10 years.

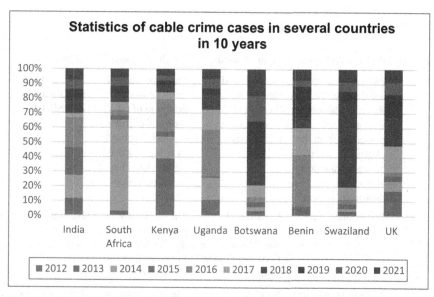

**Fig. 1.** Statistics of cable crime cases in several countries in 10 years, adapted from [2, 3].

Figure 1 illustrates an increase in cable crime in several countries between 2020/2021 and 2021/2022, which declined in 2019/2020, but started increasing again in 2020/2021, reaching 7 793 incidents. In continent, South Africa, Kenya, Uganda, Benin, Botswana, UK, India, and Swaziland contributed a great percentage of the high incidence of stolen, vandalised or damaged cable infrastructure. These incidents and attacks have a huge negative effect on railway cable infrastructure [2, 3].

In [4], it was indicated that the existing techniques address attacks and destruction to critical railway infrastructure through normal control measures and do not cover the broader boundaries of the railway environment. These methods entail vision-based detection, lidar-based detection, WSN. Yet, these approaches have weaknesses in defending critical railway infrastructure. The weaknesses in the existing techniques comprise of high false alarm rates, an unstable detection rate, requires more internal hardware resources, requires more monitoring, inadequate coverage, inability to detect cable sagging and inability to give swift alerts.

To deal with weaknesses in the existing approaches, this paper outlines a cross-sectional explorative analytics detection approach that addresses the gaps experienced by existing approaches. This approach outlines three (3) stages: Stage 1: Visual analysis on cable attributes' frequencies, Stage 2: E-M clustering process to find outliers and Stage 3: Descriptive analysis in batches on each railway sensor spot.

## 1.1  Research Questions and Contributions

This research addresses the above problem by formulating the following research question:

**Table 1.** Summary of related works on infrastructure protection methods.

| Citations | Problem addressed | Methods used | Result obtained | Limitations |
|---|---|---|---|---|
| [5] | Passenger safety accidents in the Railway Platform such as fallen passenger from the platform, caught between train doors and occurred disastrous fire etc. | Vision-based Detection: Stereo and thermal vision technology | The experimental result shows that detection of train state and object is conducted robustly by using proposed stereo-vision based object detection algorithm | Inability to deal with threat immediately |
| [6] | Power line safety distance detection | Lidar-based Detection: Small multi-rotor UAV and LOAM algorithm | The results achieved power line safety distance detection functions and cost reduction | No emphasis on real-time performance |
| [7] | The use of Wireless Sensor Network (WSN) to monitor the vegetation in the various stretches of the transmission and distribution lines, making this information available to the network operators | Wireless Sensor Network: XBee-PRO S2 radio frequency module | The experimental results demonstrated that the system is viable and cost-effective in monitoring of transmission lines | No enough work on rapid decision support system |

How can a strategy be established to monitor railway cable infrastructure protection (RCIP) threats proactively? Recently, organisations have experienced the necessity to safeguard railway critical infrastructure against attacks and destructions. This paper makes the following major contributions:

- Development of a systematic rapid decision support system to swiftly secure railway cable infrastructure based on explorative analytics in preliminary, intermediate and advanced stages, which address the time-lag of the existing methods and enrich the literature of IoT (internet-of-things).

- Early knowledge generation at preliminary stage can give railway security personnel advance notice of suspicious cable threat and their locations where crime is likely to occur.
- Detailed experimental evaluations of the proposed smart DSS performed on real-life cable behaviour from multiple sensors and cable behaviour analytics benchmarked using publicly available sensor data.

This paper is organised in the following order: Sect. 2 describes a theory of expectation maximisation (EM) clustering, Sect. 3 explains the proposed methodology, Sect. 4 provides evaluations to simulate and conduct explorative analytics on the instability of railway cable behaviours across sensor locations and determine possible cable sagging or theft and the concluding remarks are outlined in Sect. 5.

## 2 Introduction

### 2.1 Railway Infrastructure Protection Overview

According to [4], Railway infrastructure contributes significantly in people's lives worldwide. This infrastructure has several mechanisms, such as railway lines, signalling state-of-art, power systems and others. Railway infrastructure empowers people to travel throughout the world, as shared by [4], but this infrastructure requires to ensure the protection of commuters against human error, destruction and other threats that could cause loss of lives. The types of classes currently used in safeguarding railway cable infrastructure are vision-based detection and WSN [4].

### 2.2 State-of-the-Art

Several literature papers have examined protection techniques for railway cable infrastructure. Countless downsides in the current works have been discovered. In one of the research papers [7], the authors have cited the current protection techniques of the railway cable infrastructure. The summary of related infrastructure protection methods in terms of the problem addressed, method used, result obtained, and their limitations are presented in Table 1.

Although state-of-the-art techniques on infrastructure protection have been investigated in the literature. However, no enough work has considered early decision making with IoT systems on cable behaviours as a quick warning for the railway security personnel. This research develops a rapid DSS incrementally inclined on IoT to address the time-lag of existing monitoring systems.

### 2.3 Expectation Maximisation Clustering

Expectation maximisation (EM) clustering can be defined as a repetitive technique for finding maximum-likelihood estimation of parameters where the data are not complete or rely on unobserved hidden variables [8].

The EM clustering algorithm can be described as a repetitive procedure and consists of four (4) steps that include Step 1: Initialisation, Step 2: E-Step, Step 3: M-Step and

Step 4: Convergence step. Step 1: Initialisation indicates the EM parameters such as W is the current number of Gaussians; $\sigma$ is represented as the standard deviation; $\theta^0$ is the estimate at 0th iteration; $\mu$ can be defined as the mean; t can be indicated by the number of the iteration. The Gaussian can be defined as a parameter that has a bell-shaped curve and describes the distribution of the value of a variable. $\theta$ is referred to as the mode which highlights is the most regular score in the sensor dataset. Step 1: Initialisation of EM can be outlined by Eq. (1) [8]:

$$\theta^t = (\mu_1^t, \mu_2^t, \ldots, \mu_w^t) \tag{1}$$

According to [8], Step 2: E-Step indicates the estimate and expected values for a hidden variable. In Eq. (2), $H(V_{aw})$ is the expected value of the hidden variables.

$$H(V_{aw}) = \frac{\exp[-\frac{(y_a - \mu_w^{(t)})^2}{2\sigma^2}]}{\sum_{b=1}^{w} \exp[-\frac{(y_a - \mu_w^{(t)})^2}{2\sigma^2}]} \tag{2}$$

In Eq. (3), Step 3: M-step indicates the new estimate of the parameters is computed. Step 4: Convergence step indicates the stop condition checkup and if $||\theta^{(r+1)} - \theta^{(r)}|| < \varepsilon$ then stop; else, return to Step 2 [8].

$$\mu_w^{r+1} = \frac{\sum_{a=1}^{n} E(V_{aw}) y_a}{\sum_{a=1}^{n} E(V_{aw})} \tag{3}$$

## 3 Proposed Smart Cross-Sectional Explorative Analytics for RCIP

This paper presents a proposed smart cross-sectional explorative analytics detection approach for RCIP as methodology, which addresses the gaps found in the current methods. This approach explores powerful capabilities, which include visual analysis on cable attributes' frequencies, E-M clustering process to find outliers and descriptive analysis in batches on each railway sensor spot. The next section will describe the problem definition and demonstrate the real-world railway cable infrastructure.

### 3.1 Problem Definition

The reality here requires that railway critical infrastructure protection (RCIP) must have more secured physical, operational, and smart infrastructures, which can assist in preventing and minimizing railway cable vulnerabilities. This refers to cable threats such as theft, sagging, vandalism, damage, etc. on railway cable lines. Figure 2 shows the real-world railway infrastructure.

The real-world railway infrastructure consists of the cables which require to be secured against attacks or threats. This problem is strongly supported in [9]. Hence, the need for RCIPs to continually adopt a smart explorative analytics detection approach is evident, which would assist in creating early warning to securities at the operational level responsible for protecting railway cables.

**Fig. 2.** Real-world railway cable infrastructure.

## 3.2   Smart Cross-Sectional Explorative Analytics Detection Model

### Cable Sensors Description

The Libelium smart security Waspmote plug and sense passive infra-red (PIR) presence sensor is used to observe continuous vibration on the cable and high voltage or temperature. Its technical information is provided as follows: height of 22 mm, a diameter of 20.2 mm, a consumption of 170 $\mu$A and a circuit stability time of 30 s. The sensor is focused straight to the point where the railway cable infrastructure needs to be protected. Figure 3 shows PIR presence sensor.

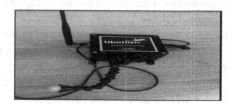

**Fig. 3.**   PIR presence sensor.

### Smart Cross-Sectional Analytics Model

Cross-sectional data can be defined as sensor data captured under situations where time is irrelevant. Explorative analysis of cross-sectional sensors cable data includes Stage 1: Visual analysis on cable attributes' frequencies, Stage 2: E-M clustering process to find outliers and Stage 3: Descriptive analysis in batches on each railway sensor spot. Figure 4 illustrates the explorative analysis of cross-sectional sensors cable data.

**Stage** 1: Visual analysis shows a frequency distributions of each sensor attributes and how sensors function situated at various locations such as sensor 1 mounted on location 1, sensor 2 mounted on location 2 and sensor 3 mounted on location 3. A frequency distribution can be defined as a count of the number of occurrences of a particular quantity of sensors located on the railway spots. The visual analysis provides a preliminary understanding of the behaviour of the cable based on the distributions of data attributes, eg. Revealing concentration on temperature values, amplitude, displacement, and variations of other attributes distributions. Quick deductions can be made at this

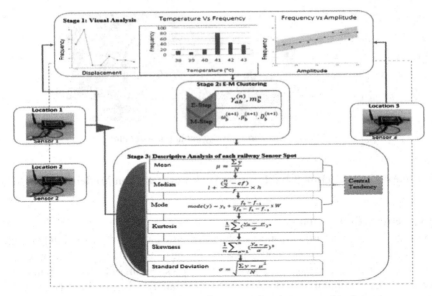

**Fig. 4.** Explorative analysis of cross-sectional sensors cable data.

point by the railway operators based on this preliminary outputs. Formalised Stages 2 and 3 are required to substantiate Stage 1.

**Stage** 2: E-M clustering operates on a set of cable data points captured from each sensor, partition them into segments, which contain similar data points. E-Step provides the estimate and expected values for sensors data, while the M-Step provides calculations of a new estimate of the parameters of the hidden variable for sensors data. Once the clustering model is built, a newly observed cable data is predicted and classified into a cluster or appears as an outlier. This research model this result as partial inference for the railway operatives.

**Stage** 3: Quantitative analysis provides statistical views of Stage 2 on the detection of outliers and relations among sensor attributes of each railway sensor spot. The descriptive analysis includes central tendency, standard deviation, mode of distribution, kurtosis, skewness and a mean of each railway sensor spot. The standard deviation on sensor observations helps in understanding how the sensor data is distributed to various locations of data points. Hence, since Stage 3 provides quantitative information on Stage 2 results, a conclusive inference on normal or suspicious cable behaviour is made on a new instance of cable parameters observed.

### Central Tendency: Skewness and Kurtosis

This section focuses on the descriptive analytics with central tendency such as skewness and kurtosis, where skewness refers to the positive or negative that deviates from the normal distribution on sensor datasets. On the other hand, Kurtosis is defined as the measurement of the flatness or peakedness of a distribution of sensor attributes. #Base1 and #Base2 highlights skewness and kurtosis conditions respectively.

**ALGORITHM 1: #Base 1 –**

| **Skewness_Sensor [1:2:3] (Voltage: Displacement): adapted from [10]** |
| --- |
| If Skewness_Sensor [1:2:3] (Voltage: Displacement) = 0 Then<br>    This implies that the railway cable is balanced and not skewed.<br>ElseIf Skewness_Sensor [1:2:3] (Voltage: Displacement) < -1 or > 1 Then<br>    This suggests that the railway cable is highly skewed.<br>ElseIf Skewness_Sensor [1:2:3] (Voltage: Displacement) (> -1 and < -0.5) or (1 and < 0.5)<br>    Then<br>    This infers that the railway cable is moderately skewed.<br>ElseIf Skewness_Sensor [1:2:3] (Voltage: Displacement) > (-0.5 and 0.5) Then<br>    This deduces that the railway cable is closer to normal.<br>EndIf |

**ALGORITHM 2: #Base 2 –**

| **Kurtosis_Sensor [1:2:3] (Voltage: Displacement): adapted from [10]** |
| --- |
| If Kurtosis_Sensor [1:2:3] (Voltage: Displacement) = 0 or 3 Then<br>    This implies that the railway cable is normal.<br>ElseIf Kurtosis_Sensor [1:2:3] (Voltage: Displacement) <3 Then<br>    This suggests railway cable is in a good state.<br>ElseIf Kurtosis_Sensor [1:2:3] (Voltage: Displacement) >3 Then<br>    This infers railway cable is in a bad state.<br>EndIf |

# 4   Experimental Evaluations

## 4.1   Experimental Setup

The experimental prototype setup in Fig. 5 was conducted to simulate the railway cable infrastructure environment. In this experimental prototype, three sensors were set up, configured and mounted at different locations connected to a base station. The sensor devices read data and transmit the signal to the base station through a wireless connection. In Fig. 5, sensor 3, mounted at location 3, illustrates the cable cut, while sensor 2, placed at location 2, demonstrates cable sag and sensor 1, positioned at location 1, signifies the normal cable state. For the purpose of testing, the cable cut and sag were deliberately introduced. The data from all sensor devices are captured and immediately analysed at the base station.

**Fig. 5.** Experimental prototype setup as railway cable infrastructure.

The data samples were collected from multiple sensors located at simulated hotspots. SPSS and WEKA statistical tools have been designed for training and testing sensor data.

## 4.2    Experiment 1: Real-Life Cable Behaviour from Multiple Sensors

The objective here is to conduct explorative analytics on the instability of railway cable behaviours across sensor location and determine possible cable sagging or theft.

### Stage 1: Cross-Sectional Visual Analytics

To assess the hypothesis, stage 1 shows visual analysis results obtained from the sensors' data with frequency distributions of the sensors' parameters - temperature, amplitude, voltage and displacement. Figure 6(a)–(d) generated from SPSS show the frequency distributions obtained from the railway cables to reveal their behaviour.

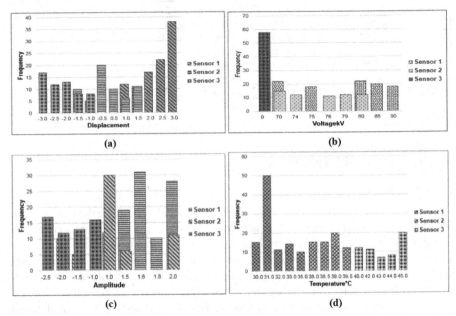

**Fig. 6.** (a) Displacement from three sensors. (b) Voltage from three sensors. (c) Amplitude from three sensors. (d) Temperature from three sensors.

### Decision for Security Agents

- In location 1, Sensor 1 appears to be highly displaced on the railway cable. These are indications of cable shift or loose cable, which raises the suspicion of cable sagging, according to the literature [4]. The above null hypothesis is rejected, which implies that the cable behaviour is unstable across the sensors. Hence, cable sagging is suspected in location 1, implying that railway security agents do not need to rush to the affected site, but a higher decision support system (DSS) stage is required.
- Sensor 2 indicates mild temperature on the railway cable. These are signs temperature indicate normal cable behaviour supported by the literature [4]. Thus, normal cable behaviour is suspected in location 2, implying that railway security agents should not respond to the affected site, but a higher DSS stage is vital.

**Table 2.** Preliminary decision support systems.

| | |
|---|---|
| Findings on displacement:<br>visual inspection on Fig. 6(a) | • Displacement distributions are expressed as follows:<br>Sensor 1 values > Sensor 2 values > Sensor 3 values |
| | • Central area for most frequent displacement values is expressed as: Sensor 1 (3) > Sensor 2 (−0.5) > Sensor 3 (−3) |
| | • Ranges of the displacement values are expressed as:<br>Sensor 2 (−1.5..1.5) > Sensor 1 (1..3) = Sensor 3 (−3..−1) |
| Findings on Voltages<br>In Fig. 6(b) visual inspection: | • Voltage distributions are observed as follows:<br>Sensor 1 values > = Sensor 2 values > Sensor 3 values |
| | • Central area is outlined for most regular voltage values as: Sensor 1 (65 and 80) > = Sensor 2 (70) > Sensor 3 (0) |
| | • Ranges of voltage values are presented as:<br>Sensor 1 (65..90) > = Sensor 2 (70..80) > Sensor 3 (0..0) |
| Findings on Amplitudes<br>From visual inspection on Fig. 6(c) | • One can extract amplitude distributions as follows:<br>Sensor 1 values > = Sensor 2 values > = Sensor 3 values |
| | • For central area for most frequent amplitude values:<br>Sensor 1 (1.6) > Sensor 2 (1) > Sensor 3 (−2.5) |
| | • The ranges of the amplitude values are defined as follows: Sensor 2 (−2..2) > Sensor 1 (1..2) > Sensor 3 (−2.5..−1) |
| Findings on Temperature<br>Figure 6(d) illustrates the visual inspections of temperature as follows: | • Temperature distributions are specified as follows:<br>Sensor 1 values < Sensor 2 values < Sensor 3 values |

(continued)

**Table 2.** (*continued*)

| | |
|---|---|
| | • For the central area for most frequent temperature values: Sensor 1 (31) < Sensor 2 (39) < Sensor 3 (45) |
| | • Temperature values' ranges are shown as: Sensor 2 (38..39.5) < Sensor 1 (30..35) = Sensor 3 (40..45) |
| Preliminary decision for security agents | • Cable sagging is suspected in location 1 <br> • Cable theft is suspected in location 3 |

- Sensor 3 shows zero voltage, meaning there is no signal on the railway cable, and this is confirmed by the test in Fig. 5. In [4], the suspected stolen cables are shown through warnings of no power and no signal transmission on the railway cable. Thus, cable theft is suspected in location 3 and railway security agents need to respond to the affected site, but a higher DSS stage is essential.

**Stage-2: Cross-Sectional EM Clustering Complementing Stage 1**

Stage 2 illustrates the cross-sectional EM clustering complementing the cable behaviour monitoring in stage 1. The results from the clusters enable a partial deduction for the railway operatives. Figs. 7(a)–(b)–Figs. 9(a)–(b) show the EM and K-means [5] clustering (C1 = cluster 1, C2 = cluster 2 and C3 = cluster 3) results of voltage and displacement attributes for three sensors generated from SPSS and Orange data mining application.

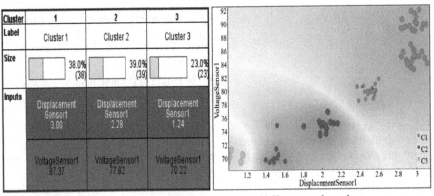

(a) Quantitative – EM clustering          (b) K-means clustering

**Fig. 7.** Sensor 1 displacement and voltage.

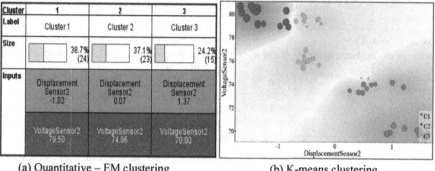

(a) Quantitative – EM clustering          (b) K-means clustering

**Fig. 8.** Sensor 2 displacement and voltage.

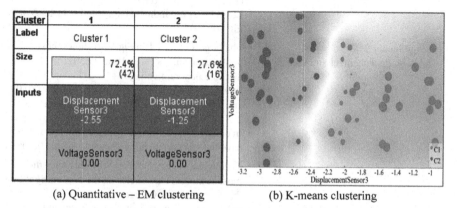

(a) Quantitative – EM clustering          (b) K-means clustering

**Fig. 9.** Sensor 3 displacement and voltage.

### Stage-3: Cross-Sectional Descriptive Analytics Complementing Stage 2

The descriptive analytics measures consist of popular central tendency such as mean, median, mode, standard deviation, skewness and kurtosis, used to understand the behaviour of sensor attributes including voltage, displacement, temperature, and amplitude over cables. The objective here is to answer the following research questions using the results obtained.

*Research Questions*

Which railway location should receive the highest priority based on the cable behaviours? Are there positive or negative correlations between any two locations on the same attribute to measure continuity?
To determine consistency, which location is reliable on the railway cable?
Is there any location that shows a stable state on the railway cable?

**Table 3.** Intermediate decision support systems.

| Location/Sensor 1: Findings on displacement and voltage | • In quantitative Fig. 7(a), one can see the dominating values (3 and 87.37) for displacement and voltage in cluster 1<br>• Minimal values (2.28 and 77.82) for displacement and voltage appear in cluster 2 in Fig. 7(a)<br>• In cluster 3, it appears that displacement and voltage have the least dominant values (1.24 and 70.22) |
|---|---|
| | • In Fig. 7(b), the most dominant blue circles are shown on the top right for displacement and voltage in cluster 1<br>• It appears that cluster 2 shows the moderate dominant of red circles<br>• Less circles are indicated in green in cluster 3 of Fig. 7(b) |
| Location/Sensor 2: Findings on displacement and voltage | • The largest cluster 1 in Fig. 8(a) shows most dominating values (−1.02 and 79.50) for displacement and voltage<br>• In Fig. 8(a), one can see the moderate values (0.07 and 74.96) for displacement and voltage in cluster 2<br>• In cluster 3, it seems that displacement and voltage have the least dominating values (1.37 and 70.00) |
| | • The largest cluster 1 in Fig. 8(b) illustrates the most dominant blue circles on the bottom right for displacement and voltage<br>• In cluster 2, it seems that displacement and voltage have the least dominant red circles on the top left of Fig. 8(b)<br>• In Fig. 8(b), one can see the moderate dominant green circles on the bottom left towards the top right for displacement and voltage in cluster 3 |
| Location/Sensor 3: Findings on displacement and voltage | • In Fig. 9(a), one can see the dominating values (−2.55 and 0) for displacement and voltage in cluster 1. This shows zero (0) voltage, which is deliberately demonstrated in Fig. 5<br>• It appears that displacement and voltage have the least dominant values (−1.25 and 0) in cluster 2 in Fig. 9(a) |
| | • In Fig. 9(b), one can point the most dominating displacement and voltage circles in blue in cluster 1, on the other hand, cluster 2 shows the least red circles |

*(continued)*

**Table 3.** (*continued*)

| Intermediate decision for security agents | • In location 1, since displacement and voltage values are dominating with 3 and 87.37 in cluster 1 of Fig. 7(a). It is also consistent with Fig. 7(b), most dominant blue circles for displacement and voltage in Cluster 1, cable sagging is then suspected, supported in [4]. Hence, this complements the preliminary decision in Table 2, but a confirmatory DSS stage is still required<br>• Figure 8(b) illustrates the most dominant blue circles for displacement and voltage in cluster 1 in location 2. Consistently, one can see most dominating values (−1.02 and 79.50) for displacement and voltage in cluster 1 in Fig. 8(a). Complimentary to Table 2, a normal cable behaviour is suspected and this is supported in [4], but additional DSS stage may be required<br>• In location 3, since dominating and minority clusters 1 and 2 respectively indicate zero voltages in Fig. 9(a) and also consistent with Fig. 9(b), cable theft/vandalism is suspected according to the literature [4]. This is deliberately demonstrated in Fig. 5, the security team should rush to the affected site but an advanced DSS may be necessary |
|---|---|

Figures 10 and 11 show the descriptive analytics results from SPSS implementation for advanced decision making in Table 4 to guide railway security agents.

| Descriptive Statistics | | | | | | | | |
|---|---|---|---|---|---|---|---|---|
| | Minimum | Maximum | Mean | Std. Deviation | Skewness | Std. Error | Kurtosis | Std. Error |
| VoltageSensor1 | 70 | 90 | 79.70 | 7.065 | .020 | .241 | -1.275 | .478 |
| VoltageSensor2 | 70 | 80 | 75.52 | 3.784 | -.303 | .304 | -1.339 | .599 |
| VoltageSensor3 | 0 | 1 | .02 | .131 | 7.616 | .314 | 56.000 | .618 |
| DisplacementSensor1 | 1.0 | 3.0 | 2.315 | .6987 | -.647 | .241 | -.873 | .478 |
| DisplacementSensor2 | -1.5 | 1.5 | -.040 | 1.0374 | .168 | .304 | -1.274 | .599 |
| DisplacementSensor3 | -3.0 | -1.0 | -2.190 | .6998 | .359 | .314 | -1.109 | .618 |
| TemperatureSensor1 | 30.0 | 35.0 | 31.640 | 1.4250 | 1.196 | .241 | .653 | .478 |
| TemperatureSensor2 | 38.0 | 39.5 | 38.734 | .5335 | -.039 | .304 | -1.226 | .599 |
| TemperatureSensor3 | 40.0 | 45.0 | 43.017 | 1.9056 | -.467 | .314 | -1.209 | .618 |

**Fig. 10.** Comprehensive cross-sectional descriptive analytics.

| Correlations | | | | | | | | | |
|---|---|---|---|---|---|---|---|---|---|
| | Displacement Sensor1 | Displacement Sensor2 | Displacement Sensor3 | Voltage Sensor1 | Voltage Sensor2 | Voltage Sensor3 | Temperature Sensor1 | Temperature Sensor2 | Temperature Sensor3 |
| DisplacementSensor1 | 1 | -.849** | -.826** | .945** | .832** | -.183 | -.915** | -.765** | -.761** |
| DisplacementSensor2 | -.849** | 1 | .944** | -.915** | -.955** | .223 | .710** | .943** | .906** |
| DisplacementSensor3 | -.826** | .944** | 1 | -.918** | -.949** | .227 | .690** | .957** | .897** |
| VoltageSensor1 | .945** | -.915** | -.918** | 1 | .899** | -.158 | -.849** | -.898** | -.925** |
| VoltageSensor2 | .832** | -.955** | -.949** | .899** | 1 | -.218 | -.645** | -.930** | -.895** |
| VoltageSensor3 | -.183 | .223 | .227 | -.158 | -.218 | 1 | .078 | .214 | .139 |
| TemperatureSensor1 | -.915** | .710** | .690** | -.849** | -.645** | .078 | 1 | .784** | .818** |
| TemperatureSensor2 | -.765** | .943** | .957** | -.898** | -.930** | .214 | .784** | 1 | .926** |
| TemperatureSensor3 | -.761** | .906** | .897** | -.925** | -.895** | .139 | .818** | .926** | 1 |

IBM SPSS

**Fig. 11.** Correlational descriptive analytics for sensors 1, 2 and 3.

**Table 4.** Advanced decision support systems.

| | |
|---|---|
| Standard deviation on Voltage locations/sensors 1–3 | • On Fig. 10, location 1 indicates the highest standard deviation with the values of 7.065<br>• This suggests that the highest priority for security agents in location 1 [answers question 1] |
| Correlations of Temperature between locations/sensors 1 and 2 | • On Fig. 11, the result of the correlation between temperature in locations 1 and 2 shows: Correlation (TemperatureSensor1: TemperatureSensor2) = 0.784<br>• This correlation infers that there is stronger positive continuity in locations 1 and 2 [answers question 2] |
| Skewness on displacement locations/sensors 1–3 | • On Fig. 10, location 2 indicates values 0.168<br>• Algorithm 1 infers that there is reliability in location 2 [answers question 3] |
| Kurtosis on Displacement locations/sensors 1–3 | • On Fig. 10, since location 2 shows values of −1.274<br>• Algorithm 2 implies a stable state in location 2 [answers question 4] |

## 4.3 Experiment 2: Cable Behaviour Analytics Using Publicly Available Sensors Data

### Stage-3: Cross-Sectional Descriptive Analytics

Stage 3 is only considered in this benchmark due to space. The central tendency such as mean, median, mode, standard deviation, skewness and kurtosis are descriptive analytics common measures used for the behaviour of sensor attributes temperature, current and thermal conductor. Here are research questions necessary to address the aim using the results attained.

*Research Questions*

1. Which substation should be given the lowest priority based on the cable behaviours?
2. Are there positive or negative correlations amongst any two substations on the same attribute to quantify power stability?
3. Is there continuous power flowing through the cable in the substation?
4. Which substation shows a dependable power that passes through the cable?

Figures 12 and 13 show the descriptive analytics results from SPSS implementation for advanced decision making in Table 5 to guide railway security agents. Using publicly available sensor data, three (3) locations have been selected, including Rise Carr substation (RCD), Darlington Melrose (DM), and Rise Carr Ianson (RCI).

| Descriptive Statistics | | | | | | | | |
|---|---|---|---|---|---|---|---|---|
| | Minimum | Maximum | Mean | Std. Deviation | Skewness | Std. Error | Kurtosis | Std. Error |
| Temp(°C)-DM | ,0 | 14,7 | 4,283 | 6,3906 | ,825 | ,012 | -1,311 | ,024 |
| Temp(°C)-RCI | ,0 | 17,6 | 5,218 | 7,8844 | ,850 | ,012 | -1,277 | ,024 |
| Temp(°C)-RCD | ,0 | 14,8 | 4,319 | 6,5273 | ,852 | ,012 | -1,269 | ,024 |
| Current (A)-DM | 18,5 | 382,8 | 109,750 | 46,4550 | ,768 | ,012 | ,039 | ,024 |
| Current (A)-RCI | ,0 | 186,1 | 33,367 | 53,0851 | 1,147 | ,012 | -,343 | ,024 |
| Current (A)-RCD | 0 | 87 | 15,10 | 24,175 | 1,162 | ,012 | -,344 | ,024 |
| ThermalCon-DM | ,00 | 1,58 | ,0830 | ,32323 | 3,653 | ,012 | 11,400 | ,024 |
| ThermalCon-RCI | ,00 | ,54 | ,0153 | ,08614 | 5,475 | ,012 | 28,024 | ,024 |
| ThermalCon-RCD | ,00 | ,54 | ,0153 | ,08614 | 5,475 | ,012 | 28,024 | ,024 |

IBM SPSS

**Fig. 12.** Comprehensive cross-sectional descriptive analytics using publicly available data.

| Correlations | | | | | | | | | |
|---|---|---|---|---|---|---|---|---|---|
| | Temp(°C)-DM | Temp(°C)-RCI | Temp(°C)-RCD | Current (A)-DM | Current (A)-RCI | Current (A)-RCD | ThermalCon-DM | ThermalCon-RCI | ThermalCon-RCD |
| Temp(°C)-DM | 1 | .889" | .890" | .001 | .840" | .832" | .382" | .238" | .238" |
| Temp(°C)-RCI | .889" | 1 | .999" | -.009 | .948" | .943" | .343" | .268" | .268" |
| Temp(°C)-RCD | .890" | .999" | 1 | -.010' | .947" | .941" | .343" | .268" | .268" |
| Current (A)-DM | .001 | -.009 | -.010' | 1 | .063" | .056" | .001 | -.002 | -.002 |
| Current (A)-RCI | .840" | .948" | .947" | .063" | 1 | .988" | .325" | .254" | .254" |
| Current (A)-RCD | .832" | .943" | .941" | .056" | .988" | 1 | .322" | .252" | .252" |
| ThermalCon-DM | .382" | .343" | .343" | .001 | .325" | .322" | 1 | .641" | .641" |
| ThermalCon-RCI | .238" | .268" | .268" | -.002 | .254" | .252" | .641" | 1 | 1.000" |
| ThermalCon-RCD | .238" | .268" | .268" | -.002 | .254" | .252" | .641" | 1.000" | 1 |

IBM SPSS

**Fig. 13.** Correlational descriptive analytics for three (3) locations (RCD, DM and RCI).

## 4.4 Performance Evaluations

This section explores the performance of the smart explorative analytics. The experiments were conducted to simulate the railway cable infrastructure environment.
*Time Saved through 3-stage Early Decision Making*
Time is an important resource to address cable theft, vandalism, etc. This smart explorative analytic can help to automate and operationalize daily monitoring of railway cables in order to increase the speed with which decisions are made, thereby saving time. Hence, the time saved in daily railway operations as a result of 3-stage early decision making is measured as shown in Fig. 14. Early knowledge generation at preliminary

**Table 5.** Advanced decision support systems.

| | |
|---|---|
| Standard deviation on temperature Rise Carr substation (RCD), Darlington Melrose (DM) substation and Rise Carr Ianson (RCI) | • On Fig. 12, the lowest standard deviation is found in Darlington Melrose substation with value of 6.3906<br>• This implies Darlington Melrose substation receives the lowest priority for security agents [answers question 1] |
| Correlations of current (A) between Rise Carr substation (RCD) and Darlington Melrose substation | • On Fig. 13, the result of the correlation between current in Rise Carr substation and Darlington Melrose substation shows: Correlation (Current_Rise Carr substation: Current_Darlington Melrose substation) = 0.56<br>• This correlation infers that there is positive continuity in Rise Carr substation and Darlington Melrose substation [answers question 2] |
| Correlations of current (A) between Rise Rise Carr Ianson (RCI) and Darlington Melrose substation | • On Fig. 13, the result of the correlation between current in Rise Rise Carr Ianson substation and Darlington Melrose substation shows: Correlation (Current_ Rise Carr Ianson substation: Current_Darlington Melrose substation) = 0.63<br>• This correlation infers that there are signs of continuous power flow in Rise Carr Ianson substation and Darlington Melrose substation [answers question 3] |
| Correlations of current (A) between Rise Carr substation (RCD) and Rise Carr Ianson (RCI) substation | • On Fig. 13, the result of the correlation between current in Rise Carr substation and Rise Carr Ianson substation shows: Correlation (Current_Rise Carr substation: Current_ Rise Carr Ianson (RCI) = 0.988<br>• This correlation suggests that Rise Carr substation and Rise Carr Ianson has a dependable power [answers question 4] |

stage 1 can give railway security personnel advance notice of suspicious cable threat and their locations where crime is likely to occur.

Figure 14 shows the data sizes of sensor data that were obtained from an experimental prototype setup used at three (3) locations in railway cable infrastructure using PowerBI tool. Three (3) BARS depict sensor data sizes for three (3) stages at three (3) locations. In order to save time, three stages should be accomplished depending on the central processing machine, i.e. high-spec machines yield faster processing response times, while low-spec machines yield slower processing responses. Since no other activity or signal is performed besides the responses, all stages receive stationary data sizes.

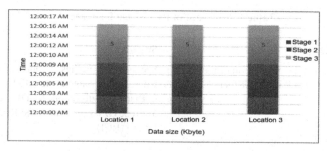

**Fig. 14.** Result of 3-stage early decision making.

During stage 1, time is measured from the start of the first abnormal event to the end of the third abnormal event, which takes three seconds and produces 1Kbyte of data. Stage 2 begins upon the end of stage 1 and ends on the sixth event, which takes approximately six (6) seconds and generates two (2) Kbytes of data. Stage 3 occurs after stage 2 and comprises continuous sensor data transmission for seven (7) seconds with a data size of five (5) Kbytes.

*Performance Accuracy*

While time saved is important, speed without accuracy can make railway cable protection processes even costlier. Errors in the cable protection after introducing this proposed system in the daily decision-making processes along with the speed of time saving should be measured. The performance of this CRIP model is a direct way to assess its efficiency as shown in Table 6.

**Table 6.** Evidence of detecting cable vandalism/theft in stage 1 confirmed in stages 2 and 3.

| Errors introduced | Stage 1 | Stage 2 | Stage 3 |
|---|---|---|---|
| Deliberate cable cut in Fig. 5 | See the result in Fig. 6(a)–(d) or Table 2 where cable sagging is suspected in location 1 and cable theft is suspected in location 3 | See the result in Figs. 7(a)–(b)–Figs. 9(a)–(b) or Table 3 where Cable theft is suspected in location 3, while the cable sagging is suspected in locations 1 | See the result in Figs. 10 and 11 or Table 4 where cable theft in location 3 and sag in location 1 |

One can see that early knowledge generation at preliminary stage 1 can give railway security personnel advance notice of suspicious cable threat.

For one to obtain more accurate and better results of the underground cable RTTR, it is necessary to increase visibility and monitoring at the targeted sites or locations supported in [11].

In comparing Table 6 which offers evidence of cable theft and vandalism detection to Table 7, which represents evidence of experiment 2, one can say that accuracy and

**Table 7.** Evidence of Experiment 2 for stage 3

| Errors introduced | Stage 3 |
|---|---|
| An error found in the calculation of the 30 min Real Time Thermal Rating (RTTR) calculations | See the result in Figs. 12 and 13 or Table 5 where underground cable RTTR is monitored in Rise Carr substation (RCD), Darlington Melrose (DM) substation and Rise Carr Ianson (RCI) |

speed play a crucial role in decision making. At the preliminary stage 1, Table 6 shows proactiveness which allows for swift action to be taken. In contrast, as shown in Table 7, the accuracy of the results from the initial stage is not guaranteed since errors can be experienced at various locations [11].

## 5   Concluding Remarks

This paper research provides a rapid decision support to detect cable threats, as well as sending timely alerts by integrating a smart cross-sectional explorative analytics of 3-stage detection approach into a sensor network used on the railway cable infrastructure protection, which addresses the delayed gaps in the existing approaches. In comparison with other methods, this method is safe and easy to use, and more adaptable to changing environments. The researchers have answered the research question which pertains to monitoring of railway cable infrastructure protection (RCIP) threats proactively. As shown in Sects. 4 (4.2 and 4.3), the experiments were performed on real-life data and publicly available data of railway cable behaviours across sensor locations and to determine possible cable sagging or theft. Researchers have deliberately introduced the cable cut and sag for the purpose of testing. Experimental results from real-life sensor and publicly available data show that the proposed detection approach is adequately robust and reliable for the detection of suspicious behaviour on railway cable infrastructure in terms of speed and accuracy. The results of this paper can be used as a reference guide to understand the railway cable infrastructure protection gaps and enrich the literature of IoT (internet-of-things) in securing railway cable infrastructure environment. Future work should explore time series sensors data to predict future suspicious cable behaviours on the railway cable infrastructure.

## References

1. Merabti, M., Kennedy, M., Hurst, W.: Critical infrastructure protection: a 21st century challenge. In: International Conference on Communications and Information Technology (ICCIT), pp. 1–6. IEEE (2011). https://doi.org/10.1109/ICCITECHNOL.2011.5762681. ISBN 978-1-4577-0402-4
2. Network Rail: National performance affecting cable theft summary. https://www.networkrail.co.uk/wp-content/uploads/2022/01/Cable-Theft-Report-P10-2021-22.pdf. Accessed 25 May 2022

3. Transnet: Weekly report: Cable Theft Statistics. https://www.transnet.net/Media/Pages/Cable-Theft-Stats.aspx. Accessed 25 May 2022
4. Lelong, A., Carrion, M.O.: On Line Wire Diagnosis using Multcarrier Time Domain Reflectometry for Fault Location (2009)
5. Oh, S.C., Kim, G.D, Jeong, W.T., Park, Y.T.: Vision-based object detection for passenger's safety in railway platform. In: International Conference on Control, Automation and Systems, COEX, Seoul, Korea. IEEE (2008)
6. Qian, J., Mai, X., Yuwen, X.: Real-time Power Line Safety Distance Detection System Based on LOAM SLAM. IEEE (2018)
7. Carvalho, F.B.S., Medeiros, T.I.O., Rodriguez, Y.P.M.: Monitoring System for Vegetation Encroachment Detection in Power Lines Based on Wireless Sensor Networks. IEEE (2018)
8. Dogdas, T., Akyokus, S.: Document clustering using GIS visualizing and EM clustering method. In: IEEE INISTA. IEEE (2013). https://doi.org/10.1109/INISTA.2013.6577647. ISBN 978-1-4799-0661-1
9. Schlehuber, C., Heinrich, M., Vateva-Gurova, T., Katzenbeisser, S., Suri, N.: Challenges and approaches in securing safety-relevant railway signalling. In: IEEE European Symposium on Security and Privacy Workshops (EuroS&PW) (2017)
10. Galvao, A.F., Montes-Rojas, G., Sosa-Escudero, W., Wang, L.: Tests for skewness and kurtosis in the one-way error component model. J. Multivar. Anal. **122**, 35–52 (2013)
11. Wang, Y., Ash, R., Gillie, M., Cross, J., Lloyd, I., Webster, A.: Lessons Learned Report Real Time Thermal Rating (2014). http://www.networkrevolution.co.uk/wp-content/uploads/2014/12/CLNR-RTTR-Lessons-Learned-Report.pdf. Accessed 21 Feb 2022

# Designing an Intangible Tele-Interaction for Point-to-Point Robot Control Using Coercive Gesture Filtering

Aditya Kiran Pal[✉], Akash Acharjee, Aniket Das, Arghajit Bhowmik, and Suman Deb

NIT Agartala, Agartala, India
adityakiran.cs@gmail.com, sumandeb.cse@nita.ac.in

**Abstract.** Teleoperation of Kinematics has been a significant research in computation which can extend the interaction from tangible to intangible devices over a large distance. This work highlights the introduction of Leap Motion along with systematic algorithmic control of a remote device. Teleoperation can be tangible or intangible, the later being fast and efficient, provided the base for all data collection methodologies. In remote operation and kinematic controls from normal to hazardous condition, teleoperation can be significant technology used for maneuvering and controlling different devices. Optical sensors primarily Leap Motion have been incorporated for achieving it. It requires agile, algorithmic and optical control, so that robust control can be transmitted for remote operation along with the feedback. This proposed system uses a Leap Motion control interface for gesture classification along with the elimination of involuntary inputs. An involuntary gesture filter has been implemented to reduce ambiguity in the captured data. Data collection through intangible approaches is preferred and different ways are widely researched. Moreover, the intercommunication during the teleoperation process needs to be scaled for data transmission at high rates. For teleoperation-based tasks, information needs to be conveyed without data loss and with minimum response delay. So this study tries to establish optimum results for data transmission by testing several criteria on different inter-network communication protocols and then selecting the most suitable method of transmission of kinematics along with the elimination of unintended gestures.

**Keywords:** Teleoperation · Intangible · Leap motion · Gesture filtering · Communication protocols

A. R. Molla et al. (Eds.): ICDCIT 2023, LNCS 13776, pp. 290–302, 2023.
https://doi.org/10.1007/978-3-031-24848-1_20

# 1 Introduction

## 1.1 Intangible Teleoperated Control System

Teleoperation, also called telerobotics is the technical term for remote control of a robot. In such a system, a human operator controls the movements of the target robot from some distance away. Since the robot is usually absent in the field of view of the operator, he/she must rely on feedback from the robot's worksite. In this world of progressive technological innovations, new ideas and methods are emerging to make peoples' lives easier. The development of computer vision tasks helps us to automate the robots accomplishing the unknown problems of complex figures. Earlier in some cases, we can recall the fact that the prominent decision-making for all sorts of tasks and problems, the toddler steps of the robots, were guided by a guardian human for necessity. However, the environment can be highly unstructured, and unfamiliar with the object's shape, and the relevant motion can be unknown. The control of robots needs human intelligence, especially in complex and dangerous environments (Fig. 1).

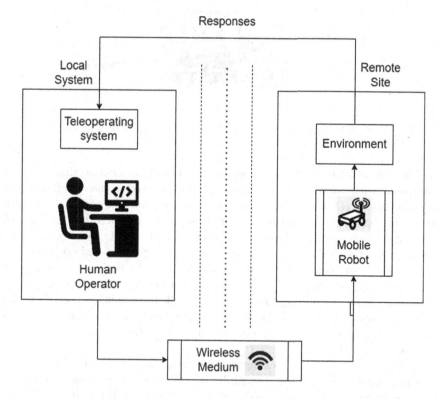

**Fig. 1.** Schematic of a teleoperated system.

Various developments have been witnessed in the field of human-computer interaction with sophisticated systems enhancing the involvement of the device with humans minimizing latency, reducing error rates in the overall protocol, and many more. There are some ugly teeth behind the soft tongue. The devices can be uncomfortable for some users with tedious processes involving that operation. The commonly used human-robot interface for remote operation includes various equipment, electromagnetic or motion capture devices, like exoskeleton equipment, or EMG devices. However, as these devices are attached to the human body, they may affect the comfort and flexibility of human operation. The ideas of gesture recognition or indicative verbal features are based on non-contact interaction and are relatively easier to deal with small tasks, such as moving, rotating, stopping and running, etc.

## 1.2    Leap Motion

**Fig. 2.** (a) Schematic (b) Working (c) Coorinate system of leap motion.

The Leap Motion Controller is an optical hand-tracking module that uses two monochromatic infrared cameras and three LEDs to detect human hand gestures. Hand Gesture detected includes the orientation, direction and position of fingers, palm, wrist, hand, etc. It captures the orientation and position data and stretch state of each finger to identify unique gestures. The Leap Motion system uses a right-hand Cartesian coordinate system with the origin centered on top of the Leap Motion Controller. The x-axis and z-axis lie in the horizontal plane, with the x-axis being parallel to the long edge of the device. The y-axis is vertical with positive values increasing toward the top, while the z-axis has positive values increasing toward the user. This allows the Leap motion to measure distances from a few centimeters to almost a meter using the Cartesian coordinate system mentioned above (Fig. 2).

The utilization of infrared cameras is advantageous when detecting gestures under ranging lighting conditions with high-performance capturing of frames as compared to computer vision-based techniques. Thus, much attention has been drawn to Leap Motion for being used as the preferred device for Teleoperation. Hence, it was adopted to recognize the gestures and employed the recognized gestures to perform the Teleoperation of the robot in this paper.

## 2   Related Works

Numerous types of research regarding teleoperation are being carried out, implementing various modern techniques. [1] highlights a Bilateral haptic teleoperation systems which allow humans to perform complex tasks in a remote or inaccessible environment, while providing haptic feedback to the human operator. At the same time, new and new devices with different functionalities are being introduced to facilitate intangible teleoperation [2]. A Leap Motion sensor based intangible teleoperation system with a virtual reality interface for mobile robots. It allows the user to teleoperate a mobile robot using bare hands. Autonomous Vehicular operations are a matter of global research, teleoperation complements autonomous features of ground vehicles in a major way [3].

Authors in [4] have used Leap Motion for Dynamic Hand Gesture recognition. Independent researches regarding Leap motion in various fields are there. Hand gesture-based assistance aids for the disabled are quite prominent in this field [5]. Moreover, due to its worldwide acceptance as an optimal tool for hand gesture recognition, it is used in gesture detection as an alternative to deep convolution networks [6]. These gesture-based models are used to implement virtual environments for various purposes, like creating learning environments [7,8], interaction with virtual reality, augmented user interfaces [9] etc. Also, gesture-to-text translation models are quite a popular field in this scenario [10,11]. In the field of Robotic movements, leap motion has introduced multiple types of research like humanoid robot integrated Exer-Learning-Interaction [12], robotic arm control [13] etc.

Teleoperation is a widely discussed topic in the current scenario [14,15]. Various data transfer protocols regarding Teleoperation are being researched incorporating features of various sensors [16]. In [17], authors have developed a robot teleoperation system using Leap Motion and Holo Lens. Similarly [18] outlines teleoperated system with Leap Motion has been designed to remotely operate on a mobile fish robot. Thus, this paper strengthens the requirement of a robust and agile transmission of control over the media over networking protocols.

## 3   Methodologies

### 3.1   Technical Set-Up

Experiments have been made on a three-wheel Holonomic Differential drive. This work involving Leap Motion based virtual control interface is mainly to complement its autonomous behaviour in case of failure situations, in doing so it should be ensured that data transfer is done securely, expending minimum bandwidth and at a higher rate. So three main data transfer architecture has been implemented in its system.

The whole interaction architecture can be divided into three major parts. Data fetching, Data transmission, and serial communication with the microcontroller. Since leap motion is built to be more time efficient in the aspect of data fetching and preprocessing, the major problem that stays is successful intercommunication at high rates. There are many Internet-based data transfer protocols. But the main issue is the security and latency of communication to operate remote robots without third-party interventions (Fig. 3).

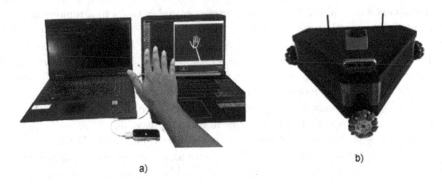

a)                                                          b)

**Fig. 3.** (a) Capturing hand gestures (b) Three-wheel holonomic drive.

After successful internet transmission from the client side, the Holonomic Drive recieves the information via the network card inside it which is relayed to the microcontroller over a USB connection. This is achieved via connecting it to the TX/RX pin of the same, which internally uses UART communication protocol to interpret the serial data. The baud rate of the serial communication is needed to be carefully set so as to keep the data fetching and interpretaion of the recieved data synchronized. Python's pyserial library provides an internal implementation of the same and writes the leap data interpreted instructions into the serial buffer. The microcontroller reads the data in the serial buffer and sends a PWM signal to the motors for movement.

### 3.2 Data Fetching

Five distinguishable effortless hand gestures were chosen and have been mapped with five different basic movement patterns. This includes ideal state, forward, backward, clockwise, and anticlockwise motion. The forward and backward actions are performed based on the pitch of the hand, whereas the other two actions require the roll of the same. The current action space includes 5 states and an acceleration range from 0–255 which is embedded in proposed custom instruction protocol as 'S [0–4] A [000–255];'. The integer following the 'S' byte is the action value whereas the 3-digit integer following the 'A' byte is the value of acceleration measured by the y- coordinate of the wrist from the sensor (150–350) and is mapped to a value between 0–255. When wrist distance exceeds the prespecified value of 150–350, acceleration is set to 0. The mapping function is (Table 1):

$$f(y_{wrist}) = \begin{cases} \frac{y_{wrist}-150}{200} \times 255, & \text{if } y_{wrist} \geq 150 \text{ or } y_{wrist} \leq 350 \\ 0, & \text{otherwise} \end{cases}$$

**Table 1.** Gesture recognition and Action mapping

| Action value | Gesture | Instruction | Mapped value |
|---|---|---|---|
| 0 | | Idle | S0A000 |
| 1 | | Forward | S1A125 |
| 2 | | Backward | S2A125 |
| 3 | | Clockwise | S3A127 |
| 4 | | Anti-clockwise | S4A127 |

The adoption of such a protocol where the hand distance and relative position both signify unique values provides us with a possible way of interpreting single-handed gestures in terms of both acceleration and direction of motion. This reduces the extra complexity of creating another complex two-handed control interface. There is also a significant reduction in the amount of noise as only single-handed operations are performed.

### 3.3 Data Transmission

**TCP Client-Server Architecture.** A communication service between an application program and the Internet Protocol is provided by the Transmission Control Protocol. It offers host-to-host communication at the Internet model's transport layer. The specific methods for transferring data over a link to another site, such as the necessary IP fragmentation to support the transmission medium's maximum transmission unit, are not necessary for an application to understand. TCP/UDP manages all handshaking and transmission details at the transport layer and provides the application with an abstraction of the network connection, generally via a network socket interface. This provides a basic interface for feedback based kinematic control of mobile robots over internet (Fig. 4).

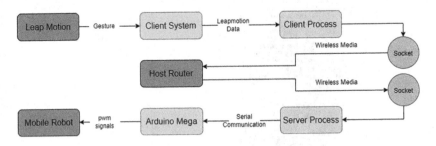

**Fig. 4.** Network schematic

**Websocket Implementation.** TCP with many of its advantages comes with the disadvantage of open unsecured ports on the server side. This can lead to unfriendly attacks and can cause harm to the server. So web sockets come into the picture. For client authentication, rather than TCP's three-way handshake, it adopts a more secure TLS/SSL-based handshake followed by normal TCP for full-duplex communication. Due to the implementation of an encrypted authentication protocol, it becomes more secure. Moreover, since it uses default browser ports, the WebSocket architecture comes with additional advantages over a normal TCP connection [19].

**ROS Implementation.** ROS is an open-source, meta-operating system for your robot [20]. It provides the services you would expect from an operating system, including hardware abstraction, low-level device control, implementation of commonly-used functionality, message-passing between processes, and package management. It also provides tools and libraries for obtaining, building, writing, and running code across multiple computers. Two systems were used for the overall control system, which was described in the aforementioned earlier methods. The systems must initially be linked via an SSH connection. The leap motion controller was one system, and the mobile holonomic drive was the other system. Both systems run roscore daemon with the mobile robot set as the ROS Master. Data is published in the leap motion/leap data topic by a leap motion talker node. The coordinates and angular coordinates of various hand parts make up the data published by the talker node. Data is published in the form of messages. ROS uses a simplified messages description language for describing the data values (aka messages) that ROS nodes publish. This description makes it easy for ROS tools to automatically generate source code for the message type in several target languages. This information is sent via the ros channel from the client to the ros master. The server system operates a listener node. This listener node keeps watching for any fresh data to arrive. When new information is discovered, the listener retrieves it from the channel, preprocesses it, and then the message is prepared to be sent using serial communication.

ROS provides very robust and efficient data transmission over SSH [21]. In case the talker node fails and shuts done. The listener remains unaffected waiting for any further messages published in the data channel. The nodes are loosely coupled which provides a safe mechanism in case of connection failure. The data transmission speed is nearly 100 times faster than TCP/IP Socket implementation.

### 3.4 Filtering Coercive Gestures

In the context of continuous data sending from client to server, one of the vital issues in leap motion is that it tends to interpret involuntary inputs to some valuable intermediary instructions. One can unintentionally place their hand over the leap motion, generating a rapid, abrupt movement. When one is operating the leap motion, interference of another hand can cause ambiguity. In a system where single-handed operation is being used, the interference of another hand could cause misinterpretation. The leap motion device may abruptly switch between the hands. Moreover even when a hand is placed into the field of view of the device, that movement can also be an involuntary one. When we examine technologies with such accuracy, such quick responses to the environment might be quite harmful (Fig. 5).

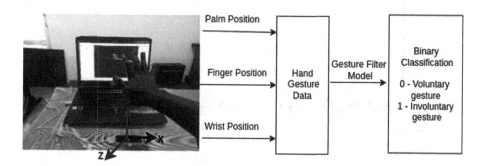

**Fig. 5.** Working of coercive gesture filter

To tackle such scenarios, a neural network-based approach has been implemented on the client side for real-time hand tracking and thus rejecting unintended gestures by the user. For the same purpose, a custom dataset was pre-pared using leap motion. leap motion APIs have been used to capture different data relating to linear coordinates, velocity, angular coordinates, etc. The dataset created was fed into a Multi-Layer Perceptron. The model was created using the Tensorflow framework which was implemented in python. The output of the model is a binary classification of Voluntary Movement.

$$L_{binarycrossentropy} = -((y\log(p) + (1 - y)\log(1 - p)))$$

The output is 0 when the gestures are classified as voluntary and 1 for invol-untary movement. The model was saved and then run on the client side in a multi-threaded environment. It continuously classifies the gestures and filters out only the voluntary movements to the server side thus, solving the problem of coercive gestures.

## 4    Experiment and Result Analysis

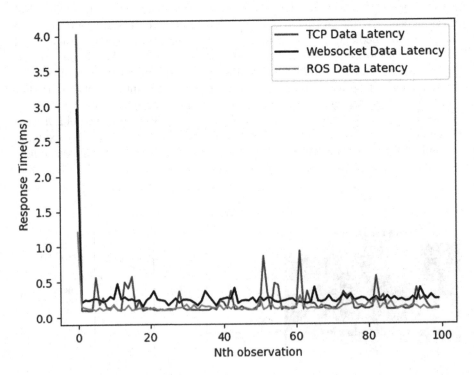

**Fig. 6.** Response time per packet over different protocols

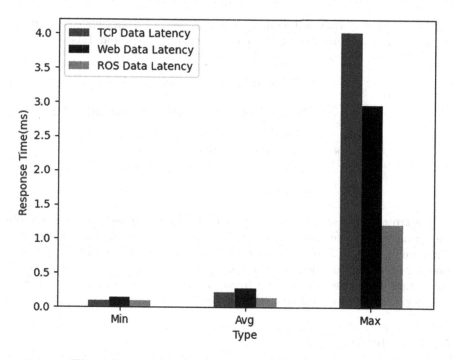

**Fig. 7.** Comparison of response time of different protocols

As mentioned above, three major networking protocols have been implemented in the architecture of mobile teleoperating robot, and results are observed. As seen in the graph of the overall latency of communication, TCP incorporates a huge amount of delay as the three-way handshake requires a considerable amount of round-trip time. Moreover, the fluctuations in ping may cause several issues in systems running with high precision. The average round trip time observed in the case of basic TCP architecture is 0.224 ms for devices in LAN. This will be amplified for devices operating at high distances. Also as open ports are observed in a bare TCP connection, there arises the problem of security and prevention of data breaching. One of the possible solutions is to adopt a more secure architecture that uses a more sophisticated handshake method that too in a minimum latency. This is where WebSockets come into the picture. Since it uses TLS/SSL-based encrypted handshake methodologies, thus provides more security. but it is observed that the average round trip time is somewhat slower than basic TCP architecture which is about 0.2707 ms. Ros on the other hand uses a secure shell (ssh) connection. SSH uses a more secure and faster key exchange algorithm as observed from the initial round trip latency graph. Moreover, the average round trip time is also less and fluctuations in latency are the minimum which is about 0.1398 ms. All this value of latency data is being calculated in the same LAN, over a hundred packets of data, to compare their performance keeping other conditions constant. Following are the various data drawn from doing latency analysis (Figs. 6, 7 and Table 2):

**Table 2.** Latency analysis.

| Protocol | Min. (in ms) | Avg. (in ms) | Max. (in ms) |
|---|---|---|---|
| TCP | 0.0893 | 0.224 | 4.013 |
| Websockets | 0.14457 | 0.2708 | 2.9570 |
| ROS | 0.09015 | 0.13989 | 1.2095 |

Regarding the speed at which leap motion data is processed, there is a significant time saving because leap motion is capable of detecting hand gestures at a maximum rate of 300 fps and with accuracy up to 0.01 mm in fingertip position, far exceeding complex neural net models as they necessitate intensive GPU computation to achieve maximum frame per second. Additionally, accuracy is also degraded when neural networks are trained.

## 5  Conclusion

The above-found results conclude that Leap motion as a sensor can be easily used to replace other gesture based models as the machine's response needs to be optimized on a real-time basis. In doing so it is possible to eliminate other time consuming processes such as preprocessing and data fetching delays. There are instances when user un-intentionally introduces some inputs to the leap device. But the proposed gesture filter can detect such anamolus inputs accurately thus eliminating possibilites of abrupt unintended motion. Moreover, ROS-provided SSH based client-server architecture has a huge advantage over other networking protocols of data transfer without compromising security and other internetwork communication issues. Proposed system also suggests The incorporation of one handed gesture for manipulating both the direction and acceleration thus reducing the noise factor even further. This Research provides a good base to try out several other intangible modes of teleoperation incorporating other more complex and fast networking protocols. In some cases, bandwidth and signal availability may be an issue. This can be optimized via carrying out research over several other advanced networking protocols that may or may not require internet bandwidth for communication.

This work incorporates using Leap Motion for Teleoperating mobile robots such a three wheeled Holonomic drive and implementing a noise filter to reduce uncontrolled kinematic behaviour. But this Teleoperation can easily be applied to interact with remote environments and its uses can be extensively generalized to operate other humanoid robots, industrial machinery, surgical robotic systems etc.

# References

1. Passenberg, C., Peer, A., Buss, M.: A survey of environment-, operator-, and task-adapted controllers for teleoperation systems. Mechatronics **20**(7), 787–801 (2010). Special Issue on Design and Control Methodologies in Telerobotics
2. Su, Y., Ahmadi, M., Bartneck, C., Steinicke, F., Chen, X.: Development of an optical tracking based teleoperation system with virtual reality. In: 2019 14th IEEE Conference on Industrial Electronics and Applications (ICIEA), pp. 1606–1611 (2019)
3. Ray, S., Sivasangari, A.: Design of mobile robot teleportation system using virtual reality. In: 2020 International Conference on Communication and Signal Processing (ICCSP), pp. 1173–1175. IEEE (2020)
4. Wei, L., Tong, Z., Chu, J.: Dynamic hand gesture recognition with leap motion controller. IEEE Signal Process. Lett. **23**(9), 1188–1192 (2016)
5. Shinde, S.S., Autee, R.M., Bhosale, V.K.: Real time two way communication approach for hearing impaired and dumb person based on image processing. In: 2016 IEEE International Conference on Computational Intelligence and Computing Research (ICCIC), pp. 1–5. IEEE (2016)
6. Shao, L.: Hand movement and gesture recognition using leap motion controller. Virtual Reality, Course Report (2016)
7. Koo, B., Kim, J., Cho, J.: Leap motion gesture based interface for learning environment by using leap motion. In: Proceedings of HCI Korea, HCIK 2015, Seoul, Korea, pp. 209–214. Hanbit Media Inc. (2014)
8. Păvăloiu, I.B.: Leap motion technology in learning (2017)
9. Hendrik, B., Masril, M., Wijaya, Y.F., Andini, S., et al.: Implementation and design user interface layout use leap motion controller with hand gesture recognition. In: Journal of Physics: Conference Series, vol. 1339, p. 012058. IOP Publishing (2019)
10. Karthick, P., Prathiba, N., Rekha, V.B., Thanalaxmi, S.: Transforming Indian sign language into text using leap motion. Int. J. Innov. Res. Sci. Eng. Technol. **3**(4), 5 (2014)
11. Koul, M., Patil, P., Nandurkar, V., Patil, S.: Sign language recognition using leap motion sensor. Int. Res. J. Eng. Technol. (IRJET) **3**(11), 322–325 (2016)
12. Nama, T., Deb, S., Debnath, B., Kumari, P.: Designing a humanoid robot integrated exer-learning-interaction (ELI). Proc. Comput. Sci. **167**, 1524–1532 (2020)
13. Yu, H., Chen, X., Liu, X.: Robotic arm control with human interactionl
14. Sheridan, T.B.: Teleoperation, telerobotics and telepresence: a progress report. Control Eng. Pract. **3**(2), 205–214 (1995)
15. Barua, H.B., Sau, A., et al.: A perspective on robotic telepresence and teleoperation using cognition: are we there yet? arXiv preprint arXiv:2203.02959 (2022)
16. Naceri, A., et al.: Towards a virtual reality interface for remote robotic teleoperation. In: 2019 19th International Conference on Advanced Robotics (ICAR), pp. 284–289. IEEE (2019)
17. Liang, C., Liu, C., Liu, X., Cheng, L., Yang, C.: Robot teleoperation system based on mixed reality. In: 2019 IEEE 4th International Conference on Advanced Robotics and Mechatronics (ICARM), pp. 384–389. IEEE (2019)
18. Mi, J., et al.: Gesture recognition based teleoperation framework of robotic fish. In: 2016 IEEE International Conference on Robotics and Biomimetics (ROBIO), pp. 137–142 (2016)

19. Skvorc, D., Horvat, M., Srbljic, S.: Performance evaluation of Websocket protocol for implementation of full-duplex web streams. In: 2014 37th International Convention on Information and Communication Technology, Electronics and Microelectronics (MIPRO), pp. 1003–1008. IEEE (2014)

20. Drumheller, W.R., Conner, D.C.: Documentation and modeling of ROS systems. In: SoutheastCon 2021, pp. 1–7. IEEE (2021)

21. Toris, R., Shue, C., Chernova, S.: Message authentication codes for secure remote non-native client connections to ROS enabled robots. In: 2014 IEEE International Conference on Technologies for Practical Robot Applications (TePRA), pp. 1–6. IEEE (2014)

# Sentiment Analytics for Crypto Pre and Post Covid: Topic Modeling

DwijendraNath Dwivedi[1]([✉]) and Anilkumar Vemareddy[2]

[1] Krakow University of Economics, Rakowicka 27, 31-510 Kraków, Poland
dwivedy@gmail.com
[2] University of Agricultural Sciences, Bangalore, India

**Abstract.** Sentiment analysis for Bitcoins and Cryptocurrency is an excellent way to understand how to make smart investment decisions. It provides broad market insights that can be useful for forming trading strategies. It is important to remember that markets are highly impacted by psychology. As such, investors should monitor market sentiment barometers to get better information on potential opportunities in the cryptocurrency market. Using these barometers is easy and can assist you in making better investment decisions. In the cryptographic market, a sentiment is a helpful tool for traders. It is because it combines the opinions, attitudes, moods, and perspectives of the public. Topic extraction was conducted to uncover keywords in feelings that capture the recurring theme of the text. This practice is often used to analyze a wide range of feelings to quickly and effectively identify the most common subjects. We used a few months of Twitter data for pre and post covid and applied the principle of latent semantic analysis and decomposition of singular values to group key water quality questions that impact people's lives. The study contributes to the literature on text exploration by providing a context to analyze the public's sense of bitcoin and cryptocurrency before and after COVID. This can help to understand key themes in negative feelings related to crypto-trading and key public concerns could be highlighted and shared with a broader community.

**Keywords:** Bitcoin · Sentiment analytics · Topic modeling · Text mining · Cryptocurrency

## 1 Introduction

Although the news is often informative, it is not an ideal source for sentiment analysis. While well-written news is a useful source of information, it can also be inaccurate. Good news will have a neutral sentiment. Social media is much different, however. Unlike news, social media threads about cryptocurrencies are driven by public material and tend to be noisy and subjective. As a result, sentiment analysis for cryptocurrencies should be conducted with care. When looking at market sentiment for cryptocurrencies, it is important to remember that the market does not consider fundamentals. Rather, it is based on the emotional state of the people with an interest in the project. This is especially true in volatile markets, which is why it is critical to conduct research on

A. R. Molla et al. (Eds.): ICDCIT 2023, LNCS 13776, pp. 303–315, 2023.
https://doi.org/10.1007/978-3-031-24848-1_21

market sentiment before investing. The sentiment of traders is essential when deciding on which cryptocurrency to buy or sell.

As Bitcoin prices are influenced by the sentiment of investors in the equity market, cryptocurrency prices also tend to rise when equity market investors are feeling bearish. In addition, since it has a small trading volume, this kind of research can help find out if there is a correlation between sentiment and cryptocurrency prices. The findings will be useful for both investors and traders. The research will help to identify trends and predict price movements for other cryptocurrencies.

As Bitcoin prices increase, traders use a market sentiment as an indicator of price changes. Traders will borrow USD in a bullish market to buy cryptocurrency. During a bearish market, they will borrow USD from an investor. If the cryptocurrency price is falling, traders will sell. Similarly, if it is rising, it will fall. This means that cryptocurrency is highly correlated with the price of the dollar.

Despite the volatility in the market, the cryptocurrency market is still a great place to make smart investment decisions. A few experts say the market is bullish, while others claim it is a bearish one. This is a fact, many factors affect the value of a particular crypto. In addition to market sentiment, many investors also use real-time tweets and other data sources to evaluate a crypto's value.

Bitcoins and Cryptocurrency have become popular topics in the finance world. Despite its popularity, it is largely unregulated. Therefore, the price of a crypto-currency may be driven by public sentiment. This could mean the difference between a successful investment and a complete failure. A thriving cryptocurrency market will be based on the positive and negative opinions of investors. The sentiment of an individual coin can be influenced by many factors.

Positive or negative sentiment can lead to a trend in cryptocurrency. For example, a bullish sentiment can influence a currency's price. A bullish sentiment may lead to a downward trend. A bearish sentiment may lead to a bullish market. A strong VIX indicates high levels of fear. The same is true for a low VIX. A rising VIX will promote the price of bearish crypto.

## 2   Literature Survey

Gupta et al. (2021) attempted to use contextual analysis of the text to determine what characteristics of the product or service stimulate user sentiment. Dwivedi et al. (2021) refined sentiment analysis and thematic modeling of the government's response and documented the post-COVID situation by comparing the UAE and the Kingdom of Saudi Arabia. Dwivedi et al. (2022) performed topic and sentiment analysis using Twitter data to identify key concerns regarding data quality and data impurity. Dwivedi et al. (2022) have attempted to use text contextual analysis to categorize Twitter data to understand positive and negative feelings about COVID-19 vaccination and would like to highlight key concerns. Dwivedi et al. (2022) analyzed medical research in the United Arab Emirates vs. World Health Organization for COVID-19. The objective was to identify the key themes for both organizations. Dwivedy et al. (2022) used context analysis of texts to categorize Twitter data based on positive and negative feelings. This was linked to the ethical challenges of AI and key concerns were emphasized.

Alghamdi and Alfalqi (2015) found that new techniques or tools were required to organize, search, index, and review extensive data. As they observed the explosion of electronic documents and archives. Hofmann (2001) presented two key approaches to natural language processing (NLP) and statistical programs such as thematic modeling for such analysis. In contrast to NLP methods that mark parts of speech and grammatical structure, statistical models and thematic models are based mainly on the assumption of the "word bag" (BoW). In BoW templates, the collection of textual documents is quantified in a document-term matrix (DTM). This takes into account the incidence of each word (columns) for each document (rows). Deerwester et al. (1990) were one of the earlier researchers who presented one of the first topic models. He leveraged semantic latency analysis (LSA) and decomposition of singular values (SVD). Asmussen and Moller (2019) presented a unique framework that took benefit of the topic modeling techniques to conduct a review of the exploratory literature of a large collection of articles. The framework offered by them makes it possible to review a large number of documents in a transparent, effective, and reproducible manner using the LDA method. In general, there are two approaches to automatic document processing: supervised learning and unsupervised learning. In supervised learning, the manual coding of a collection of documents is being done. This takes a long time to achieve the result. On the other hand, unsupervised learning methods, such as topic modeling, do not have the prerequisites to manually code the documents, saving a lot of time for an exploratory review of the extensive collection of papers. Gotipati et al. (2018) Leveraged subject modeling and data visualization methods to understand the student's feedback from seven post-graduate courses that were taught at the Singapore University of Management. They evaluated rules-based methods and statistical classifiers to extract topics. Al-Obeidat et al. (2018) then proposed a sandbox for extracting opinions on the subject and analyzing feelings for extracting questions and their associated feelings from a database. LDA was used for theme extraction and the "bag-of-words" sentiment analysis algorithm. Polarity was established using the frequency of the positive/negative words in the document. Benedetto and Tedeschi (2016) outline standard approaches to social media sentiment analysis and related cloud-based issues. Ajeet Ram Pathak et al. (2021) the proposed approach is that it functions at the sentence level to extract the subject using online latent semantic indexing with regulation constraint. Md. Mokhlesur Rahman et al. (2021) explored the features related to positive and negative sentiments of the people about reviving the economy. This was done for the United States (US) during the COVID-19 global crisis. It took into consideration the situational uncertainties (i.e., changes in work and travel patterns due to lockdown policies), economic slowdown and associated trauma, as well as emotional factors like depression. Jikyung (Jeanne) Kim et al. (FY22) showed that the excessive reaction of consumers to negative news and the negative feeling of the takers intensifies this excessive reaction, leading to negative livestock farming.

(Dwijendranath Dwivedi & Anilkumar Vemareddy) attempted the data quality text mining analysis using twitter data on sentiment analysis to know the public opinions on the data quality issues and what are all the main topics which people are talked about the data quality issues against positive and negative sentiments. The text preprocessing methods are followed like cleaning the stop words and removing the special characters and applied the sentiment algorithm to classify text in to positive and negative words. On the topic modeling used different algorithms and got the better results after comparison.

This papers shows the different algorithms for topic modeling they have used and compared the results and we followed the same methodology and improved the results.

## 3   Data and Methodology

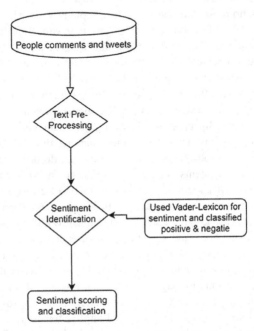

**Fig. 1.** Process flow for topic modeling (Dwivedi et al. 2021)

Preprocessing text: This step is required for text analysis to transform human language into a machine-readable format for subsequent processing and analysis. There are some mandatory steps to request clean-up, which are listed below (Fig. 1).

- Convert all the text to lowercase
- Removing stop words, sparse terms, and particular words
- Convert numbers into words or remove the numbers
- Removing white spaces (leading and ending spaces)
- Removing punctuation (all types of special characters or symbols)

First, we have started to eliminate duplication of rows, and it is essential to delete duplicate data or rows to avoid unbiased results. Convert all text to lowercase to prevent more than one copy of the same word. For example ("drinking water" is considered to be two different words).

*We were deleting punctuation because it adds more information* handling text data. In addition, this will shrink the size of the training dataset. We eliminate keywords that often appear in the text or we create a list of keywords, or we use predefined libraries. We used stopword and text, blob libraries that will deal with stopwords. For stewards, we have deleted common words in the general scenario, but we can also delete naturally occurring comments from our textual data. We can therefore check the ten words that occur frequently, and then decide which to delete.

*Spelling correction* of the text: We have seen tweets with many spelling mistakes, or short words will be used. In this situation, the spell-checking step is useful to reduce the number of copies of the word. For this, we have a Textblob library it will handle spelling mistakes.

**Tokenization** is the process of dividing the text into a series of words or phrases. In our example, we used the text blob library to transform our tweets into blob and convert them into a group of words.

**Stemming** refers to the removal of suffices, like "ing," "ly," "s," etc., through a straightforward rules-based approach. For that, we will use Porter Stemmer of the NLTK library.

Lemmatization is a more suitable method than stemming because it converts the term to its root term, rather than just stripping it enough. it uses vocabulary and proceeds to a morphological analysis to obtain the root word. Hence, we generally prefer to use lemmatization instead of stamping.

We've done all the basic preprocessing steps to clear the text, and now we need to extract the characteristics using natural language techniques.

**N-grams are defined as a combination of several words used in combination.** N-grams, bigrams, and trigrams were used. Unigrams will not have a great deal of information compared to bigrams and trigrams. We use these bigrams or trigrams to grasp the structure of the language, such as which letter or word is likely to follow that given. Those recommendations are going to depend on the implementation of our study. Sometimes, if we use low grams and do not grasp the essential differences or if we sometimes take long grams, it will not capture the overall sense of the expression.

## Part-of-Speech Tagging (POS)

The marking of a part of the speech assigns mostly speeches to each word of the text according to its context and its definition (nouns, verbs, adjectives, and others).

As presented in the Fig. 2A, firstly started removing the duplication of rows to avoid unbiased results. Further, converted all the text into lower cases to prevent multiple copies of the same word. For example ("Crypro Currency" crypto currency" will be considered as two different words). Followed by removing punctuation as it might add any extra information or reduce the size of the training dataset while handling text data. Also, eliminated stop words that are frequently occurring words in the text by using text blob library in python. The tweets with many spelling mistakes, or short words were observed in the twitter data, hence spelling correction step are performed with the help of text blob library.

After the above steps, tokenization was done to divide the text into a sequence of words or sentences, transforming our tweets into a blob and then by converting them into a series of words. Followed by Stemming refers to the removal of suffices, like

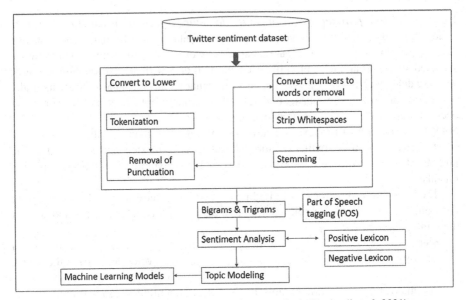

**Fig. 2.** Preprocessing process for sentiment analysis (Dwivedi et al. 2021)

"ing," "ly," "s," etc., by a simple rule-based approach by using PorterStemmer from the NLTK library of python. Some may use Lemmatization as it is more practical option than stemming because it converts the word into its root word, rather than just stripping the suffices. It makes use of the vocabulary and does a morphological analysis to obtain the root word. Therefore, researchers usually prefer lemmatization over stemming.

After basic preprocessing steps of cleaning the text extracted the features using the following natural language techniques. N-Grams which identifies the combination of multiple words used together. We have used N-grams, bigrams, and trigrams. Unigrams has not captured much information as compared to bigrams and trigrams. Hence used bigrams or trigrams to capture the language's structure, like what letter or word likely to follow the given one. Further, part-of-speech tagging mainly assigns speeches to each word of the text based on its context and definition (nouns, verbs, adjectives, and others).

Secondly, **topic modelling,** is the process of extracting or obtaining required features from the bag of words. This is an important technique since each word present in the corpus has considered as a feature in natural language processing. This feature reduction will help us to focus on the right content instead of going through the entire text in the training data. There are many methods used for topic modeling, Latent Dirichlet Allocation (LDA) is one of method which is used to analyze the topic modeling in the present study.

LDA is a statistical and graphical model used to obtain relationships between multiple documents in a corpus. It is developed using the variational exception maximization (VEM) algorithm for obtaining the maximum likelihood estimate from the whole corpus of text. Traditionally, this can be solved by picking out the top few words in the bag of words. However, this completely lacks the semantics in the sentence. This model follows the concept that the probabilistic distribution of topics can describe each document, and the probabilistic distribution of words can explain each topic. Thus, it helps to get a much clearer vision of how the topics are connected. It considers all corpus of entire documents in the data. After preprocessing of the corpus, each bag of words consists of common words. Using LDA model, the topics related to each document has been derived and can group all corpus into a particular group for further usage. The flow chart below details the process of topic modeling.

## 4   Results

The adoption of topic modeling using the Latent Dirichlet Allocation (LDA) method of Twitter data about Bitcoins/cryptocurrencies resulted in the statistical distribution of several topics on the trending of Bitcoins/Cryptocurrencies. The topics were studied separately for the pre and post-Covid periods to understand the most prevalent topics by the people on the trending of bitcoins/cryptocurrencies before and after the Covid period.

**Pre-covid Period – Topics**
At the initial stage, the twits were categorized into four topics as seen in Fig. 3. Most people are probably opting for cryptocurrencies as it is shown in topic 0, most popular topic is to buy, follow and list which depicts that people are showing favoritism towards cryptocurrencies. Further, people are aware of and probably show interest in cryptocurrencies due to the metaverse and blockchain facilities which are probably the influencing factors towards cryptocurrencies as may be seen from topic 2 of Fig. 3. The metaverse, block, get ready, and programmed were high-probability words in topic 2. Subsequently, topic 2 and 3, depicts that people are using weekly, close, and project as the most popular words concerning cryptocurrencies.

**Fig. 3.** Topics

In Fig. 4, the topics are shown in two dimensional scatter plot, wherein each topic is represented by the bubble. The larger the bubble, the more frequent is that topic. Topic

1 is the most popular and topic 4 is the least important topic. The distance between the two bubbles depicts the approximate similarity between the topics. The bar chart shows the top 30 most popular topics, the bar represents the term frequency within the selected topic (Fig. 5).

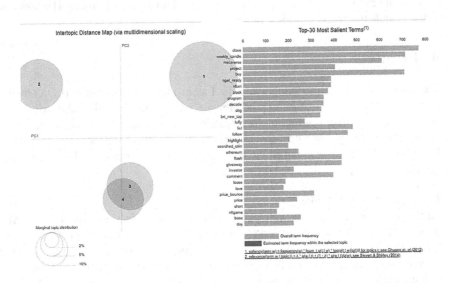

**Fig. 4.** Two-dimensional scatter plot – topics

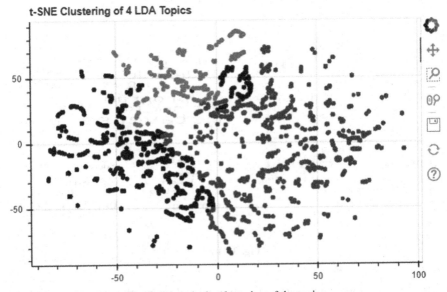

**Fig. 5.** Reveals the clustering of the topics

## Post-covid Period – Topics

During the post-Covid period, cryptocurrencies have gained popularity. The majority of people are writing/commenting on cryptocurrencies. As seen from Fig. 6, the most prevalent comments are crypto trading, trade, and bullish, which reveal that most people may be interested in trading cryptocurrencies. Topic 1, comprises the words buy, price, project, and invest which also depicts that people were commenting on the cryptocurrency trading and which is a positive sign towards the cryptocurrencies. Further, topics 2 and 3, highlights currency, bit, week, crypto trade, and liquidate are may be forecasting the positive signs of peoples engaging in cryptocurrency trading.

**Fig. 6.** Topic categorization

Figure 6, represents the topics derived from LDA, the larger bubble is the most prevalent topic and the smaller bubble is the least popular topic. The distance between the bubbles in the two-dimensional scatter plot represents the similarity between the topics (Fig. 7).

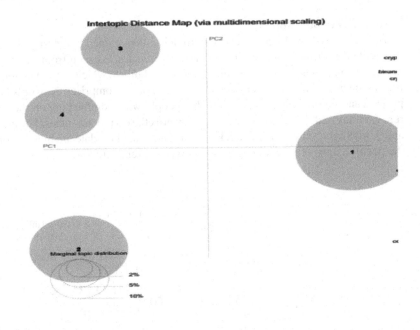

**Fig. 7.** Two-dimensional scatter plot

Topic modeling results on Pre-Covid period.

| Topic 1 | 0.055 * "buy" + 0.037 * "list" + 0.036 * "follow" + 0.034 * "flash" + 0.034 * "giveaway" |
|---|---|
| Topic 2 | 0.073 * "metaverse" + 0.046 * "nget_ready" + 0.046 * "nftart" + 0.044 * "block" + 0.041*" |
| Topic 3 | 0.122 * "close" + 0.113 * "weekly_candle" + 0.032 * "highlight" + 0.031 * "searched_coin" |
| Topic 4 | 0.039 * "program" + 0.039 * "decade" + 0.022 * "money" + 0.019 * "future" + 0.009 * "soon" |
| Topic 5 | 0.036 * "luffy" + 0.015 * "support" + 0.014 * "market" + 0.013 * "live" + 0.012 * "rest" |
| Topic 6 | 0.025 * "short" + 0.022 * "move" + 0.021 * "pump" + 0.015 * "coin" + 0.012 * "fear" |
| Topic 7 | 0.035 * "price" + 0.021 * "big" + 0.019 * "time" + 0.018 * "point" + 0.015 * "drop" |
| Topic 8 | 0.053 * "project" + 0.012 * "great" + 0.012 * "claim" + 0.012 * "way" + 0.011 * "faucet" |

Topic models play an essential role in exploring text data, especially with a large volume of text data, to understand the structures and groups of interest. The LDA was used for the topic modeling wherein the key topics were extracted from the bag of words. As the model follows the concept of the probabilistic distribution of topics that describes each document, the probabilistic distribution of words can explain each topic to get a clearer vision of how the topics are connected.

Bitcoin has the most popular cryptocurrency, has become a highly valued asset of Pre-covid. Take a close look at the historical prices of Bitcoin; you will see that it has tremendously outperformed itself in value by rising over 500% in six months. Other cryptocurrencies such as Ethereum, Ripple, Dogecoin etc., have also performed well. The crypto market is continuing to witness its longest bull run to date 2019 was a big year for cryptocurrencies, as more and more investors jumped from the stock market to enjoy the crypto frenzy. As per Coinmarket cap, the total crypto-market capitalization in January 2019 was around $130 billion.

The rising interest in cryptocurrencies drove the crypto-market capitalization to a whopping $180 billion, by the end of July. In December 2019, Bitcoin pushed all its barriers and wider adoption of cryptos lead the market tad under the $200 billion threshold. It is worth noting that the pre-covid era witnessed volatility with usual movements and a few spikes in crypto prices. A correction of the 15 to 50 percent range could be seen in the crypto assets.

Topic modeling results on Post-Covid period.

| Topic 1 | 0.027 * "cryptotrading" + 0.021 * "project" + 0.009 * "new" + 0.008 * "money" + 0.008 * "start" |
|---|---|
| Topic 2 | 0.022 * "buy" + 0.018 * "cryptotrade" + 0.013 * "price" + 0.013 * "ethereum" + 0.011 * "look" |
| Topic 3 | 0.015 * "invest" + 0.013 * "trade" + 0.012 * "week" + 0.010 * "nft" + 0.009 * "currency" |
| Topic 4 | 0.031 * "trading" + 0.022 * "great" + 0.018 * "coin" + 0.013 * "gain" + 0.010 * "check" |
| Topic 5 | 0.021 * "bcheck" + 0.013 * "list" + 0.012 * "opensea" + 0.009 * "offer" + 0.008 * "platform" |
| Topic 6 | 0.039 * "liquidate" + 0.030 * "binance_future" + 0.023 * "binance" + 0.014 * "bybit" + 0.009 * "trader" |
| Topic 7 | 0.012 * "learn" + 0.011 * "strategy" + 0.009 * "outperform" + 0.009 * "chat" + 0.007 * "nhttp" |
| Topic 8 | 0.008 * "drop" + 0.008 * "unus_se" + 0.006 * "unussedleo" + 0.005 * "bthe_road" + 0.005 * "speed_ahead" |

The covid outbreak caused havoc and every financial asset lost its value. However, one sector was particularly booming—the crypto sector. January 2020 was a good month for crypto assets with the trading volume spiking from $200 billion to $255 billion.

Investors had two favorite coins—Bitcoin and Ethereum. As per the Coin market cap tracker, Bitcoin was in the dominant category alluring over 65 percent of investors. Ethereum also garnered quite a bit of investor for cryptos. Little did investors know about the upcoming crash.

# 5 Conclusion and Discussion

The study has been carried out in two phases, namely, i) the pre-Covid period – to understand the prevalent topics by people on cryptocurrencies and ii) the Post-Covid period – the people's interactions on cryptocurrencies. The LDA is used as it is the most popular approach in NLP for topic modeling. The twits before and after Covid were analyzed using the LDA approach by categorizing them into several topics. The topics revealed that during the pre-Covid period, people are commenting on cryptocurrencies concerning knowing the trading of cryptocurrencies, especially for ease of entering into trade or understanding the cryptocurrencies. The topics after the Covid period reveal that people may be showing interest in entering into the trading of cryptocurrencies. It may be revealed from the study that Covid is having a positive impact on the trading of cryptocurrencies.

# References

Gupta, A., et al.: Understanding consumer product sentiments through super-vised models on cloud: pre and post COVID. Webology 18(1), 406–415 (2021). https://doi.org/10.14704/web/v18i1/web18097

Dwivedi, D.N., Anand, A.: The text mining of public policy documents in response to COVID-19: a comparison of the United Arab Emirates and the Kingdom of Saudi Arabia. Public Gov./Zarządzanie Publiczne 55(1), 8–22 (2021). https://doi.org/10.15678/ZP.2021.55.1.02

Dwivedi, D.N., Pathak, S.: Sentiment analysis for COVID vaccinations using Twitter: text clustering of positive and negative sentiments. In: Hassan, S.A., Mohamed, A.W., Alnowibet, K.A. (eds.) Decision Sciences for COVID-19. ISOR, vol. 320, pp. 195–203. Springer, Cham (2022). https://doi.org/10.1007/978-3-030-87019-5_12

Dwivedi, D.N., Anand, A.: A Comparative study of key themes of scientific research post COVID-19 in the United Arab Emirates and WHO using text mining approach. In: Tiwari, S., Trivedi, M.C., Kolhe, M.L., Mishra, K.K., Singh, B.K. (eds.) Advances in Data and Information Sciences: Proceedings of ICDIS 2021. LNNS, vol. 318, pp. 341–350. Springer, Singapore (2022). https://doi.org/10.1007/978-981-16-5689-7_30

Dwivedi, D.N., Wójcik, K., Vemareddyb, A.: Identification of key concerns and sentiments towards data quality and data strategy challenges using sentiment analysis and topic modeling. In: Jajuga, K., Dehnel, G., Walesiak, M. (eds.) Modern Classification and Data Analysis: Methodology and Applications to Micro- and Macroeconomic Problems. STUDIES CLASS, pp. 19–29. Springer, Cham (2022). https://doi.org/10.1007/978-3-031-10190-8_2

Alghamdi, R., Alfalqi, K.: A survey of topic modeling in text mining. Int. J. Adv. Comput. Sci. Appl. 6(1), 147–153 (2015). https://doi.org/10.14569/ijacsa.2015.060121

Hofmann, T.: Unsupervised learning by probabilistic latent semantic analysis. Mach. Learn. 42(1–2), 177–196 (2001). https://doi.org/10.1023/A:1007617005950

Deerwester, S., Dumais, S.T., et al.: Indexing by latent semantic analysis. J. Am. Soc. Inf. Sci. 41(6), 391–407 (1990)

Asmussen, C.B., Møller, C.: Smart literature review: a practical topic modelling approach to exploratory literature review. J. Big Data 6(1) (2019). https://doi.org/10.1186/s40537-019-0255-7

Gotipati, S., et al.: Text analytics approach to extract course improvement suggestions from students' feedback. Res. Pract. Technol. Enhanced Learn. 13(1) (2018). https://doi.org/10.1186/s41039-018-0073-0

Al-Obeidat, F., Kafeza, E., Spencer, B.: Opinions sandbox: turning emotions on topics into actionable analytics. In: Belqasmi, F., Harroud, H., Agueh, M., Dssouli, R., Kamoun, F. (eds.) AFRICATEK 2017. LNICSSITE, vol. 206, pp. 110–119. Springer, Cham (2017). https://doi.org/10.1007/978-3-319-67837-5_11

Benedetto, F., Tedeschi, A.: Big data sentiment analysis for brand monitoring in social media streams by cloud computing. In: Pedrycz, W., Chen, S.-M. (eds.) Sentiment Analysis and Ontology Engineering. SCi, vol. 639, pp. 341–377. Springer, Cham (2016). https://doi.org/10.1007/978-3-319-30319-2_14

Kim, M.M., et al.: Socioeconomic factors analysis for COVID-19 US reopening sentiment with Twitter and census data. Heliyon 7(2), e06200 (2021). https://doi.org/10.1016/j.heliyon.2021.e06200

Dwivedi, D.N., et al.: How responsible is AI?: Identification of key public concerns using sentiment analysis and topic modeling. IJIRR 12(1), 1–14 (2022). https://doi.org/10.4018/IJIRR.298646

# A Rough Set Based Approach to Compute Impact of Non Academic Parameters on Academic Performance

Ahan Chatterjee, Swagatam Roy, and Aniruddha Mandal[✉]

Department of Computer Science and Engineering, The Neotia University, Diamond Harbour, South 24-Pargana, Kolkata 743368, India

**Abstract.** Our academic performance can differ due various to non-academic factors namely, self-motivation and mind stability where peace of mind plays a pivotal role; self-confidence after any kind of setback, environment and financial pressure, family support, and capability of an individual to make a combat. These non-academic factors may lead to stress affecting our academic performance. In this paper, we aim to assess the mentioned non-academic factors affect our academic performance. The paper is modeled on survey performed on college students and depending upon their responses related to these non-academic factors. We have carried out the work through statistical analysis and machine learning algorithms. Density bases Clustering algorithm is being used to cluster similar kind of performance and analyze how the factors differ from each other. Moreover, we also presented a comparative study between academic parameters such as attendance, interest in subject and travelling time in comparison to such non-academic performances. A rough set is also being created comprising the parameters which gives the best result among the students. The entire hypotheses are being tested on the various non-academic factors and how they impact academic performances and statistical calculations. Conclusion is drawn to check if non-academic factors really impact academic performances?

**Keywords:** Rough Set Theory · Descriptive Analysis · Regression · DBSCAN · Multicollinearity · Clustering

## 1 Introduction

In the recent decades, many new institutions of higher education have sprouted in various corners of the country. However, students who enter college do not get decent grades and academic performance, and the responsibility lies with the faculty. A better understanding of this is that studies of non-academic performance can have important consequences. Our data were collected primarily from students at Neotia University. Our approach of collection is an online survey, to prevent data loss through offline methods.

The paper is segmented into various sections. Section 2 contains the literature review, Sect. 3, contains various descriptive statistical analyses that were carried out over our collected data, with verification of our hypothesis and graphical visualization. In Sect. 4,

© The Author(s), under exclusive license to Springer Nature Switzerland AG 2023
A. R. Molla et al. (Eds.): ICDCIT 2023, LNCS 13776, pp. 316–327, 2023.
https://doi.org/10.1007/978-3-031-24848-1_22

we used a rough set model that examines the target variables, the key features that influence CGPA. Thus using multinomial regression algorithm, we optimize the maximum accuracy of the model with least *RMSE* value to state factors is mostly influencing. Sect. 5, contains density based clustering where we classify the traits that high and low-performing students using DBSCAN.

## 2 Literature Review

Non-Academic factors are innumerable, and they do affect the academic performances as proved by previous research works. Non-Academic comprises sleeping pattern, attendance, confidence, social integration, self-motivation and many more. Research work by Mulugeta Tesfay and Dawit Zekiros [1] states environment, family support, distance travelled to attend college, financial support impact academic performance negatively or positively through multiple linear regression and descriptive analysis. Sarah M.S. Shathele and Anitha Oommen [2] research work suggests anxiety, stress and lack of sleep negatively and family support and awareness of the course affects positively the academic performance through the Chi-Square test. Correlation (r = 0.549) between career interest, aptitude, mental health, happiness and their impact on academic performance is stated by Maryam Mehria et al. [3]. Ayça Akdil Sönmez and Ali Talip Akpınar [4] suggest that sigma value of variance (ANOVA) is bigger than 0.05 and there is no relationship and effect in between academic performance and physical environmental factors. Andrew Lepp et al. [5] investigated cell phone use and texting have negative correlation on academic performance and positive correlation on anxiety. The study by Trevor G. Bennett & Simon M.Yalams [6] shows positive correlation of 0.656 between attendance and performance in the class and the obtained p-value of 0.05 shows that the results can be replicated. Prima Vitasaria et al. [7] investigates that anxiety has a negative correlation (r = −0.264, n = 205) with academic performance. Hasan Afzal et al. [8] research work states that motivation has positive variance on academic performance. Academic performance has large effects on likelihood of retention, academic self-discipline, commitment to college and social connectedness have direct effects on retention as investigated by Jeff Allen, Steven B. Robbins, et al. [9]. Increasing students' connections benefits students experiencing mental and behavioral concerns but also aids in suicide prevention initiatives and improves academic outcomes as researched by Susan M. De Luca et al. [10]. Hysenbegas et al. [11] that students with depression reported a decrease in approximately half of a letter grade compared to non-depressed peers. It has been found to have a negative association between depression and GPAs, as suggested by Eisenber et al. [12] the linking of behavioural health functioning with academic performance.

## 3 Descriptive Statistical Analysis and Hypothesis Testing

The dataset is prepared from The Neotia University, across 230 students from various streams of education. All the students are selected in random basis, with high mix of students coming from different socio-economic background, with different cultures. We aim to study all those factors.

Null Hypothesis, $H_0$ = Higher anxiety tendency among students do not affect academic performance,

$$pvalue = 0.000057$$

As p value is < 0.05 thus Alternate Hypothesis stands which conclude that higher anxiety tendencies among students affect academic performance (Fig. 1).

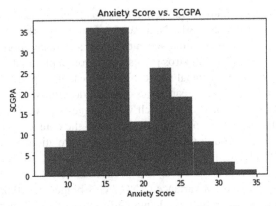

**Fig. 1.** Histogram plot between anxiety and CGPA. Source: Created by Author, based on dataset

Null Hypothesis = $H_0$ = lower family support and environment among students do not affect academic performance,

$$pvalue = 0.000045$$

As the p value is < 0.05 thus Alternate Hypothesis stands which conclude that higher anxiety tendencies among students affect academic performance (Fig. 2).

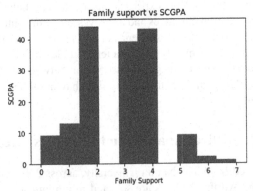

**Fig. 2.** Histogram plot between family support and CGPA. Source: Created by Author, based on dataset

Null Hypothesis = $H_0$ = interest to attend classes among students do not affect academic performance,

$$pvalue = 0.000070$$

As the p value is $< 0.05$ thus Alternate Hypothesis stands which conclude that attendance among students affect academic performance (Fig. 3).

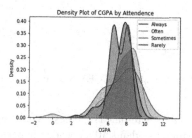

**Fig. 3.** Density Plot between interest to attend classes and CGPA. Source: Created by Author, based on dataset

Null Hypothesis = $H_0$ = higher financial pressure among students do not affect academic performance,

$$pvalue = 0.000063$$

As the p value is $< 0.05$ thus Alternate Hypothesis stands which conclude that higher financial pressure among students affect academic performance (Fig. 4).

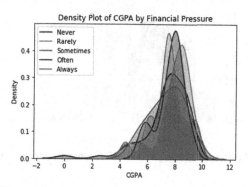

**Fig. 4.** Density plot of financial pressure and CGPA. Source: Created by Author, based on dataset

Null Hypothesis = $H_0$ = daily lower sleeping tendency among students do not affect academic performance

$$pvalue = 0.000053$$

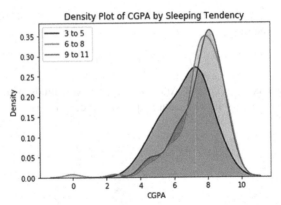

**Fig. 5.** Density plot of financial pressure and CGPA. Source: Created by Author, based on dataset

As the p value is $< 0.05$ thus Alternate Hypothesis stands which concludes that daily lower sleeping tendency among students affect academic performance (Fig. 5).

Null Hypothesis $= H_0 =$ lower motivation and confidence do not affect academic performance.

$$pvalue = 0.000056$$

As the p value is $< 0.05$ thus Alternate Hypothesis stands which concludes that lower motivation and confidence do affect the academic performance (Fig. 6).

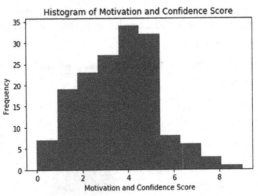

**Fig. 6.** Histogram of Motivation in Study and CGPA. Source: Created by Author, based on dataset

Null Hypothesis $= H_0 =$ higher social integration among the students do not affect academic performance

$$pvalue = 0.000049$$

As per the obtained p value, p value $< 0.05$ the Alternate Hypothesis is accepted (Fig. 7).

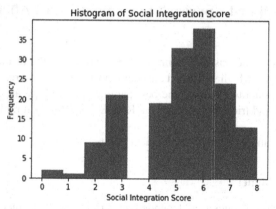

**Fig. 7.** Histogram of social integration and CGPA. Source: Created by Author, based on dataset

Null Hypothesis = $H_0$ = higher cell phone using tendency among students do not affect academic performance

$$pvalue = 0.000053$$

Since the p value < 0.05, then the alternate hypothesis is accepted is rejected which concludes that higher cell phone usage affects academic performance (Fig. 8).

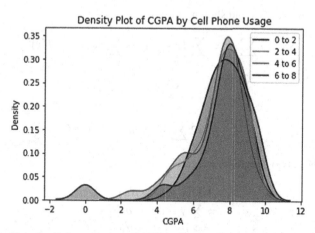

**Fig. 8.** Density plot of cell phone usage and CGPA. Source: Created by Author, based on dataset

In some cases, we have only categorical variables, thus we have opted for density plot. From the descriptive analysis portion, we can understand that our entire taken hypothesis stands true for alternative hypothesis. Thus, every non-academic factor taken here influences the target variable CGPA. Moreover, we have plotted a median curve through which we can analyze how the peers are better respect to the median marks under various stated circumstances.

## 4  Rough Set Based Approach to Compute Most Affecting Parameters

We have collected numerous data points, but few will affect the result directly. So we take those parameters which will affect the result most. Thus for this reason we have used rough set based model to calculate the best parameters. We have applied multinomial regression model and tried to minimize the RMSE value for each rank to find the best suitable rank.

Taking the parameters that affect the data points.

### 4.1  Regression Model

Here the error term is shown as u, depended on variable as y, x as independent

$$y = a + bx + u \tag{1}$$

The error is assumed that it is uncorrelated with the explanatory variables

$$y' = a' + b'x \tag{2}$$

The error comes as,

$$u' = y - y' = y - (a' - b'x) \tag{3}$$

We will optimize SSR and shown in Eq. 4.

$$SSR = \sum u'^2 = \sum (y - y')^2 = \sum (y - a' - b'x)^2 \tag{4}$$

$$\frac{\partial SSR}{\partial a'} = -2 \sum (y - a' - b'x) = 0 \tag{5}$$

$$\frac{\partial SSR}{\partial b'} = -2 \sum x(y - a' - b'x) = 0 \tag{6}$$

Solving these two equations we get,

$$\begin{aligned} -2 \sum (y - a' - b'x) &= 0 \\ \Rightarrow \sum y - na' - b' \sum x \\ \Rightarrow a' &= y'' - b'x'' \end{aligned} \tag{7}$$

Substituting the required variable,

$$b' = \frac{\sum xy - y'' \sum x}{\sum x^2 - x'' \sum x} \tag{8}$$

## 4.2  Rough Set Model

The RST model is taken when there is an inconsistent data. This is mostly adapted to find hidden prevalent patterns in the dataset, where other algorithm fails to trace the patterns. This model works on approximation model where there is 2 bounds the upper and lower approximation bound (Fig. 9).

2 features are there namely decisional and conditional their relation is shown as,

$$A = C \cup D \tag{9}$$

$$I(P) = \left\{ (x, y) \in U^2 | \forall a \in P, a(x) = a(y) \right\} \tag{10}$$

Here a(x) and a(y) are values of feature for object x and object y.

The lower approximation of X wrt. P is defined by the Eq. 11.

$$P_L X = \{x | [x_P] \underline{C} X\} \tag{11}$$

The upper approximation of X wrt. P is defined by the Eq. 12.

$$P_U X = \{x | [x_P] \cap X \neq \emptyset \tag{12}$$

$$m_{ij} = \left\{ a \in C : a(x_i) \neq a\left(x_j\right) \Delta\left(d \in D, \left(d(x_i) \neq d\left(x_j\right)\right)\right) \right\}, \forall i, j = 1, 2, 3, \ldots n \tag{13}$$

$d(x_i) and d\left(x_j\right)$ represent the class labels of object. The core set is represented as shown in Eq. 9.

$$CoreSet = \cup \left\{ m_{ij} | |m_{ij}| = 1 \right\}, \forall i, j = 1, 2, 3, \ldots n \tag{14}$$

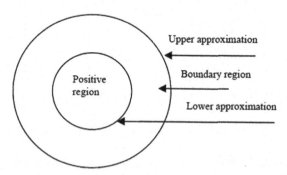

**Fig. 9.**  Rough set theory

### 4.3 Proposed Algorithm for Feature Reduction

**Algorithm:** Feature Sorting and Best-Case Scenario
**Input:** All Features (AF)
**Output:** Minimized Feature
Begin
    Apply CF on (AF)
    ST CF (AF) in decreasing order of correlation = CFS
    Name R1 to highest order and R8 to lowest
    Apply RSTA (CFS)
End

### 4.4 Results

From the result it is evident that our reduced dataset is giving a lower RMSE value subsequently we can state that reduced dataset has higher accuracy. So, we can say that reducing the dataset is giving more accurate results. The set with 6 most correlated parameters is giving the best result (Table 1 and Fig. 10).

**Table 1.** Accuracy table

| RMSE with Whole Dataset | RMSE with Reduced Dataset |
|---|---|
| **1.05** | 1.001 |

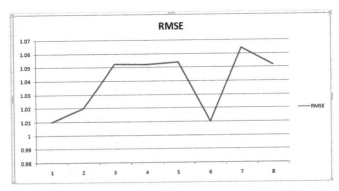

**Fig. 10.** RMSE change for different ranks. Source: Created by Author, based on dataset

The OLS Equation is represented as:

$$
\begin{aligned}
AcademicPerformance \\
&= 6.72 + 0.44 * SleeplingTendency + (-0.02) \\
&* MotivationandConfidenceScore + 0.14 \\
&* Social\ Integration + (-0.18) * Cell\ phone\ usage + 0.05 \\
&* Anxiety + (-0.08) * FamilySupportandEnviroment \\
&+ 0.10 * Attendance + 0.02 * FinancialPressure
\end{aligned}
$$

Linear Regression Model gives the above equation, which is computed on the collected dataset.

- The coefficient of sleeping tendency is 0.44 which means that if the student sleeps increase by unit (1) then performance will be decreased by 0.44%.
- The coefficient of motivation and confidence score is −0.02 which simplifies that if the student has decreases unit (1) of motivation and confidence then performance will decrease by 0.02%.
- The coefficient of social integration is 0.14 which means that if the student has an increase in social integration by unit (1) then performance will be decreased by 0.14%.
- The coefficient of cell phone usage is −0.18 which simplifies that if the student has increase in cell phone usage by unit (1) then performance will decreased by 0.18%.
- The coefficient of anxiety is 0.05 which means that if the student anxiety increases by unit (1) then performance will be decreased by 0.05%.
- The coefficient of family support and environment is −0.08 which simplifies that if the student decreases unit (1) of family support and environment then performance will decrease by 0.08%.
- The coefficient of attendance is 0.10 which means that if the student attendance increases by unit (1) then performance will be increased by 0.10%.
- The coefficient of financial pressure is 0.02 which simplifies that if the student decreases unit (1) of financial pressure then performance will increase by 0.02%.

## 5 Density Based Clustering Approach

In this section we have converted the continuous variable, CGPA into categorical form to create clusters. In this section we have created 3 clusters, high, moderate, low marks achieved. Now we will try to analyze that which factors influence most the high marked students, and will try to inject those traits into others for comparatively better performance.

Let $U$ the whole dataset and each object $a$ in $U$ is characterized by a set $F = \{F_1, F_2, ..., Fn\}$ of $n$ extracted features. Then we normalize the data into $[0, 1]$. Then all the values are being are formulated by $(x, x + 0.1]$. In this step all the values are being discrete.

Then all the similar points are being partitionedæ.

$$
PFi = \{P_{i_1}, P_{i_2}, ..., P_{i_{10}}\}
$$

Let 2 partitions obtained using features $F_i$ and $F_j$ are $P_{Fi} = \{P_{i_1}, P_{i_2}, \ldots, P_{i_{10}}\}$ and $P_{Fj} = \{P_{j_1}, P_{j_2}, \ldots, P_{j_{10}}\}$ respectively. Similarity of $F_i$ to $F_j$ is computed using Eq. (15).

$$S_{ij} = \frac{1}{10} \sum_{k=1}^{10} \max_{1 \leq l \leq 10} \left\{ \frac{|P_{i_k} \cap P_{j_l}|}{|P_{i_k} \cup P_{j_l}|} \right\} \tag{15}$$

Then graphs are being drawn with $G = (F, E, W)$ where F is the features, E is the nodes of the graph and W is being the set of edges. Weights in the graph is assigned as

$$w_{ij} = \left( S_{ij} + S_{ji} \right) / 2$$

Here 3 clusters are formed, for 3 classes (Fig. 11).

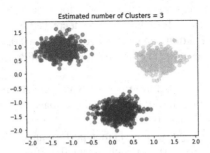

**Fig. 11.** Density based clustering for 3 classes of CGPA. Source: Created by Author, based on dataset

It's seen that using lower mobile, less financial stress, proper sleeping tendency, and regular attending class leads to higher marks among the students.

## 6  Conclusions

We have carried out the descriptive analysis at first, which shows that non-academic factor do affect the performance of a student academically. It shows that large number of students using mobile or less social integration, less sleeping tendency is suffering from poor grades. In feature reduction section it's observed that there is a need of reducing the features to get best results. We have achieved such in this section. Finally, we have categorized the target variable to 3 classes to get which trait is being most counted for students with better result.

## References

1. Vitasaria, P., Wahabb, M.N.A., Othmanc, A., Herawand, Sinnadurai, S.K.: The Relationship between Study Anxiety and Academic Performance among Engineering Students (2010)
2. De Luca, S., Franklin, C., Yueqi, Y., Johnson, S., Brownson, C.: The Relationship Between Suicide Ideation, Behavioral Health, and College Academic Performance (2016)

3. Peugeot, M.A.: Impact of academic and nonacademic support structures on third grade reading achievement (2017)
4. Chatterjee, A., Sinha, T.: Correlation between absence, interest in the field and grades in an organization using regression model. Int. J. Eng. Adv. Technol. **8**, 1436–1441 (2019). https://doi.org/10.35940/ijeat.F8118.088619
5. Chatterjee, A., Roy, S., Das, S.: Feature selection using rough set theory from infected rice plant images. In: Das, A.K., Nayak, J., Naik, B., Dutta, S., Pelusi, D. (eds.) Computational Intelligence in Pattern Recognition. AISC, vol. 1120, pp. 417–427. Springer, Singapore (2020). https://doi.org/10.1007/978-981-15-2449-3_36

# *Varta Rasa* - A Simple and Accurate System for Emotion Recognition in Conversations

Rosalin Parida[2(✉)], Sai Babu Udayagiri[3], Bhushan Jagyasi[1], Surajit Sen[3], Aditi Debsharma[1], Pallavi Gawade[1], and Gopali Contractor[1]

[1] Artificial Intelligence Practice, Accenture, Mumbai, India
{bhushan.jagyasi,aditi.debsharma,
pallavi.s.gawade,gopali.contractor}@accenture.com
[2] Artificial Intelligence Practice, Accenture, Kolkata, India
rosalin.parida@accenture.com
[3] Artificial Intelligence Practice, Accenture, Bangalore, India
surajit.sen@accenture.com

**Abstract.** Emotion Recognition in Conversation (ERC) has become one of the most explored topics in Speech Technology. Unlike Emotion Recognition in a single utterance, context plays an important role in Emotion Recognition in Conversation. In literature complex architectures are proposed for ERC which makes the developement of real-time applications difficult. In this paper, we propose a simple architecture called *Varta Rasa* (in Sanskrit, **Varta** means conversation and **Rasa** means emotions) for ERC, which is a stacked ensemble model. Extensive experiments on the IEMOCAP dataset generated results comparable to state-of-the-art models while providing significantly lower complexity.

**Keywords:** Emotion recognition in conversation · Speech processing

## 1 Introduction

Emotion Recognition in Conversation (ERC) is becoming important for numerous applications such as opinion mining, human-robot interaction, call centre data analytics in healthcare, finance, retail and other industries. In ERC, context plays an important role in determining the emotion of an utterance. Previous works on ERC [2,4,6–8,10] have shown that context and fusion of modalities, such as speech, text and video, have significant impact on model performance. To capture context and use different modalities previous researchers have focused on designing the right architecture. Most of these architectures have complicated design elements. Various fusion strategies such as attention [7], concatenation [10] and gates [8] have been employed. Poria *et al.* [10] captures only self dependency, Hazaraika *et al.* [4] captures both self dependency and inter-personal dependencies and Mao *et al.* [8] focuses on interaction between modalities. These complex architectures are not easy to productionize as they need more resources and their latency is higher.

A. R. Molla et al. (Eds.): ICDCIT 2023, LNCS 13776, pp. 328–334, 2023.
https://doi.org/10.1007/978-3-031-24848-1_23

In this work, we propose **Varta Rasa**, a light-weight architecture for recognition of emotion from conversations. *Varta Rasa* is based on a multi-level stacked ensemble architecture which includes context and multi-modal features. In this architecture we have introduced two new concepts. First, we have introduced a tree-based model to capture non-linearities better, instead of sequence-based models. Second, we introduced bagging through ensemble learning which would capture nuances of the emotions well. We have also used Bidirectional Encoder Representations from Transformers (BERT) embeddings for text encoding.

Overall, in this paper, we make the following contributions:

- We propose a simpler tree based ERC architecture which is more pragmatic for the development of practical applications
- We show that multi-level stacking along with superior embeddings can produce very good results in ERC
- Our experiments with a bench mark dataset (IEMOCAP dataset) show that our method performs comparable to the state-of-the-art architectures while requiring significantly less number of computations.

## 2   Proposed *Varta Rasa* Architecture

For emotion recognition in conversation we propose *Varta Rasa* architecture as depicted in Fig. 1. Given a video consisting of the input sequence of N number of utterances $[(u_1, p_1), (u_2, p_2), ....., (u_N, p_N)]$, where $u_t$ is a utterance spoken by party (speaker) $p_t$ at time $t$, the task is to predict the emotion label $e_t$ at time $t$. When classifying emotions for any utterance in conversation, previous and future utterances can provide important contextual information [10]. Hence,

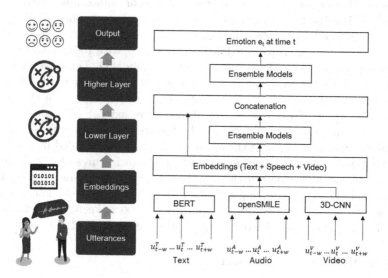

**Fig. 1.** *Varta Rasa*: A simple and accurate Multi-level Stacked Ensemble Architecture for ERC

to predict the emotion of an utterance $u_t$ at time $t$, utterances $u_{t-w}, \ldots u_{t-1}$, $u_t, u_{t+1} \ldots, u_{t+w}$ are considered, where $w$ is the context window.

In comparison to the unimodal system, fusing textual clues along with audio and visual modalities for the multimodal system yields higher accuracy in ERC [4,10].Different embeddings extraction methods are used as described in Sect. 3 for each of the modalities. Concatenated vector of these texts, speech, and visual embeddings with context is given as input to predict the emotion of an utterance.

The concatenated vector is fed to multi-level stacked ensemble model which consists of two layers of LightGBM models with extremely randomized trees. The architecture consists of 10 base learners (LightGBMs with extra trees) in both the layers together. The reason for the choice of the architecture is to maintain simplicity and lightness. An AutoML framework called Autogluon is used to build the model. The output of the second layer of the ensembled model will generate the target emotion $e_t$ at time $t$.

## 3    Experimental Setting

We have used the Interactive Emotional Dyadic Motion Capture (IEMOCAP) dataset [1] for carrying out experiments on Emotion recognition in conversation. The IEMOCAP data set is the most widely used open source conversational dataset which contains approximately 12 h of audio, transcriptions, video, and motion-capture recordings. Following Hazarika *et al.* [4], out of 5 sessions of IEMO-CAP dataset, we used first four sessions of transcripts as the training set, and the last one as the testing set. The validation set is extracted from the randomly shuffled training set with the ratio of 80:20. Also, we considered six emotion classes: Happy, Sad, Neutral, Angry, Excited, and Frustrated for the classification task.

In this work, we have used text, speech and visual modalities to perform ERC task. For **Text**, we have used three types of embeddings: BERT, CNN and Glove. For BERT embeddings [3] we have used pre-trained BERT models from huggingface transformers library. For CNN embeddings, we have used the feature vectors for text as provided by Poria *et al.* [10] with 100 dimensions. Glove embeddings are extracted from pretrained Glove model [9]. For **Speech** features, we have used the features as provided by Poria *et al.* [10]. They extracted audio features from voice data 30 Hz frame-rate and a sliding window of 100 ms. The open source openSMILE is used to automatically extract 6373 audio features which has been reduced to 100 features by feature selection methods. Again, for **Visual** features we have used the extracted features as provided by Poria *et al.* [10]. They have used 3D-CNN for extracting visual features of 512 dimensions for each utterance. We have further reduced these 512 dimensional vector to 200 dimensional vector by applying Principal Component Analysis. For initial and last few utterances, zero padding is applied as context to keep the length of every example same. The final size of the embedding vector is 4272 dimensions.

## 4    Results and Discussion

In this section, we present the outcomes obtained from the experiments on the IEMOCAP dataset using *Varta Rasa* architecture. We have included a comparison of our system with previous works on the same dataset. We also compare the model complexity of *Varta Rasa* with the baseline architecture of bc-LSTM. At last, we present the effect of Modality, Text Embedding, and Ensemble learning on the performance of the model.

**Table 1.** Comparison of $F_1$-score of *Varta Rasa* with baseline methods

| Methods | Happy | Sad | Neutral | Angry | Excited | Frustrated | Wt.Avg |
|---------|-------|-----|---------|-------|---------|------------|--------|
| bc-LSTM [10] | 35.63 | 62.90 | 53.00 | 59.24 | 58.85 | 59.41 | 56.19 |
| Dialogue TRM [8] | 48.7 | 77.52 | 74.12 | 66.27 | 70.24 | 67.23 | 69.23 |
| ICON [4] | 35.6 | 69.2 | 53.5 | 66.3 | 61.1 | 62.4 | 59.0 |
| DialogueRNN [7] | 33.18 | 78.80 | 59.21 | 65.28 | 71.86 | 58.91 | 62.75 |
| DialogueGCN [2] | 42.75 | 84.54 | 63.54 | 64.19 | 63.08 | 66.99 | 64.18 |
| BiERU-1c [6] | 31.56 | 84.13 | 59.66 | 65.25 | 74.32 | 61.54 | 64.59 |
| *Varta Rasa* | **39.64** | **75.62** | **64.46** | **67.5** | **71.19** | **61.48** | **64.8** |

Multiple experiments are performed by varying modalities (text, audio and video), text embeddings (BERT, CNN and Glove) and context windows (3, 4 and 5). Best model from our experiments is stacked ensemble model using bidirectional embeddings of text, audio and video features with 3 context window size. Our best model is compared with baseline models from literature in Table 1. In terms of $F_1$ score our model outperforms bc-LSTM, ICON, DialougeRNN and shows marginally better performance compared to DialougeGCN and Bi-ERU, however shows lesser performance compared to DialogueTRM.

**Table 2.** Model complexity of *Varta Rasa* and baseline methods

| Methods | Memory Occupied | Learnable Parameters |
|---------|-----------------|----------------------|
| bc-LSTM [10] | 8 MB | 2,038,057 |
| ICON [4] | 24.14 MB | 680,306 |
| DialogueRNN [7] | 34.70 MB | 8,880,406 |
| DialogueGCN [2] | 5.28 MB | 1,348,046 |
| *Varta Rasa* | **300 KB** | **37,255** |

**Complexity of Model.** We compare the memory complexity and time complexity of our model in Table 2. *Varta Rasa* occupied $300KB$ which is 25 times lower than the memory occupied by the bc-LSTM which is $8MB$. Most of the baseline models will occupy memory comparable or larger to that of the bc-LSTM.

When it comes to time complexity, LightGBM has a time complexity of $O(0.5 \times feature \times bin)$ [5]. In our case, number of features are 7455 and number of bins are 10. So time complexity is $O(37255)$. Time complexity of LSTM model or any deep learning model is approximately equal to the order of learnable parameters. For instance, bc-LSTM model has $2,038,057$ learnable parameters which is approximately 55 times more than *Varta Rasa*.

**Ablation Studies.** We have shown that it is possible to achieve comparable or better results even with simpler models in ERC as compared to complex architectures in literature. The obvious question is that why would a simpler model perform so well? In this section we present our findings that F1 score in ERC task is mainly determined by: good text embeddings (BERT) and bagging methods (Ensemble learning).

**Table 3.** Effect of modality on the performance ($F_1$ score) of *Varta Rasa*

| Modality | Happy | Sad | Neutral | Angry | Excited | Frustrated | Wt.Avg |
|---|---|---|---|---|---|---|---|
| V | 26.05 | 32.63 | 28.0 | 23.69 | 44.57 | 37.51 | 33.36 |
| A | 2.67 | 65.31 | 41.69 | 49.39 | 42.49 | 42.92 | 43.04 |
| T | 28.87 | 67.42 | 57.49 | 64.21 | 60.37 | 63.68 | 59.14 |
| T+A | 32.35 | 74.49 | 59.20 | 60.67 | 67.01 | 56.50 | 60.08 |
| T+A+V | 39.64 | 75.62 | 64.46 | 67.5 | 71.19 | 61.48 | 64.8 |

**V** - Video, **A** - Audio, **T** - Text

In Table 3, we show that the best performance is achieved when each of the modalities among text, video, and audio are considered. A good choice of text embeddings can represent the semantics of utterance in a better way and will have a huge influence on the accuracy of ERC. As we can see from Fig. 2(a), BERT embeddings show superior accuracy over 1D-CNN and Glove embeddings. This is due to the superior nature of BERT embeddings in capturing the context of the utterance accurately. Also, from Fig. 2(b) we can see that stacked ensembling improves the F1 score by 6.74% points over the XGBoost model with everything remaining the same. The nuances between the emotions such as happy & excited; and sad & frustrated are better captured using stacked ensemble learning.

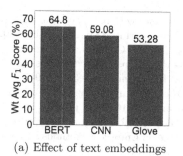

(a) Effect of text embeddings

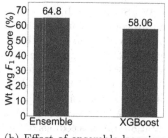

(b) Effect of ensemble learning

**Fig. 2.** Effect of text embeddings and model ensembling on performance

## 5    Conclusion

In this paper, we have proposed a simple yet accurate model to perform the ERC task. We have shown that use of ensemble learning and BERT embeddings can improve the accuracy. We also argue that our simpler architecture facilitates the development of practical applications. Using the IEMOCAP dataset and extensive comparison with numerous previously proposed methods, we show that our model achieves results comparable to the state-of-the-art models with an additional advantage of being lightweight.

## References

1. Busso, C., et al.: Iemocap: interactive emotional dyadic motion capture database. Lang. Resour. Eval. **42**, 335–359 (2008). https://doi.org/10.1007/s10579-008-9076-6
2. Deepanway, G., Navonil, M., Soujanya, P., Niyati, C., Alexander, G.: DialogueGCN: a graph convolutional neural network for emotion recognition in conversation. arXiv preprint arXiv:1908.11540 (2019)
3. Devlin, J., Chang, M.W., Lee, K., Toutanova, K.: BERT: pre-training of deep bidirectional transformers for language understanding. arXiv preprint arXiv:1810.04805 (2018)
4. Hazarika, D., Poria, S., Mihalcea, R., Cambria, E., Zimmermann, R.: Icon: interactive conversational memory network for multimodal emotion detection. In: Proceedings of the 2018 Conference on Empirical Methods in Natural Language Processing, pp. 2594–2604 (2018)
5. Ke, G., et al.: LightGBM: a highly efficient gradient boosting decision tree. Adv. Neural. Inf. Process. Syst. **30**, 3146–3154 (2017)

6. Li, W., Shao, W., Ji, S., Cambria, E.: BiERU: bidirectional emotional recurrent unit for conversational sentiment analysis. arXiv preprint arXiv:2006.00492 (2020)
7. Majumder, N., Poria, S., Hazarika, D., Mihalcea, R., Gelbukh, A., Cambria, E.: DialogueRNN: an attentive RNN for emotion detection in conversations. In: Proceedings of the AAAI Conference on Artificial Intelligence, vol. 33 (2019)
8. Mao, Y., et al.: DialogueTRM: exploring the intra-and inter-modal emotional behaviors in the conversation. arXiv preprint arXiv:2010.07637 (2020)
9. Pennington, J., Socher, R., Manning, C.D.: GloVe: global vectors for word representation. In: Proceedings of the 2014 Conference on Empirical Methods in Natural Language Processing (EMNLP), pp. 1532–1543 (2014)
10. Poria, S., Cambria, E., Hazarika, D., Majumder, N., Zadeh, A., Morency, L.P.: Context-dependent sentiment analysis in user-generated videos. In: Proceedings of the 55th Annual Meeting of the Association for Computational Linguistics (volume 1: Long papers), pp. 873–883 (2017)

# An Optimal Approach for Multi-class Object Detection

Ankit Deb$^{(\boxtimes)}$ , Rapti Chaudhuri , and Suman Deb

NIT, Agartala, India
ankitdeb98@gmail.com, {rapti.ai,sumandeb.cse}@nita.ac.in

**Abstract.** Object detection in an image aims to point out all the desired objects in the target image and considers the positional enlightenment to obtain machine perspective knowledge. In object detection vanishing gradient has been the most significant problem that can occur, it results in a model with many layers being specified to unable to learn on a specific dataset. It could cause models to a low-grade solution. To get grip on the obstacles, this paper has analyzed certain algorithms for precise object detection from a customized dataset. In this approach, multiple objects in an image have been classified and localized with the help of Mask Region Colvolutional Neural Network (R-CNN). This work is pulled off using a regional convolutional neural network, pixellib, OpenCV, and TensorFlow, resulting in a preferable desired output. Implementing this technique and algorithm, based on deep learning, which is also based on machine learning requires framework understanding. The experimental result and analysis are done with multiple classes of an image to verify the efficiency of the mask R-CNN technique. The algorithm has been compared with other existing object detection strategies to establish the accuracy and reliability of the concerned determined algorithm. The systematic execution enabled clarification of the suggested technique to be precise in analyzing multi-class object in an image.

**Keywords:** Mask-RCNN · Object detection · Image classification · Image segmentation · Deep learning

## 1 Introduction

Achieving the best idea about an image is to classify the objects within the image by object detection, the problem here is challenging and interesting [1]. With the help of image segmentation more intelligent and accurate robots can be built for better knowledge about the classification of objects [2]. To use mask R-CNN we need to go through some of the layers of the convolutional neural network namely, the Convolutional Layer, Pooling Layer, Fully Connected Layer, and Softmax. Regional convolutional neural network(R-CNN) for object segmentation is a challenging piece of work that desired both, object localization to pin point and sketch boundary box around each object in an image and classification of the object to anticipate the accurate class of object that was pinpointed.

© The Author(s), under exclusive license to Springer Nature Switzerland AG 2023
A. R. Molla et al. (Eds.): ICDCIT 2023, LNCS 13776, pp. 335–340, 2023.
https://doi.org/10.1007/978-3-031-24848-1_24

Mask R-CNN uses a boundary box to detect multiple objects and objects that are overlapped in an image. The architecture of the Mask R-CNN model is shown in Fig. 1. Some of the object detection techniques which can be used in the work includes Reigion-based Convolutional Neural Network(R-CNN), Fast R-CNN, Faster R-CNN, Histogram-based object detecting algorithm, Single Shot Detector(SSD), and different version of You Only Look Once(YOLO) models. Image segmentation is the technique of identifying object present in images or videos. This can be achieve by using Mask R-CNN. The prime focus of the experiment is to teach machine about objects and their classes, identify, understand the specification of an image just like humans do. Image segmentation can be used for many applications namely, Medical imaging [3], Self driven cars [4], Face recognition [5], Satellite imaging [6]. In this work it is shown that the object can be identified in largely varying lighting and background condition with high accuracy and it is better in comparison to other existing models.

This paper is arranged in major sections describing the related work that has been done on image segmentation, discussion of methodology and finally experimental result and analysis.

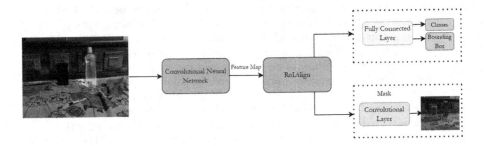

**Fig. 1.** Architecture of Mask R-CNN.

## 2   Related Work

In this section of the work, various algorithms are discussed which are used in object detection, mainly the Mask R-CNN technique.

The presented work in reference [7] shows the active use of the Mask regional convolutional neural network technique in the field of object detection. State-of-the-art methods for detecting objects of general classes are primarily rooted in the deep convolutional neural network. In the research paper of Girshick et al. [8], he proposed a multi-stage pipeline designated as region-based convolutional neural networks(R-CNN), for training deep convolutional networks to analyze region proposals for object detection. Later, Fast R-CNN [9] was proposed where the accuracy of object detection in an image can be more precise. In Faster R-CNN [10] the region proposals were give rise to region proposal network(RPN).

Apart from framework that consist of region proposal, procedure that straight away perform position regression and classification have also been put forward to object detection. Single shot detection(SSD) [11] give rise to collective fixed size anchor boxes on each one of the location in order to forecast classification score. Another algorithm that is widely used in object detection in an image is You Only Look Once(YOLO). Yolo has been used in many applications that needs to do object detection. YOLO [12] divides the image into different grids and at the same time predicts the bounding boxes and classification score for each one of the grid. In this work, study the behavior of different optimization methods distinguish itself for image segmentation and object classification using modern deep learning based object detection strategy. This paper discussed about the complexities that has been faced in autonomous driving. In autonomous driving object detection can provide valuable contextual information about the vehicles surroundings.

## 3   Methodology

This section gives a brief description of the entire procedure carried out for object detection dividing into subsequent subsections discussed below:

R-CNN is a pretrained region proposal model. It is a first generation computational neural network. Region proposals are used to localize objects within an image. Mask R-CNN is also works with two stage process, Region Proposal Network(RPN), in the second stage, Mask R-CNN works side by side to predicting the class of each object and giving it bounding box, it also gives binary mask for each output (Fig. 2).

In neural network, more layer suggests that the network will perform good, but the weights of the starting layer of neural network won't be updated correctly through back propagation. Due to this error gradient is back propagated to the previous layers and the frequent multiplication makes the gradient small, resulting a low level output which will be not precise for the model. Residual Network(Res-Net) resolve this complication by using identity matrix [13]. ResNet make use of a skip connection in which native input is also added to the output of the convolutional network [14]. For the neural network block X is our input and the model will learn true distribution(H(X). The remaining between the input and output can be denoted as Eq. 1

$$R(X) = Output - Input = H(X) - X \tag{1}$$

The Region Proposed Network(RPN) [15] takes the input image and generate candidate boxes which is known as anchor boxes. RoIPooling losses a lot of data in the process and to overcome the problem RoIAlign has been used in the assignment to tackle the problems and to bring down all the object proposals to their actual size.

Region Proposal Model has a cost function to train and it can be written as:

$$Loss_{RPN} = Loss(\{p_i\}, \{t_i\}) + Loss_{featuremap} \tag{2}$$

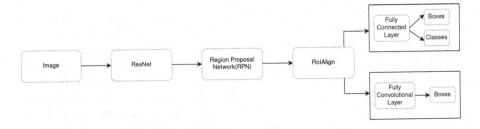

**Fig. 2.** Detailed workflow of Mask R-CNN.

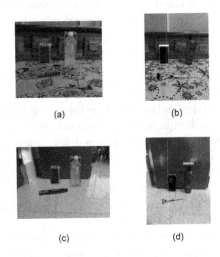

**Fig. 3.** Object detection has been done in different conditions using different algorithms. In figure (a) and (c) object detection has been done using Mask R-CNN. In figure (b) and (d) object detection has been done using YOLO V5

## 4 Result and Analysis

The Mask R-CNN and YOLO V5 algorithm is able to identify the different class of object present in the image. Figure 3 shows the sample test image that is obtain by two of the algorithms. Mean average precision of both of the models is shown in the Fig. 4. Accuracy of Mask R-CNN and YOLO V5 is given in the Table 1

**Table 1.** Accuracy of the model

| Mask R-CNN | YOLO V5 |
|------------|---------|
| 99.77%     | 97.7%   |
| 99.89%     | 98.99%  |

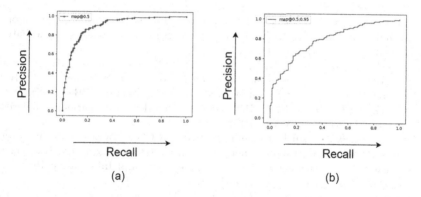

**Fig. 4.** Mean average precision for both of the models.

## 5   Conclusion

The segmentation of multi class object from a data set has been successfully demonstrated using Mask R-CNN and YOLO V5. A good precision of the required function will be important for object detection. The main goal of this project is to determine a object in an image with high accuracy and it has been achieved in a exact manner. The training model shows that the accuracy of the obtained output is high. For future work, the mask rcnn model will be implemented for the real time object detection in a faster approach to overcome the challenges that we faced in this work.

## References

1. Saad, A., Mohamed, A.A.: An integrated human computer interaction scheme for object detection using deep learning. Comput. Electr. Eng. **96**, 107475 (2021)
2. Brandenburg, S., Machado, P., Shinde, P., Ferreira, J.F., McGinnity, T.M.: Object classification for robotic platforms. In: Silva, M.F., Luís Lima, J., Reis, L.P., Sanfeliu, A., Tardioli, D. (eds.) ROBOT 2019. AISC, vol. 1093, pp. 199–210. Springer, Cham (2020). https://doi.org/10.1007/978-3-030-36150-1_17
3. Chaudhuri, R., Deb, S.: Machine learning approaches for microscopic image analysis and microbial object detection (MOD) as a decision support system. In: 2022 First International Conference on Artificial Intelligence Trends and Pattern Recognition (ICAITPR), pp. 1–6. IEEE (2022)
4. Prajwal, P., Prajwal, D., Harish, D.H., Gajanana, R., Jayasri, B.S., Lokesh, S.: Object detection in self driving cars using deep learning. In: 2021 International Conference on Innovative Computing, Intelligent Communication and Smart Electrical Systems (ICSES), pp. 1–7 (2021)
5. Coşkun, M., Uçar, A., Yildirim, Ö., Demir, Y.: Face recognition based on convolutional neural network. In: 2017 International Conference on Modern Electrical and Energy Systems (MEES), pp. 376–379 (2017)
6. Ji, S., Wei, S., Meng, L.: Fully convolutional networks for multisource building extraction from an open aerial and satellite imagery data set. IEEE Trans. Geosci. Remote Sens. **57**(1), 574–586 (2019)

7. Sun, P., et al.: Sparse R-CNN: end-to-end object detection with learnable proposals. In: Proceedings of the IEEE/CVF Conference on Computer Vision and Pattern Recognition (CVPR), pp. 14454–14463 (2021)
8. Girshick, R., Donahue, J., Darrell, T., Malik, J.: Rich feature hierarchies for accurate object detection and semantic segmentation. In: Proceedings of the IEEE Conference on Computer Vision and Pattern Recognition (CVPR) (2014)
9. Girshick, R.: Fast R-CNN. In: Proceedings of the IEEE International Conference on Computer Vision (ICCV) (2015)
10. Ren, S., He, K., Girshick, R., Sun, J.: Faster R-CNN: towards real-time object detection with region proposal networks. In: Cortes, C., Lawrence, N., Lee, D., Sugiyama, M., Garnett, R., (eds.), Advances in Neural Information Processing Systems, vol. 28. Curran Associates Inc (2015)
11. Kanimozhi, S., Gayathri, G., Mala, T.: Multiple real-time object identification using single shot multi-box detection. In: 2019 International Conference on Computational Intelligence in Data Science (ICCIDS), pp. 1–5 (2019)
12. Han, X., Chang, J., Wang, K.: You only look once: unified, real-time object detection. Procedia Comput. Sci. **183**, 61–72 (2021)
13. Lu, X., Kang, X., Nishide, S., Ren, F.: Object detection based on SSD-ResNet. In: 2019 IEEE 6th International Conference on Cloud Computing and Intelligence Systems (CCIS), pp. 89–92 (2019)
14. Li, Z., Peng, C., Yu, G., Zhang, X., Deng, Y., Sun, J.: Detnet: design backbone for object detection. In: Proceedings of the European Conference on Computer Vision (ECCV) (2018)
15. Nabati, R., Qi, H.: Rrpn: radar region proposal network for object detection in autonomous vehicles. In: 2019 IEEE International Conference on Image Processing (ICIP), pp. 3093–3097 (2019)

# Performance Analysis of Routing Protocol for Low Power and Lossy Networks (RPL) for IoT Environment

Subodh Manvi, K. R. Shobha[✉], and Soumya Vastrad

Department of Electronics and Telecommunication Engineering, M S Ramaiah Institute of Technology, Bangalore, India
subodhmanvi97@gmail.com, shobha_shankar@msrit.edu, soumyav4@gmail.com

**Abstract.** The Internet of Things (IoT) has emerged over the last few years as an intriguing and promising paradigm that is expanding quickly and opening a wide range of applications for humanity. Due to the low power, lossy, and resource-limited nature of IoT devices, an effective routing protocol is very essential for which the Routing Protocol for Low-Power and Lossy Network (RPL) was standardized by the Internet Engineering Task Force (IETF) as RFC6550 in 2012. A path-constructing method termed as objective function (OF) is used in RPL to optimize the routing paths considering different metrics. The two standard objective functions used in RPL are Minimum Rank with Hysteresis Objective Function (MRHOF) and Objective Function Zero (OF0). To determine if the algorithm is appropriate for a variety of dynamic scenarios in IOT, this work focuses on performance analysis of RPL utilising the two-objective function. Network Convergence Time, Power Consumption, Control Overhead, Packet Delivery Ratio (PDR), and Latency are the measures used to evaluate the protocol's performance. According to the findings, MRHOF provides superior network quality performance than OF0.

**Keywords:** IoT · RPL · 6LoWPAN · Static networks · Dynamic networks · Homogenous and heterogeneous traffic

## 1 Introduction

Internet of Things (IoT) has become a popular term in the twenty-first century. The constantly developing IoT is regarded as a groundbreaking technology that connects common products to the internet. Wireless Sensor Networks (WSNs) are crucial to the development and expansion of IoT.

WSN is recognized as a cutting-edge data collection network that significantly raises the dependability and effectiveness of infrastructure systems. Low power and lossy networks are characterized by communication links having high packet loss and low throughput. LLNs are basically networks consisting of many embedded devices with limited memory, power and processing resources interconnected by a variety of links such as IEEE 802.15.4 or low power Wi-Fi. One of the key standards supporting LLNs is

A. R. Molla et al. (Eds.): ICDCIT 2023, LNCS 13776, pp. 341–348, 2023.
https://doi.org/10.1007/978-3-031-24848-1_25

IEEE 802.15.4 which forms the backbone of WSNs.There is a wide scope of application areas for LLNs, including industrial monitoring, building automation and smart cities projects.

IoT devices have less memory, slow transfer rate, andlow energy consumption. 6LoWPAN was developed to transport IPv6 datagrams over restricted networks (Fig. 1) as IPV6 does not consider the characteristics of constrained devices.RPL is intended to work over a variety of connection layers, including those that are confined, may be lossy, or are frequently used in conjunction with host or router devices that are extremely constrained. Moreover, it is used to respond to LLNs requirements such as load balancing, unbalanced energy consumption and network traffic. The RPL routing protocol for low power and lossy networks builds a Destination Oriented Directed Acyclic Graph (DODAG) based on a set of metrics and constraints. The best parent or the most efficient route to the destination is chosen and specified using the OF as indicated in Fig. 2.

The objective function MRHOF uses the Expected Transmission Count (ETX) as a metric whereas OF0 uses hop count to calculate the optimal root. The choice of path selection mechanism is indicated in Fig. 3.

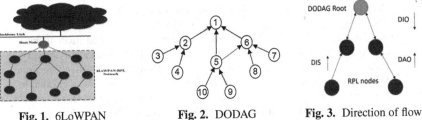

**Fig. 1.** 6LoWPAN based IoT network     **Fig. 2.** DODAG topology construction     **Fig. 3.** Direction of flow of control messages

**Minimum Rank with Hysteresis Objective Function (MRHOF):** IT is an objective function aimed at selecting routes which minimize a metric, while using hysteresis to decrease churn corresponding to small changes in metric values.The objective function is achieved in two steps. First, the minimum path cost is found out. Second, it switches to that minimum rank path if and only if it is shorter than the already in use current path by at least a given threshold. The said mechanism is called 'Hysterisis'.

**OF0:** IT is an objective function that seeks to link a DODAG Version that enables adequate connectivity to a particular group of nodes, regardless of whether the optimal way will be made available in accordance with a particular metric. The metric used is minimum hop count, and the path with the lowest hop count is given the highest priority to be chosen by OF0. If one is available, OF0 chooses a plausible backup successor in addition to a preferred parent.

## 2    Related Work

LLNs are a class of networks consisting of both routers and their interconnects which are in turn constrained [1]. LLN routers regularly work with constraints on processing power, memory, and energy (battery power). Their interconnects are described by high loss rates, low information rates, and instability.RPL was created in 2012 by the IETF as the routing protocol for LLNs [2] and was later standardized as RFC6550 [1]. Lot of work has been carried out on RPL-based conventions related to security, vigor, dependability, energy efficiency and future possibilities of RPL in connection to IoT systems [3, 4].

MRHOF [5] picks routes to minimize a metric while utilizing hysteresis to reduce churn in reaction to modest metric changes. RPL can find stable minimum-latency pathways from the nodes to a root in the Directed Acyclic Graph (DAG) instance [RFC6550] by combining MRHOF with the latency metric.RPL specification [6] describes a general Distance Vector protocol that can be applied to numerous different network types by using certain OFs. The RPL specification specifies limitations on how nodes choose a parent set from among their neighbors.

Different performance metrics like Network convergence time, ETX, power consumption, Packet Delivery Ratio (PDR), hop, latency, and Packet Delivery Ratio in Mobility Nodes are considered for the evaluation of MRHOF and OF0 [7]. For analyzing the performance of RPL operating systems (OS) like Contiki, TinyOS, and FreeRTOS are popular. Among these Contiki OS is popular as it uses less memory footprint and the code written for simulation can be ported directly to the hardware [8].

In a real time usage, IoT applications must support both homogenous and heterogeneous traffic as well as static and dynamic networks [9]. QWL-RPL uses OF based on queue size and workload for heterogeneous traffic to accomplish a dependable route with better performance.Some researchers have addressed the problems of energy consumption, signaling expense, handover delay, and route stability for mobile nodes in RPL [10]. Mobility Enhanced RPL uses a new strategy to identifymobile nodes and refresh the DODAG to improve performance of RPL for mobile nodes [11, 12].

Cross layer approaches have been tried to minimize the power consumption of RPL by using a single metric which is a combination ETX and local power required for transmission [13]. Several Physical and MAC layer metrics have been analyzed to improve the efficiency and stability of RPL Since each metric has its own drawback combination of different metrics have been tried out to improve the overall performance of RPL in real time environments [14].

RPL is essentially a routing protocol designed for data routing in static topologies.With the penetration of IOT into applications like smart cities [15] and healthcare, testing the performance of RPL by providing mobility to nodes using different mobility models [16] as well as different types of traffic has become a primary concern which is not addressed by many researchers.

## 3 Proposed Work

The study aims at exploring the performance of RPL using different objective functions in varied IOT environments and understands their performance statistics. It is proposed to study the suitability of RPL for homogeneous and heterogeneous traffic in both static and dynamic networks. It is proposed to introduce mobility to the nodes and conduct a detailed study to find out if RPL is suitable for real time dynamic IoT environment.

## 4  Simulation and Network Setup

The simulation is performed in Cooja by utilizing Unit Disk Graph Model (UDGM) as displayed in Fig. 5. The study uses the Cooja test system's example network, which includes five to thirty client hubs and one server hub as the DODAG's root (Fig. 4).

**Fig. 4.**  Cooja simulator view

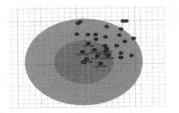

**Fig. 5.**  30 nodes traffic scenario

## 5  Results and Analysis

**Table 1.**  Simulation parameters

| Parameters | Value |
|---|---|
| Simulation time | 300000 ms |
| Number of nodes | 10,20,30 |
| PHY & MAC protocol | 802.15.4 with CSMA |
| Packet size | 127 bytes |
| uIP payload buffer size | 140 bytes |
| Buffer occupancy | 4 packets |
| Mote device model | Z1 Zolertia |
| CWmin | 0 |
| CWmax | 31 |
| Maximum back-off stage | 5 |
| Maximum retry limits | 8 |
| Traffic | 1 ppm, 2 ppm |
| Scripting Version | Python 3.7 |
| Transmission range of each node | 50m |
| Interference range | 100m |

The parameters set for simulation in Cooja simulator are as listed in Table 1. This study focuses on performance analysis of RPL using MRHOF and OF0 objective functions. The simulations performed integrate mobility of nodes which is a critical phenomenon for real time IoT applications. The simulation in homogenous environment is performed for traffic rates of 1 packet per minute (ppm) and 2 packets per minute (ppm).

**Scenario: 1: Static and Mobile Nodes in Homogeneous Environment**

Graphs in Fig. 6 show that traffic rates have very little impact on the convergence time and convergence time follows an increasing path against the increasing number of nodes, both in case of MRHOF and OF0. Power consumption in case of a homogenous environment decreases with the rise in the number of nodes while it is observed that consumption in MRHOF is observed to be more than that compared to OF0. OF0 shows a gradual increase in latency in static as well as mobile environments with the increase in traffic rates as well as the number of nodes in homogenous environments. MRHOF shows a spike in latency when the traffic and the nodes increase by a larger amount, indicating that MRHOF is suitable for heavy traffic conditions and large networks whereas OF0 is better suited for applications requiring delayed responses.Control overhead is typically more significant in dense networks than in networks with sparse population as well as Control overheads are higher in case of MRHOF when compared to OF0. This shows that a significant portion of the battery's power is lost during the transmission and receiving of control overheads. This is because the network becomes extremely unstable because of the OF0's inability to address the congestion and load balancing problems.Packet delivery ratio is not affected much in either of the OFs even with the introduction of mobility.

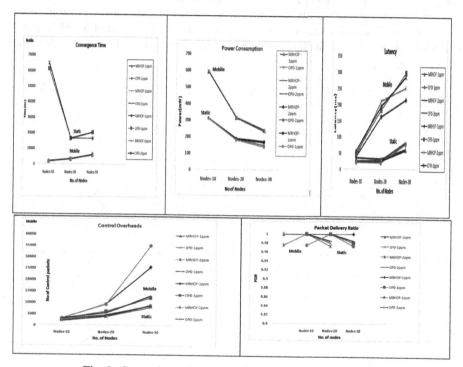

**Fig. 6.** Comparison of parameters in homogenous environment

## Scenario 2: Static Nodes and Mobile Nodes in Heterogeneous Environment

Convergence time follows a decreasing curve against the increasing number of nodes, both in the case of MRHOF and OF0 making it better suited for applications requiring faster network setup, while in the case of mobile nodes, it follows a flat path against the increasing number of nodes initially for both objective functions. But as the number of nodes increases, a sudden spur in the curve indicates sudden rise in convergence time in mobile environments. The energy consumption is observed to be more in case of OF0 than in MRHOF in both static as well as mobile environments. This makes MRHOF networks more scalable and reliable for real time implementation of IoT. The end-to-end delay in OF0 reduces as additional nodes are added. There is a longer delay because, regardless of congestion, OF0 selects the route with the fewest hops in homogenous environments and as a result, delays are increased by both link-level and load-level congestion whereas, it can be observed that latency increases gradually in case of OF0 compared to MRHOF in mobile environments.OF0 is suitable for applications that require delayed response to applications. Thereby, it can be concluded that MRHOF is better suited for real time IoT scenarios with heavy traffic conditions and large networks. Control overheads for MRHOF are larger compared to OF0 in both static as well as mobile environments which can cause a significant loss of the battery power in the transmission and reception of control overheads. This is due to the inability of OF0 to address the congestion and load balancing problems. Packet delivery ratio has not varied much for both OFs in static environments but follows a decreasing trend in mobile heterogeneous environments. The PDR is observed to fall more significantly in case of MRHOF when compared to OF0.This indicates OF0 is better suited for IoT applications in dynamic network (Fig. 7).

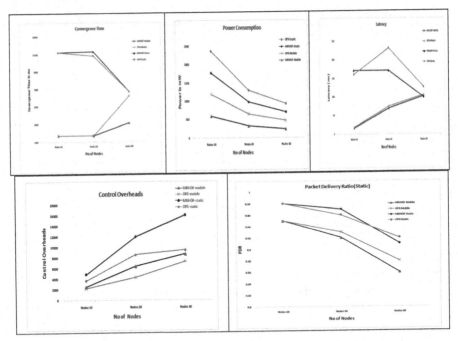

**Fig. 7.** Comparison of parameters in heterogeneous environments.

## 6 Conclusion and Future Scope

Analysis has been carried out with homogenous and heterogeneous traffic for both dynamic and static networks. MRHOF outperforms OF0 in terms of network dependability, however OF0 has a faster rate of network convergence and consumes lesser power during the DODAG convergence time. OF0 is appropriate for usage in networks with mobile nodes and networks with power limits.

Machine learning can be used as a method to provide a self-learning, self-adaptive RPL protocol to suit different traffics loads and applications.

## References

1. Accettura, N., Grieco, L.A., Boggia, G., Camarda, P.:Performance analysis of the RPL routing protocol. In: 2011 IEEE International Conference on Mechatronics, pp. 767–772 (2011). https://doi.org/10.1109/ICMECH.2011.5971218
2. Karrufa, H., Al-Kashoash, H.A.A., Kemp, A.H.: RPL-based routing protocols in IoT applications: a review. IEEE Sens. J. 19(15), 5952–5967 (2019). https://doi.org/10.1109/JSEN.2019.2910881
3. Hadaya, N.N., Alabady, S.A.: Improved RPL protocol for low-power and lossy network for IoT environment. SN Comput. Sci. 2, 341 (2021). https://doi.org/10.1007/s42979-021-00742-1
4. Winter, T., et al.: RPL : IPv6 routing protocol for low power and lossy networks. RFC 6550, IETF ROLL WG (2012)
5. Gnawali, O., Levis, P.: The minimum rank with hysteresis objective function. RFC 6719 (Proposed Standard), Internet Engineering Task Force (2012)
6. Thubert, P.: Objective Function Zero for the Routing Protocol for Low-Power and Lossy Networks (RPL). RFC 6552 (Proposed Standard) (2012)
7. Pradeska, N., Widyawan, W.N., Kusumawardani, S.S.: Performance analysis of objective function MRHOF and OF0 in routing protocol RPL IPV6 over low power wireless personal area networks (6LoWPAN). In: 2016 8th International Conference on Information Technology and Electrical Engineering (ICITEE), pp. 1–6 (2016). https://doi.org/10.1109/ICITEED.2016.7863270
8. Mosaddeq, A., Zikria, Y.B., Hahm, O., Yu, H., Bashir, A.K., Kim, S.W.: A survey on resource management in IoT operating systems. IEEE Access 6, 8459–8482 (2018)
9. Musaddiq, A., Zikria, Y.B., Zulqarnain, S.W., et al.: Routing protocol for Low-Power and Lossy Networks for heterogeneous traffic networks. J. Wirel. Comput. Netw. 2020, 21 (2020). https://doi.org/10.1186/s13638-020-1645-4
10. Yadav, R.K., Awasthi, N.: A survey on enhanced RPL: addressing the mobility in RPL. In: 2020 Fourth International Conference on I-SMAC (IoT in Social, Mobile, Analytics and Cloud) (I-SMAC), pp. 1189–1195 (2020). https://doi.org/10.1109/I-SMAC49090.2020.9243405
11. Korbi, E., Ben Brahim, M., Adjih, C., Saidane, L.A.: Mobility enhanced RPL for wireless sensor networks. In: 2012 Third International Conference on The Network of the Future (NOF), pp. 1–8 (2012). https://doi.org/10.1109/NOF.2012.6463993
12. Lindsey, S., Raghavendra, C.S.: PEGASIS: power-efficient gathering in sensor information systems. In: Proceedings, IEEE Aerospace Conference, p. 3 (2002). https://doi.org/10.1109/AERO.2002.1035242
13. Estepa, R., Estepa, A., Madinabeitia, G., García, E.: RPL cross-layer scheme for IEEE 802.15.4 IoT devices with adjustable transmit power. IEEE Access 9, 120689–120703 (2021). https://doi.org/10.1109/ACCESS.2021.3107981
14. Iova, O., Theoleyre, F., Noel, T.: Stability and efficiency of RPL under realistic conditions in wireless sensor networks. In: 2013 IEEE 24th Annual International Symposium on Personal, Indoor, and Mobile Radio Communications (PIMRC), pp. 2098–2102 (2013). https://doi.org/10.1109/PIMRC.2013.6666490

15. Sterbenz, J.P.G.: Smart city and IoT resilience, survivability, and disruption tolerance: challenges, modelling, and a survey of research opportunities. In: 2017 9th International Workshop on Resilient Networks Design and Modeling (RNDM), pp. 1–6 (2017). https://doi.org/10.1109/RNDM.2017.8093025
16. Safaei, B., et al.: Impacts of mobility models on RPL-based mobile IoT infrastructures: an evaluative comparison and survey. IEEE Access **8**, 167779–167829 (2020). https://doi.org/10.1109/ACCESS.2020.3022793

# A Novel Image Steganography Technique Using AES Encryption in DCT Domain

Aman Sahu[✉] and Chittaranjan Pradhan

School of Computer Engineering, KIIT University, KIIT Road,
Bhubaneshwar 751024, Odisha, India
aman.sahu0520@gmail.com

**Abstract.** Steganography is a process to conceal the existence of a message. Image Steganography involves hiding the message onto a cover image to obtain a resultant stego image. This paper discusses a procedure for hiding coloured message images onto coloured cover images using a lossy Image Steganography technique. The secret images undergo Discrete Cosine Transformation, quantization and AES encryption before embedding using the Least Significant Bit embedding (LSB) technique onto the cover image. Finally, Peak signal-to-noise ratio (PSNR), Structural Similarity Index measurement (SSIM), zero normalized cross-correlation (ZNCC) are used for comparing the cover and stego images as well as the extracted and original images.

**Keywords:** Steganography · Image steganography · DCT · LSB embedding · Encryption · AES

## 1 Introduction

Although the internet has made communication easier, secure transmission requires the use of either cryptography or steganography. Steganography is concerned with hiding the existence of a secret message whereas cryptography is the technique of protecting the message's contents. This paper discusses a lossy approach to hide image content using image steganography for different message image sizes of $64 \times 64$, $128 \times 128$, $192 \times 192$, $256 \times 256$ and $512 \times 512$ pixels wherein the preprocessing consists of discrete cosine transformation, quantization of the message image, encryption using the AES 128 bit encryption and embedding onto the cover image using the least significant bit (LSB) embedding technique. We also compare the cover image to the stego-image as well as the original message image to the extracted message image using the mean squared error, peak signal-to-noise ratio, structural similarity index measurements, zero normalized cross-correlation parameters.

### 1.1 Discrete Cosine Transform

DCT is a quick computing type of Fourier transform which converts a signal or image to the frequency domain from the spatial domain. It aids to split the image

A. R. Molla et al. (Eds.): ICDCIT 2023, LNCS 13776, pp. 349–354, 2023.
https://doi.org/10.1007/978-3-031-24848-1_26

into subbands of the spectrum composed of high, medium, and low-frequency components. DCT separates images into different frequencies. We implement DCT as stated in [1]

The DCT equation computing the i, j$^{th}$ entry of the DCT of an image is

$$DCT(i,j) = \left(\tfrac{1}{2M}\right)^{1/2} G(i).G(j) \sum_{x=0}^{M-1}\sum_{y=0}^{M-1} pixel(x,y) \cos\left[\tfrac{\pi.i}{2.M}(2x+1)\right]\cos\left[\tfrac{\pi.j}{2.M}(2y+1)\right] \tag{1}$$

where

$$G(k) = \begin{cases} \tfrac{1}{\sqrt{2}}, & \text{if } k=0 \\ 1, & k \geq 0 \end{cases} \tag{2}$$

pixel(x,y) represents the x, y$^{th}$ pixel of an image. M is the block size on which the DCT is applied.

## 2   Relevant Work

In 2019, Khalaf et al. [2] compare standard LSB embedding (no encryption no compression) steganography to their proposed technique. The proposal is to encrypt the secret message followed by LSB embedding and DCT transformation. The proposed algorithm works on the cover image, breaking it into blocks, applying DCT transformation to it, followed by LSB embedding of the LSB of the secret file, replacing the LSB of each DCT coefficient. Similarly, extraction is performed in reverse to retrieve the secret file. In 2020, Nadish et al. [3] proposed a technique of image steganography wherein the data is embedded in the edge pixels of the carrier image in existing image steganography techniques with data hiding in the DCT domain algorithm. In 2020, Baziyad et al. [4] a precise segmentation process using the region-growing segmentation method is applied to maximize the homogeneity level between pixels in a segment, maximizing the hiding capacity while achieving improved stego quality. The cover image is segmented into homogenous segments in order to utilize the energy compaction property of the DCT. In 2020, Hongzhu et al. [5] proposed an algorithm that utilizes cover image scrambling and an optimized modified quantization table in the DCT domain to get different embedding effects.

## 3   Proposed Technique

The proposed embedding technique works on the message image, transforms it using Discrete Cosine Transformation (DCT), quantization [6], Advanced Encryption Standard (AES) 128-bit encryption using a secret key, and Least Significant Bit [7] embedding onto a cover image.

### 3.1   Embedding

Input: Message Image, Cover Image, Key

- Split message image to RGB channels
- For each channel, extract an 8 × 8 block (total three blocks)
- For each block
    - Apply DCT transform
    - Apply Quantization
    - Apply Zig zag ordering [1]
    - Encrypt data with AES encryption using key
- Concatenate processed block data together
- Embed concatenated data onto the cover image using LSB embedding

Output: Stego image as depicted in Fig. 1

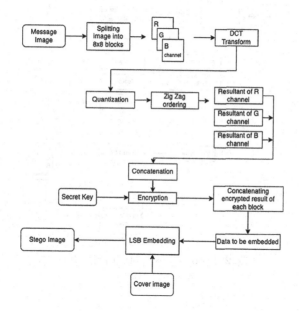

**Fig. 1.** Proposed embedding procedure

### 3.2   Extraction

Input: Stego image, Key

- Read Stego image
- LSB extraction of embedded data
- Split extracted data into blocks of 192 elements (8 × 8 block x3 channels)
- Decrypt each block of data with AES 128-bit technique using key
- For each decrypted block

- Split the 192 elements into 3 subblocks, each having $8 \times 8 = 64$ elements. These correspond to the r,g,b channels for each $8 \times 8$ block to be reconstructed
- For each subblock
  * Apply Zig Zag decoding to get an $8 \times 8$ matrix [1]
  * Apply de-quantization
  * Apply IDCT
  * Each subblock then forms the corresponding $8 \times 8$ block of r, g, b channels
  * Fill the corresponding pixel values onto the new image

Output: Lossy Reconstructed Message Image As depicted in Fig. 2

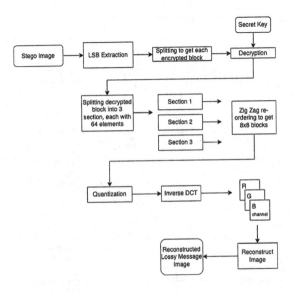

**Fig. 2.** Proposed extraction procedure

## 4    Results

We compare the cover image and the stego image using the Mean Squared Error(MSE), Peak Signal-To-Noise Ratio (PSNR) metrics to compare the image compression quality and Structural Similarity Index Measurement(SSIM) to quantify the image quality degradation due to compression in our message image. The MSE represents the cumulative squared error between the compressed and the original image, whereas PSNR computes the peak signal-to-noise ratio, in decibels, between two images. The lower the value of MSE, the lower the error, and the higher the PSNR, the better the quality of the compressed, or reconstructed image (Tables 1 and 2).

## 4.1 Tables

**Table 1.** MSE, PSNR and SSIM values

| Message image size | MSE | PSNR | SSIM |
|---|---|---|---|
| 64[a] | 0.006728262 | 69.85177443[b] | 0.99995267 |
| 128 | 0.026886606 | 63.83544385 | 0.99981275 |
| 192 | 0.060410066 | 60.31971051 | 0.99960699 |
| 256 | 0.107273259 | 57.82588887 | 0.99933051 |
| 512 | 0.428690458 | 51.80936544 | 0.99787260 |

This table shows the MSE, PSNR, and SSIM values while comparing the cover image and the stego image, for different sizes of the message image

[a] The image size is NxN pixels

[b] PSNR values are in dB

**Table 2.** SSIM - Original and Extracted Message Image

| Message image size | SSIM |
|---|---|
| 64 | 0.936997721 |
| 128 | 0.987625997 |
| 192 | 0.992307347 |
| 256 | 0.997509904 |
| 512 | 0.956731626 |

This table shows the SSIM values while comparing the original message image and extracted image, for different sizes of the message image

## 5  Conclusion

In this paper, we applied Discrete Cosine Transform and Quantisation techniques to message image, AES 128-bit encryption, and embedding in a cover image using Least Significant Bit Embedding method to implement Image Steganography. The cover image and stego-images were compared using Mean Squared Error and Pixel Strength to Noise Ratio metrics. The MSE values increase from 0.00672 for $64 \times 64$ coloured message image, to 0.42869 for $512 \times 512$ message image. PSNR values showed an inverse relation, where $64 \times 64$ message image had a PSNR value of 69.85177 dB, which decreased to 51.80937 dB for a $512 \times 512$ message image. The structural similarity index between the cover and stego-images also decreased from 0.99995 for $64 \times 64$ image to 0.99787 for a $512 \times 512$ image. When comparing the original message image to extracted image, their structural similarity index scores were 0.93699 for 64 × 64 message image increasing to 0.95673 for a 512 × 512 message image.

For future works to the proposed approach, we used a single-level DCT transform, as opposed to a Discrete Wavelet Transform which is a multi-level transform. Further, the proposed algorithm uses a lossy compression mechanism, which can be replaced by a lossless compression or similar technique so that the extracted message image is exact to the original one.

# References

1. Cabeen, K., Gent, P.: Image compression and the discrete cosine transform. In: Math 54. College of Redwood
2. Khalaf, A.A.M., Fouad, O., Hussein, A., Hamed, H., Kelash, H., Ali, H.: Hiding data in images using DCT steganography techniques with compression algorithms 17 (2019). https://doi.org/10.12928/telkomnika.v17i3
3. Ayub, N., Selwal, A.: An improved image steganography technique using edge based data hiding in DCT domain. J. Interdisc. Math. 23(2), 357–366 (2020). https://doi.org/10.1080/09720502.2020.1731949
4. Baziyad, M., Rabie, T., Kamel, I.: Achieving stronger compaction for DCT-based steganography: a region-growing approach. In: Rocha, Á., Adeli, H., Reis, L.P., Costanzo, S., Orovic, I., Moreira, F. (eds.) WorldCIST 2020. AISC, vol. 1160, pp. 251–261. Springer, Cham (2020). https://doi.org/10.1007/978-3-030-45691-7_24
5. Dai, H., Cheng, J., Li, Y.: A novel steganography algorithm based on quantization table modification and image scrambling in DCT domain. Int. J. Pattern Recognit Artif Intell. 35, 2154001 (2020). https://doi.org/10.1142/S0218001421540001X
6. Roberts, E.: Lossy data compression: JPEG. https://cs.stanford.edu/people/eroberts/courses/soco/projects/data-compression/lossy/jpeg/index.htm
7. Singh, A.K., Singh, J.: Steganography in images using LSB technique (2015)
8. Bansal, D., Chhikara, R.: An improved DCT based steganography technique. Int. J. Comput. Appl. 102, 46–49 (2014). https://doi.org/10.5120/17887-8861

# Text Classification Using Correlation Based Feature Selection on Multi-layer ELM Feature Space

Rajendra Kumar Roul[1]([⊠])(iD), Jajati Keshari Sahoo[2](iD), and Gaurav Satyanath[3](iD)

[1] Department of Computer Science, Thapar Institute of Engineering and Technology, Patiala, Punjab, India
raj.roul@thapar.edu
[2] Department of Mathematics, BITS Pilani, K.K.Birla, Goa Campus, Sancoale, India
jksahoo@goa.bits-pilani.ac.in
[3] Department of Electrical and Computer Engineering, Carnegie Mellon University, Pittsburgh, PA, USA
gsatyana@andrew.cmu.edu

**Abstract.** Text data available on the Web are generally unstructured. Text classification, a machine learning technique, has proven to be a great alternative to structure textual data in a cost-effective, faster, and scalable manner. This study examines the feature space of Multilayer ELM (ML-ELM) for the classification of text data with the help of a novel feature selection technique termed as Correlation-based Feature Selection (*CRFS*). Experimental results show that the feature space of ML-ELM is better for text classification compared to the traditional vector space.

**Keywords:** Correlation · ELM · Feature selection · ML-ELM · Term frequency

## 1 Introduction

Transferring unstructured text data into a structural form to identify meaningful patterns is known as text data mining. Text mining which uses natural language processing to extract the hidden pattern from text data has a wide range of applications such as text clustering, categorization, sentiment analysis, text retrieval, text summarization etc. [1]. Many techniques are used for text classification, and they all fall under three types of systems:- machine learning-based systems, rule-based systems, and hybrid systems. Traditional machine learning techniques for text classification are support vector machine, naive Bayes, logistic regression, maximum entropy, decision trees, etc. The literature on text classification has been dominated by deep learning techniques motivated by the outstanding results of deep neural networks in text mining, image processing, and natural language processing [2]. Although the existing machine and deep learning techniques have many advantages, there are still some limitations that cannot be ignored, such as

i. Machine learning techniques cannot capture the discriminative features automatically from the training data. Their performances heavily depend on data representation, and it is labor-intensive.

A. R. Molla et al. (Eds.): ICDCIT 2023, LNCS 13776, pp. 355–361, 2023.
https://doi.org/10.1007/978-3-031-24848-1_27

ii. Deep learning techniques have limitations such as they need large memory bandwidth, huge training time is required because of backpropagation, architecture is very complex, preserving interdependencies among the internal layers for a long time is quite difficult etc.

iii. Displaying the data becomes more complex when ample storage space is required due to the growth of the dataset size.

iv. When the input data grows exponentially on the limited dimensional space, distinguishing input features on *TF-IDF* vector space becomes challenging.

Hence it is not easy to generalize the text classification models to a new domain. An efficient deep learning classifier called Multi-layer ELM was introduced in the year 2013 by Kasun et al. [3] to address the above problems. In this vein, this research investigated the feature space of Multi-layer ELM [3,4], which extensively exploits the advantages of *ELM feature mapping* [5,6] and ELM autoencoder to address the constraints mentioned above. The goal of this study is to investigate the extended feature space of ML-ELM (*HDFS-MLELM*) and to thoroughly test this feature space for text classification in comparison to the *TF-IDF vector space* (*VS-TFIDF*).

The major contributions of the paper can be summarized as follows:

i. It is clear from the past literature that no research on classification using text data has been done on the Multi-layer ELM's enlarged feature space. As a result, in light of the benefits mentioned above, this study can be considered as a new direction in the text classification domain.

ii. A novel feature selection termed as Correlation-based Feature Selection *CRFS* is proposed for selecting the essential features from a big corpus, which enhances the performance of the classification process.

iii. This work studies *HDFS-MLELM*, and uses text data to thoroughly investigate multiple classification algorithms on *HDFS-MLELM* and on *VS-TFIDF*.

## 2 Methodology

1. Pre-processing:
Let corpus $P$ consists of $C$ classes. At the beginning of the feature engineering, all documents of each class are combined into a single set called $D_{large}$. Then lexical-analysis, stop-word deletion, HTML tag removal, and stemming [1] are done on $D_{large}$. Natural Language Toolkit[2] is used to extract index terms from $d_{large}$. After completing the basic data cleaning, the first set of features was derived from $D_{large}$, and a term-document matrix is created.

2. Correlation based feature selection (CRFS):
Using $k$-means clustering algorithm [7], the $D_{large}$ is divided into $n$ term-document clusters $td_i, i \in [1, n]$. The following steps discuss the methodology used to extract the important features from each cluster $td_i$.

[1] https://pythonprogramming.net/lemmatizing-nltk-tutorial/.
[2] https://www.nltk.org/.

i. Calculating Centroid: The centroid of $td_i$ is calculated using Eq. 1.

$$sc_i = \frac{\sum_{j=1}^{r} t_i}{r} \qquad (1)$$

Then cosine-similarity is computed between $t_j \in td_i$ and $sc_i$.

ii. Generating correlation matrix: Eq. 2 is used to find the correlation $(cr)^3$ between pair of terms $t_i$ and $t_j$.

$$cr_{t_i t_j} = \frac{C_{t_i t_j}}{\sqrt{(V_{t_i} * V_{t_j})}} \qquad (2)$$

where, $C_{t_i t_j}$ is the covariance between $t_i$ and $t_j$, and $V_{t_i}$ and $V_{t_j}$ are their variances respectively.

iii. Rejection of high correlated terms from $td_i$: Terms that are highly correlated in a cluster are generally considered as a sort of synonym, and hence they do not discriminate well in the cluster. To find those terms in $td_i$, initially, those terms with the maximum cosine-similarity score in $td_i$ get selected. Subsequently, a set of terms are identified which are highly correlated to $t_i$ ($\leq -0.87$ or $\geq 0.89)^4$ and that set of terms get removed from $td_i$. This step is repeated for the next highest cosine-similarity score term until $td_i$ gets exhausted. Finally, all highly correlated terms are removed from $td_i$.

iv. Computing Discriminating Power Measure (DPM): (DPM) [8] is a technique that measures the relevance, i.e., the importance of a term in a cluster. If the DPM score of a term inside an unbiased cluster is very high, then that term is an important term for that cluster. It is because many documents of the cluster contain that term. The cohesion or tightness of that term is very close to the cluster's center.

- For each $t_i \in td_i$, the document frequency inside ($DF_{in,t_i}$) and outside ($DF_{out,t_i}$) of $td_i$ are calculated using Eqs. 3 and 4 respectively.

$$DF_{in,t_i} = \frac{no.\ of\ documents\ \in td_i\ and\ have\ t_i}{no.\ of\ documents\ \in td_i} \qquad (3)$$

$$DF_{out,t_i} = \frac{no.\ of\ documents\ have\ t_i\ and\ \notin td_i}{no.\ of\ documents\ \notin td_i} \qquad (4)$$

- The difference between inside and outside document frequency of $t_i \in td_i$ is computed using Eq. 5.

$$DIFF_{td_i,t_i} = |DF_{in,t_i} - DF_{out,t_i}| \qquad (5)$$

- Equation 6 computes the DPM score of each term.

$$DPM(td_i, t_i) = \sum_{i=1}^{P} DIFF_{td_i,t_i} \qquad (6)$$

---

[3] https://libguides.library.kent.edu/SPSS/PearsonCorr.

[4] decided experimentally so that we will not lose more terms

v. Selection of candidate terms having High *DPM* scores: The terms of term-document cluster are arranged as per the *DPM* scores, and higher $k\%$ terms are selected as the candidate terms. This step is repeated for each $td_i$ so that every $td_i$ has top $k\%$ candidate terms in them.

3. Generating feature vector:
   To build the input feature vector, all the top $k\%$ features of each $td_i$ are merged into a list $L_{list}$.

4. ML-ELM feature mapping technique:
   ML-ELM cleverly leveraged the extended representation (i.e., $n < L$) technique of the ELM autoencoder [5], where $n$ and $L$ are the number of input and hidden layer nodes, respectively. $L_{list}$ is mapped from low-dimensional feature space to a higher-dimensional feature space of ML-ELM [9, 10]. Before the transformation, $L$ is set to a higher value than $n$. This makes all the features of $L_{list}$ linearly separable.

5. Classification on MLELM-HDFS:
   Different supervised learning algorithms employing $L_{list}$ as the input feature vector are run individually on TFIDF-VS and MLELM-HDFS respectively.

## 3   Analysis of Experimental Results

To conduct the experiment, four benchmark datasets (WebKB[5], Classic4[6], 20-Newsgroups[7], and Reuters[8] are used and the details are shown in Table 1. The proposed approach for the classification of text data is implemented using python 3.7.3 on Spyder IDE running on a system with Intel Core i11 processor, 32 GB RAM, and 24 GB GPU.

**Table 1.** Corpus statistics

| Datasets | Training docs used | Testing docs used | Terms selected for training | 10% of terms |
|---|---|---|---|---|
| 20-NG | 11292 | 7527 | 32269 | 3239 |
| DMOZ | 38000 | 31067 | 39886 | 3989 |
| Classic4 | 4256 | 2838 | 15970 | 1602 |
| Reuters | 5484 | 2188 | 13532 | 1351 |

### 3.1   Discussion

For practical reasons, six distinct classification approaches are performed on the *HDFS-MLELM* and on the *VS-TFIDF*, employing four datasets individually. The obtained accuracies and F-measures are shown in Figs. 1–4 and Figs. 5–8 respectively. Following conclusions are drawn from the findings:

---

[5] http://www.cs.cmu.edu/afs/cs/project/theo-20/www/data/.
[6] http://www.dataminingresearch.com/index.php/2010/09/classic3-classic4-datasets/.
[7] http://qwone.com/~jason/20Newsgroups/.
[8] http://www.daviddlewis.com/resources/testcollections/reuters21578/.

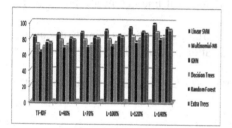

Fig. 1. 20-NG (Accuracy)

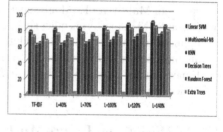

Fig. 2. Classic4 (Accuracy)

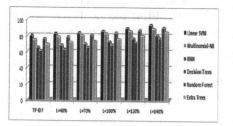

Fig. 3. Reuters (Accuracy)

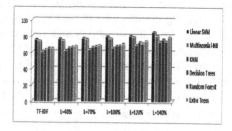

Fig. 4. DMOZ (Accuracy)

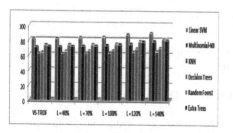

Fig. 5. 20NG (F1-measure)

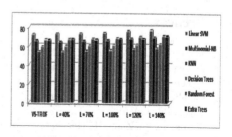

Fig. 6. Classic-4 (F1-measure)

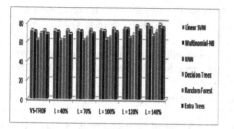

Fig. 7. Reuter (F1-measure)

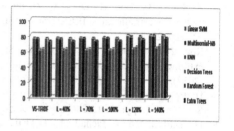

Fig. 8. DMOZ (F1-measure)

i. Compared to the *VS-TFIDF*, the obtained empirical findings in all three feature spaces of Multi-layer ELM are superior.
ii. Linear SVM outperforms other supervised learning algorithms, owing to its convex optimization property [11,12], and generalisation property [13,14], both of which are independent of feature space dimension.
iii. F-measure and accuracy are better in *HDFC-MLELM* whereas it is close on equal dimensional space.

## 4    Conclusion and Future Work

This paper studied a novel correlation-based feature selection technique that has been used for categorizing text data on the feature space of ML-ELM. Traditional machine learning algorithms are run on the feature space of ML-ELM and on TFIDF vector space to justify the suitability and importance of the proposed approach. Experimental results revealed that the feature space of ML-ELM is more suitable for text classification compared to the conventional TFIDF vector space. This work can further be extended by finding the variance of hidden layer weights to comprehend the whole operation of ML-ELM. Also, further studies are required to find the suitability of ML-ELM on a vast dataset having noise.

**Acknowledgement.** We thank Thapar Institute of Engineering and Technology for providing the seed money grant to do this research work.

## References

1. Du, J., Vong, C.-M., Chen, C.P.: Novel efficient RNN and LSTM-like architectures: recurrent and gated broad learning systems and their applications for text classification. IEEE Trans. Cybern. **51**, 1586–1597 (2020)
2. Goodfellow, I., Bengio, Y., Courville, A., Bengio, Y.: Deep Learning, vol. 1. MIT press, Cambridge (2016)
3. Kasun, L.L.C., Zhou, H., Huang, G.-B., Vong, C.M.: Representational learning with extreme learning machine for big data. IEEE Intell. Syst. **28**(6), 31–34 (2013)
4. Roul, R.K.: Impact of multilayer elm feature mapping technique on supervised and semi-supervised learning algorithms. Soft. Comput. **26**(1), 423–437 (2022). https://doi.org/10.1007/s00500-021-06387-9
5. Huang, G.B., Zhou, H., Ding, X., Zhang, R.: Extreme learning machine for regression and multiclass classification. IEEE Trans. Syst. Man Cybern. Part B (Cybernetics) **42**(2), 513–529 (2012)
6. Roul, R.K., Asthana, S.R., Kumar, G.: Study on suitability and importance of multilayer extreme learning machine for classification of text data. Soft. Comput. **21**(15), 4239–4256 (2017). https://doi.org/10.1007/s00500-016-2189-8
7. Hartigan, J.A., Wong, M.A.: Algorithm as 136: a k-means clustering algorithm. J. R. Stat. Soc. Ser. C (Appl. Stat.) **28**(1), 100–108 (1979)
8. Dreiseitl, S., Ohno-Machado, L.: Logistic regression and artificial neural network classification models: a methodology review. J. Biomed. Inform. **35**(5–6), 352–359 (2002)

9. Huang, G.-B., Chen, L., Siew, C.K., et al.: Universal approximation using incremental constructive feedforward networks with random hidden nodes. IEEE Trans. Neural Networks **17**(4), 879–892 (2006)
10. Roul, R.K., Rai, P.: A new feature selection technique combined with ELM feature space for text classification. In: Proceedings of the 13th International Conference on Natural Language Processing, pp. 285–292. ACL (2016)
11. Bengio, Y., LeCun, Y., et al.: Scaling learning algorithms towards AI. Large-scale kernel machines **34**(5), 1–41 (2007)
12. Roul, R.K., Agarwal, A.: Feature space of deep learning and its importance: comparison of clustering techniques on the extended space of ML-ELM. In: Proceedings of the 9th Annual Meeting of the Forum for Information Retrieval Evaluation, pp. 25–28 (2017)
13. Vapnik, V.N.: An overview of statistical learning theory. IEEE Trans. Neural Networks **10**(5), 988–999 (1999)
14. Roul, R.K.: Study and understanding the significance of multilayer-ELM feature space. In: Bellatreche, L., Goyal, V., Fujita, H., Mondal, A., Reddy, P.K. (eds.) BDA 2020. LNCS, vol. 12581, pp. 28–48. Springer, Cham (2020). https://doi.org/10.1007/978-3-030-66665-1_3

# Multi-objective Pelican Optimization Algorithm for Engineering Design Problems

Y. Ramu Naidu[✉]

School of Sciences, National Institute of Technology Andhra Pradesh,
Tadepalligudem, Andhra Pradesh, India
y.ramunaidu@gmail.com

**Abstract.** This work presents an efficient multi-objective version of the Pelican Optimization Algorithm (POA) which is recently proposed in the family of meta-heuristic algorithms. It is called a multi-objective Pelican Optimization Algorithm (MOPOA). From the literature, it is observed that the POA performed well on a set of unconstrained classical optimization problems as well as some engineering design problems. To extend its applicability to multi-objective engineering design models, the MOPOA has been proposed and applied for two engineering design models, four bar truss and speed reducer problems. The obtained results are compared with the literature and they proved that the MOPOA is an efficient and robust optimizer.

**Keywords:** Pelican optimization algorithm · Multi-objective problems · Engineering design problems

## 1 Introduction

Optimization means achieving the best result under a given set of circumstances. From the mathematics point of view, it is an objective method of finding the maximum or minimum value of a function under some conditions [1]. These conditions are called constraints. Optimization problems are classified into two types based on the number of objective functions. One is a single objective optimization problem, and the other is a multi-objective optimization problem (MOPs). In MOPs, there are more than one objective functions which are conflicted with each other. Unlike the traditional method with a single objective function, an optimum does not necessarily exist to optimize all objective functions simultaneously. That means, here, a set of optimal solutions is obtained, and all are incomparable. This set is called Pareto optimal set. A decision maker can choose the most compromised one from the set based on his requirements.

In recent days, multi-objective optimization has gained importance since real-world problems have been formulated as multi-objective optimization problems. To solve such problems, many multi-objective optimizers have been introduced by researchers worldwide since no optimizer exists to solve all kinds of optimization problems. For instance, Coello and Coello [2] introduced multi-objective Particle swarm optimization (MOPSO) to address the multi-objective optimization problems. In which, Particle swarm optimizer is the main algorithm, and a special mutation operator is incorporated to enrich the exploratory capabilities of the MOPSO. K. Deb and his team [3] proposed an elitist multi-objective genetic algorithm (NSGA-II). In the NSGA-II, the Genetic algorithm is the key algorithm, and the non-dominated sorting scheme is used to classify the candidate solutions into different levels (fronts). It also reduces the algorithm's computational complexity. Later, many multi-objective meta-heuristic algorithms have proposed [4–7].

## 2    Pelican Optimization Algorithm

Recently, the optimizer so-called Pelican Optimization Algorithm (POA) was proposed by Trojovský P. and Dehghani M [8], which is based on the social behavior and hunting strategy of pelicans. The mathematical model of the POA is as follows: Like swarm-based algorithms, the POA is also started with an initial population. This population is generated randomly according to the lower and upper bounds of the problem. The mathematical expression for the initialization of the population is

$$p_{i,j} = lb_j + rand * (ub_j - lb_j), \; i = 1, 2, \ldots, N \; \text{and} \; j = 1, 2, \ldots, D \qquad (1)$$

where $p_{i,j}$ is the $j$th value of the $i$th Pelican (solution). $lb_j$ and $ub_j$ are lower, and upper bounds of the $j$th variable of a problem, respectively. The value $N$ represents the population size and $D$ is the dimension of the domain of a problem. $rand$ is a uniform random number in $[0, 1]$.

The POA mimics two typical characteristics of Pelicans, social behavior and hunting strategy, to obtain new positions of candidate solutions in the search space. Especially the hunting strategy is modeled in two phases.

Phase-1: Moving towards prey (exploration phase): In this phase, the pelicans find the location of the prey and move toward the prey so that an algorithm can explore the new search areas in the domain. This phase is modeled, mathematically, as

$$p_i^{FP_1} = \begin{cases} p_i + rand \cdot (pr - I \cdot p_i), & f_{pr} < f_i \\ p_i + rand \cdot (p_i - pr), & \text{otherwise} \end{cases} \qquad (2)$$

where $p_i^{FP_1}$ is the new position of the $i$th Pelican $(p_i)$, $FP_1$ stands for the first phase. Here, $pr$ is the position of the prey and $f_{pr}$ is its fitness value, and $f_i$ is the fitness of the $i^{th}$ Pelican. The value of $I$ is either 1 or 2, taken randomly. It is noted that the prey is selected randomly from the population of Pelicans. Each Pelican updates its position according to Eq. (2).

Phase-2: Winging on the water surface (Exploitation Phase)
In the second phase, after reaching the water surface, the Pelicans spread and

beat their wings to scoop up the prey in the vicinity. This strategy is modeled, mathematically as:

$$p_i^{FP_2} = p_i + R\left(1 - \frac{it}{Maxit}\right)(2\ \text{rand}\ - 1)p_i \tag{3}$$

where $p_i^{FP_2}$ is the new position value of the $i$th Pelican. $R$ is a constant and its value is 2. The terms $it$, and $Maxit$ represent the current iteration number, and total number of iterations, respectively.

## 3 The Proposed MOPOA

This section presents the proposed multi-objective Pelican optimization algorithm (MOPOA). It combines a non-dominated sorting (NDS) approach and crowding distances (CD) method for maintaining diversity within the MOPOA algorithm. In NDS, all candidate solutions are ranked. The first-ranked candidate solutions are not dominated by any other candidate solutions, whereas the second-ranked candidate solutions are dominated by at least one of the first-ranked members and so on. The CD is utilized to maintain diversity among the same ranked candidate solutions.

---

**Algorithm 1.** Pseudo code of MOPOA

---

1: **function** MOPOA($N$, $D$, $M$, $Max - Gen, lb,\ ub$)
2:      Generate initial population ($P$)
3:      $g = 1$
4:      **while** $g < Max - Gen$ **do**
5:          Select $Pr$ from the front-1 candidates          ▷ Choose the prey randomly
6:          **for** $i = 1$ to $N$ **do**
7:              Generate the vector $I_{1 \times D}$ with the elements 1 or 2
8:              Find the new positions of Pelicans $NewP^1$ using Eqn. (2)
9:              Calculate the fitness values of $NewP_i^1$
10:          **end for**
11:          $TP = P \cup NewP^1$          ▷ Combine parent population and new population, $TP$-Total population
12:          Update the population using non-dominated rank and CD
13:          **for** $j = 1$ to $N$ **do**
14:              Generate new population $NewP_i^2$ using Eqn. (3)
15:          **end for**
16:          Update the population using non-dominated rank and CD
17:      **end while**
18: **end function**

---

It is crucial that the preponderant features of the POA (i.e. choosing the prey) be correctly outlined in order to make it into an efficient multi-objective optimization algorithm. Ordinary optimization problems have only one objective function to be optimized. In the traditional POA, the prey (solution) is selected randomly, and based on its fitness value, the new candidate solution is generated by Eq. (2). Nevertheless, in the MOO, there are more than one objective functions to be optimized simultaneously. Therefore, the process of selecting prey has to be changed in the MOPOA. Here, to select the prey, the ranking method is used. The entire process of the MOPOA is shown in Algorithm 1.

In Algorithm 1, the predefined parameters such as population size, number of objective functions, variable range, and maximum number of generations are initialized. Firstly, an initial population is created in the feasible space, and find fitness values for each candidate solution in the population. Next, use a non-dominated sorting process, assign the non-dominated rank and compute the CD for each candidate solution. Here, phase 1 starts. In this phase, select one candidate solution as prey and produce a new population by using Eq. (2). Then, combine the parent population and the new population to get the total population. Again, apply the non-dominated sorting process on the total population and assign non-dominated ranks and compute the CD. Based on these quantities, non-dominated ranks and CD choose the best candidate solutions for the second phase. In this phase, the new population is produced by Eq. (3). Now, again combine the parent population with the new population, which is obtained by Eq. (3) and assign non-dominated rank and calculate the CD for each solution. The combined population is sorted based on NDR and the CD. Lastly, select the best population of size $N$ for the next generation. This process is repeated until the termination condition is met.

## 4   Results and Discussions

In this section, two multi-objective mechanical design problems are studied from the literature to assess the performance of the proposed MOPOA. Two performance metrics, Generational Distance (GD) and metric of spacing (S) are considered in this work in order to measure the efficiency and spread of the MOPOA. The common parameter, population size, is 100.

### 4.1   Four-bar Truss Design Problem

The four-bar design problem is a well-known design problem, and many researchers have studied it [9,10]. The mathematical description is as follows:

$$\text{minimize} \begin{cases} F_1(x) = L\left(2x_1 + \sqrt{2x_2} + \sqrt{x_3} + x_4\right) \\ F_2(x) = \frac{FL}{E}\left(\frac{2}{x_2} + \frac{2\sqrt{2}}{x_2} - \frac{2\sqrt{2}}{x_3} + \frac{2}{x_4}\right) \end{cases}$$

subject to

$$\left(\frac{F}{\sigma}\right) \le x_1 \le 3 \times \left(\frac{F}{\sigma}\right), \ \sqrt{2} \times \left(\frac{F}{\sigma}\right) \le x_2 \le 3 \times \left(\frac{F}{\sigma}\right)$$
$$\sqrt{2} \times \left(\frac{F}{\sigma}\right) \le x_3 \le 3 \times \left(\frac{F}{\sigma}\right), \ \left(\frac{F}{\sigma}\right) \le x_4 \le 3 \times \left(\frac{F}{\sigma}\right)$$

where

$$F = 10\text{KN}, E = (2)10^5\text{KN/cm}^2, \ L = 200 \text{ cm}, \sigma = 10\text{KN/cm}^3.$$

The MOPOA is utilized to solve the four bass truss design. For a fair comparison, 10000 function evaluations are taken. The statistical measurements, mean value (mean), and standard deviation (SD) for 30 independent runs are taken into account, and the obtained results are reported in Table 1 and obtained Pareto fronts are shown in Fig. 1. The results are compared with that of NSGS-II [3], MOPSO [2], MWCA [10], micro-GA [11], and PAES [12]. In the case of the GD metric, the MOPOA performed very well compared to competitors. The obtained Pareto front by MOPOA is very close to the true Pareto front. It indicates that the MOPOA has a high convergence ability. From Table 1, it is observed that the mean value of $S$ metric is very less compared to other methods. It is evident that the MOPOA has good distribution among candidate solutions. From Fig. 1, we can observe that MOPOA is able to earn more Pareto optimal solutions on or near the true Pareto front. In both metrics, standard deviation values are very small. It shows the consistency of the MOPOA.

**Table 1.** Results for four-bar truss design problem

| Method | GD | | S | |
|---|---|---|---|---|
| | Mean | SD | Mean | SD |
| NSGA-II | 0.3601 | 0.0470 | 2.3635 | 0.2551 |
| MOPSO | 0.3741 | 0.0422 | 2.5303 | 0.2275 |
| Micro-GA | 0.9102 | 1.7053 | 8.2742 | 16.8311 |
| PAES | 0.9733 | 1.8211 | 3.2314 | 5.9555 |
| MOWCA | 0.2076 | 0.0055 | 2.5816 | 0.0298 |
| MOPOA | $2.57e-04$ | $1.28e-05$ | 0.0089 | 0.0014 |

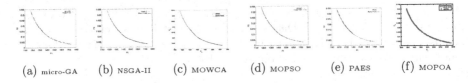

(a) micro-GA    (b) NSGA-II    (c) MOWCA    (d) MOPSO    (e) PAES    (f) MOPOA

**Fig. 1.** Best Pareto optimal fronts obtained by proposed MOPOA and competitors for the Four-bar truss design problem

### 4.2 Speed Reducer Design Problem

The second problem is speed reduced design problem, and it has been widely used in past work to assess the performance of the proposed methods. The mathematical definition of this problem is given below:

$$\text{Min } f_1(x) = 0.7854x_1x_2^2 \left(10x_3^2/3 + 14.933x_3 - 43.0934\right) - 1.508x_1\left(x_6^2 + x_7^2\right)$$
$$+ 7.477\left(x_6^3 + x_7^3\right) + 0.7854\left(x_4x_6^2 + x_5x_7^2\right)$$

$$\text{Min } f_2(x) = \frac{\sqrt{\left(745.0x_4/x_2x_3\right)^2 + 1.69 \times 10^7}}{0.1x_6^3}$$

subject to

$$g_1(x) = \frac{1.0}{x_1 x_2^2 x_3} - \frac{1.0}{27.0} \leq 0, \ g_2(x) = \frac{1.0}{x_1 x_2^2 x_3^2} - \frac{1.0}{397.5} \leq 0, \ g_3(x) = \frac{x_4^3}{x_2 x_3 x_6^4} - \frac{1.0}{1.93} \leq 0$$

$$g_4(x) = \frac{x_5^3}{x_2 x_3 x_7^4} - \frac{1.0}{1.93} \leq 0, \ g_5(x) = x_2 x_3 - 40.0 \leq 0, \ g_6(x) = \frac{x_1}{x_2} - 12 \leq 0$$

$$g_7(x) = 5 - \frac{x_1}{x_2} \leq 0, \ g_8(x) = 1.9 - x_4 + 1.5 x_6 \leq 0, \ g_9(x) = 1.9 - x_5 + 1.1 x_7 \leq 0$$

$$g_{10}(x) = \frac{\sqrt{(745 x_4 / x_2 x_3)^2 + 1.69 \times 10^7}}{0.1 x_6^3} - 1300 \leq 0$$

$$g_{11}(x) = \frac{\sqrt{(745 x_5 / x_2 x_3)^2 + 1.575 \times 10^8}}{0.1 x_7^3} - 1100 \leq 0$$

$$2.6 \leq x_1 \leq 3.6, \ 0.7 \leq x_2 \leq 0.8, \ 17 \leq x_3 \leq 28, \ 7.3 \leq x_4 \leq 8.3,$$
$$7.3 \leq x_5 \leq 8.3 \ 2.9 \leq x_6 \leq 3.9, \ 5.0 \leq x_7 \leq 5.5$$

It is solved by the MOPOA, and results are presented in Table 2 and the obtained Pareto front is shown graphically in Fig. 2. For a fair comparison, 15000 function evaluations are taken. The obtained results are compared with NSGA-II [3], MOALO [9], MOWCA [10], Micro-GA [11], and PAES [12]. From Table 2, it is confirmed that the MOPOA outperformed other competitors in terms of GD. It is observed that the MOPOA produces less spacing metric value compared to other algorithms which are considered in this work, i.e., the MOPOA achieves a good uniform distribution among Pareto optimal solutions. All our claims are supported by graphical representation. The standard deviation values of the GD and spacing metric, which are obtained by the MOPOA, are very less compared to others. It shows the robustness of the MOPOA in solving engineering design problems.

**Table 2.** Results for speed reducer design problem

| Method | GD | | S | |
|---|---|---|---|---|
| | Mean | SD | Mean | SD |
| NSGA-II | 9.843702 | 7.08103039 | 2.765449155 | 3.53493787 |
| Micro-GA | 3.117536 | 1.67810867 | 47.80098 | 32.80151572 |
| PAES | 77.99834 | 4.21026087 | 16.20129 | 4.26842769 |
| MOWCA | 0.98831 | 0.17894217 | 16.68520 | 2.69694436 |
| MOALO | 1.1767 | 0.2327 | 1.7706 | 2.769 |
| MOPOA | 0.0027 | 8.0984e − 04 | 0.0012 | 2.2174e − 04 |

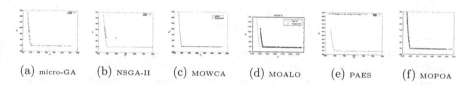

(a) micro-GA  (b) NSGA-II  (c) MOWCA  (d) MOALO  (e) PAES  (f) MOPOA

**Fig. 2.** Best Pareto optimal fronts obtained by proposed MOPOA and competitors for the Speed reducer design problem

# 5  Conclusion

This paper presents a new version of the POA called the multi-objective Pelican optimization algorithm. In the proposed method, elitist non-dominated sorting and crowding distance are adopted to enrich the convergence and diversity among solutions. To demonstrate the efficiency and robustness of the proposed multi-optimizer, two engineering problems, four bar truss design, and speed reducer problems are solved, and results are reported in terms of statistical measurements like mean value and standard deviation. Apart from these measurements, graphical representations are also reported. The obtained results are compared with the literature. From the results and graphical representations, it is concluded that the MOPOA outperformed competitors in terms of the GD and metric of spacing. In future research work, it is recommended to apply MOPOA to multi-modal multi-objective optimization.

# References

1. Rao, S.S.: Engineering Optimization, 4th edn. John Wiley & Sons Inc, New Jersey (2009)
2. Coello, C.A.C., Pulido, G.T., Lechuga, M.S.: Handling multiple objectives with particle swarm optimization. IEEE Trans. Evol. Comput. **8**, 256–279 (2004)
3. Deb, K., Pratap, A., Agarwal, S., Meyarivan, T.: A fast and elitist multiobjective genetic algorithm: NSGA-II. IEEE Trans. Evol. Comput. **6**, 182–197 (2002)
4. Naidu, Y.R., Ojha, A.K.: Solving multiobjective optimization problems using hybrid cooperative invasive weed optimization with multiple populations. IEEE Trans. Syst. Man Cybern. Syst. **48**(6), 821–832 (2018)
5. Ramu Naidu, Y., Ojha, A.K., Susheela Devi, V.: Multi-objective jaya algorithm for solving constrained multi-objective optimization problems. In: Kim, J.H., Geem, Z.W., Jung, D., Yoo, D.G., Yadav, A. (eds.) ICHSA 2019. AISC, vol. 1063, pp. 89–98. Springer, Cham (2020). https://doi.org/10.1007/978-3-030-31967-0_11
6. Abdel-Basset, M., et al.: MOEO-EED: a multi-objective equilibrium optimizer with exploration-exploitation dominance strategy. Knowl. Based Syst. **214**, 106717 (2021)
7. Got, A., Zouache, D., Moussaoui, A.: MOMRFO: multi-objective manta ray foraging optimizer for handling engineering design problems. Knowl. Based Syst. **237**, 107880 (2022)
8. Trojovský, P., Dehghani, M.: Pelican optimization algorithm: a novel nature-inspired algorithm for engineering applications. Sensors **22**, 855 (2022). https://doi.org/10.3390/s22030855
9. Mirjalili, S., Jangir, P., Saremi, S.: Multi-objective ant lion optimizer: a multi-objective optimization algorithm for solving engineering problems. Appl. Intell. **46**(1), 79–95 (2016). https://doi.org/10.1007/s10489-016-0825-8
10. Sadollah, A., Eskandar, H., Kim, J.H.: Water cycle algorithm for solving constrained multi-objective optimization problems. Appl. Soft Comput. **27**, 279–298 (2015)
11. Coello, C.C., Pulido, G.T.: Multiobjective structural optimization using a micro-genetic algorithm. Struct. Multidiscip. Optim. **30**, 388–403 (2005)
12. Knowles, J., Corne, D.: The Pareto archived evolution strategy: a new baseline algorithm for multiobjective optimization. In: Proceedings of the 1999 Congress on Evolutionary Computation, pp. 98–105. IEEE Press, Piscataway, NJ (1999)

# Prediction of Accident and Accident Severity Based on Heterogeneous Data

Sneha Kandacharam[✉] and B. Rajathilagam

Department of Computer Science and Engineering, Amrita School of Computing, Amrita Vishwa Vidyapeetham, Coimbatore, India
k_sneha@cb.students.amrita.edu, b_rajathilagam@cb.amrita.edu

**Abstract.** Studies of traffic accident analysis, as well as prediction, have traditionally relied on small-scale datasets with limited coverage, so limiting the scope and usefulness of these analyses. There is also the issue that many large-scale databases are either confidential, outdated, or missing important contextual factors like environmental stimuli (weather, points of interest, etc.). There are presently 37 million records stored in the US Accidents dataset, including crashes that occurred anywhere in the 48 contiguous states between 2016 and 2021. We were able to piece together details like date, time, place, weather, season, and landmarks from this information. Used deep neural networks that have a trainable embedding component, a fully connected network, and a recurrent network for time-sensitive data and time-insensitive data, respectively (for capturing spatial heterogeneity). Our research includes the prediction of the occurrence of accident incidents using deep neural networks and an understanding of Accident Severity against machine learning models.

**Keywords:** Accident prediction · Deep Learning. Accident Severity

## 1 Motivation

Around the globe, 1.35 million people died in vehicle collisions in 2016. A traffic accident resulted in 20 to 50 million additional injuries or disabilities. For young people and children aged 5 to 29 as well, vehicle collisions are the major cause of death. Death rates are three times greater in developing nations than in developed nations, which are gradually becoming more motorized. Both the World Bank and the WHO has said that governments must take action to minimise the number of vehicle collisions since it is much high in both developing and developed nations. There is a significant economic cost associated with traffic accidents in addition to the social cost. According to several studies, traffic collisions may cost nations 2% of their GDP. As per the World Bank "halving deaths and injuries due to road traffic could potentially add 22% to GDP per capita in Thailand, 15% in China, 14% in India, over 2014–2038." Developing countries face a substantial economic challenge as a result of traffic accidents. Therefore, both developing and developed nations must prioritize lowering traffic accidents and enhancing road safety.

A. R. Molla et al. (Eds.): ICDCIT 2023, LNCS 13776, pp. 369–374, 2023.
https://doi.org/10.1007/978-3-031-24848-1_29

## 2  Previous Work

Using deep learning algorithms is a cutting-edge method that may be used to find patterns and structures in high-dimensional data, develop learning patterns, and find correlations in the data that go beyond immediate neighbours [1]. Deep learning has been used in a variety of applications, like signal processing, speech recognition, computer vision, as well as NLP [2].

Ref [3] proposed a CNN + LSTM model, which was used to detect traffic events, such as accidents, using a labelled dataset built of traffic-related data extracted from Twitter data. This model was concerned with ensembled deep learning frameworks applied to vehicle collisions prediction and analysis.

Using a mix of CNN and LSTM networks, the authors of [4] developed a technique for real-time vehicle collisions risk prediction on urban arterials. A SdAE (Stack Denoise Convolutional Auto-Encoder) with 8 hidden layers and a batch normalising technique was suggested in Ref. [5]. An ensemble model comprising the LSTM layer, hybrid LSTM-CNN layer, and CNN layer made up the spatio-temporal convolutional LSTM ("Long short-term memory"), or STCL-Net Model, which was presented by [6]. The authors used their model to forecast vehicle collisions in New York City using a variety of spatio-temporal combinations and various time and spatial grid configurations. A deep learning model with three layers—a spatio-temporal layer, an embedding layer, and a spatial layer —known as DSTGCN or "Deep Spatio-Temporal Graph Convolutional Network" was introduced in Reference [7].

## 3  Proposed Solution

The model that is the focus of this research makes use of a variety of inputs to better represent spatial as well as temporal heterogeneity. In context of embedding representation, we are capable of extracting latent spatiotemporal characteristics. Grid-search was used to do hyper-parameter tuning to determine the ideal number of recurrent layers (options included {1, 2, 3}); the ideal recurrent cell type (options include {GRU, RNN, and LSTM}); grid-cell embedding vector size (options include {50, 100, and 150}); and activation function ({sigmoid, ReLU, or tanh}) for each fully connected layer. To train the model, we utilized the Adam optimizer [8] with a 0.001 initial learning rate. The following are available in the model.

- Recurrent Component: We employ a series of 8 vectors, each of size 24 (i.e., time-variant qualities), which may be seen as a series of such vectors (provided their temporal order), to define our prediction framework; consequently, the recurrent neural network models could be useful to us. We adopt an LSTM model [9] that consists of two recurrent layers, each with 128 LSTM cells. The result is a vector with a size of 128.
- Embedding Component: This component, when provided the grid cell index, produces a distributed representation of that cell that contains crucial data about traffic characteristics, geographic heterogeneity, and the influence of other environmental stimuli on accident incidence. As we train the whole pipeline, this distributed representation

will be generated. We use the sigmoid activation function on a feed-forward layer with a size of 128 and feed this representation to it. The embedding matrix has a dimension of $|R| \times 128$; the input dataset's grid-cell regions are collected in a set called R.

- Description-to-Vector Component: It makes use of Desc2Vec data, which is a plain language description of past traffic occurrences in a grid cell. Using the sigmoid activation function, we send the Desc2Vec of a grid cell to a feed-forward layer with a size of 128.
- Points-of-Interest Component: It makes use of points-of-interest data, which is spatial attribute representation (a vector of size 13). We apply the sigmoid activation function to a feed-forward layer with a size of 128 and a POI vector.
- Fully-connected Component: The final prediction is made by this component using the results of the previous components. Here, there are four dense layers with the corresponding sizes of 512, 256, 64, and 2. After $2^{nd}$ and $3^{rd}$ layers, we also use batch normalization [10] to accelerate the training process. The output of the final layer is subjected to Softmax after using ReLU as the activation function for the preceding three layers (Fig. 1).

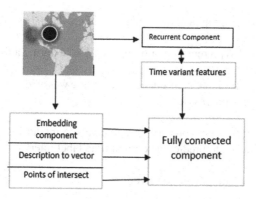

**Fig. 1.** Overview of the proposed model

## 4 Materials, Procedure

Moosavi and his colleagues [11] have created a US Accident Dataset that includes data from all 49 states. Since February 2016, data was continually gathered with various data sources, like many APIs that provide streaming traffic event data [11]. There are traffic events like Broken Vehicle, Accident, Construction, Congestion, Lane Blocked, Event, Flow incident (Table 1).

We frame the issue as follows in light of the preliminary information:

Given:

– "A spatial grid $R = \{r_1, r_2, \dots, r_n\}$, where every $r \in R$ represents a 2 km × 2 km geographical area.

**Table 1.** Shows the attributes of the dataset.

| Total Attributes | 45 |
| --- | --- |
| Traffic Attributes | Source, id, severity, TMC, starting time, starting point, ending time, end_point, description, and distance |
| Address Attributes | Street, number, city, side (right/left), state, county, country, zip-code |
| Weather Attributes | Temperature, time, humidity, wind chill, visibility, pressure, precipitation, wind speed, wind direction, and conditions (such as snow, rain, etc.) |
| POI Attributes | Give-way, Amenity, Junction, Railway, No-exit, Station, Roundabout, Traffic Calming, Stop, Turning Loop, Traffic Signal |

- A series of fixed-length time period $T = \{t_1,t_2,...,t_m\}$, where |t| is set to 10 min, for $t \in T$.
- For each $r \in R$ geographical region, a database of traffic occurrences $E_r = \{e_1, e_2,...\}$.
- For every $r \in R$ geographical area, weather observation records database $W_r = \{w_1,w_2,...\}$.
- For every $r \in R$ geographical area, points of interest database $P_r = \{p_1,p_2,...\}$.
Create:
- Using $P_r$, $E_r$, and $W_r$ a representation $F_{rt}$ for an area during a time $t \in T$ for which $r \in R$.
- a binary label $L_{rt}$ for $F_{rt}$, where 1 denotes that there was at least one collision during t in area r and 0 denotes otherwise.
Find:
- Using data from the previous six-time intervals, a model M to forecast $L_{rt}$ with $\langle F_{r\,ti-6}, F_{r\,ti-5},..., F_{r\,ti-1}\rangle$, to predict the current time label interval".
Objective:
- Attempt to reduce the prediction error.

## 5   Results

The test data used in the cross-validation procedure served as the basis for all outcomes, including performance ratings, which allowed the model to be assessed using real-world data that it had never observed before. The scikit-learn software was used to generate all graphs and metrics. Each class was equally represented in the test data provided to the model. At the ideal threshold, the accuracy scores were finally attained (taking into account other measures like recall and precision).

Figure 2 displays the ROC curve for Atlanta. The accuracy was 87% and the AUROC was 0.87. An entirely naive classifier is shown by the dotted line in the centre. The confusion matrix for the Atlanta model is demonstrated in Table 2, and the accuracy, f1 scores, as well as recall that were chosen for an ideal threshold are shown in Table 3. These results demonstrate that the scores accurately represent the model's actual performance since the support for each class is almost substantial and equal.

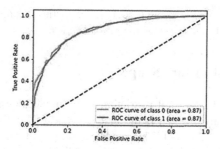

**Fig. 2.** ROC curve for Atlanta city

**Table 2.** Confusion matrix for Atlanta city

| Class | Actual 0 | Actual 1 |
| --- | --- | --- |
| Predicted 0 | 11970 | 10880 |
| Predicted 1 | 1960 | 2630 |

**Table 3.** Atlanta model class-based scores

| | Recall | Precision | F1-Score |
| --- | --- | --- | --- |
| Class 0 | 0.75 | 0.93 | 0.83 |
| Class 1 | 0.94 | 0.8 | 0.86 |

**Table 4.** When considering the Accident Severity attribute, with different baseline models, we found the following accuracies.

| Model | Accuracy |
| --- | --- |
| Logistic Regression | 95.4% |
| K Nearest Neighbours | 93.5% |
| Decision Tree (gini) | 96.8% |
| Random Forest | 97.4% |

Random forest tells us about the important features of the model as well.

## 6   Conclusion

Before accidents ever occur, real-time accident prediction algorithms may assist in allocating the appropriate emergency resources. We deployed a deep-learning algorithm to anticipate traffic accidents. We discovered that this model is superior for the accident classification task. As a consequence, the prediction model was able to provide more reliable and objective findings due to all of the attributes the dataset has. We compared different traditional machine learning models for severities that are caused by accidents and found that random forest provides the best accuracy.

## 7 Limitations and Future Scope

The use of this model has several restrictions. The model doesn't take into account driver and vehicle features, which is the first drawback. This three-dimensional data is not compatible with the model. This model doesn't train depending on severity of accident, which is another drawback. It considers all accidents equally, irrespective of the number of deaths or the level of damage. Another extension would be to develop a comprehensive system that would enable municipal authorities to collect information, train, and forecast vehicle collisions with a few simple clicks.

## References

1. Bengio, Y., Courville, A., Vincent, P.: Representation learning: a review and new perspectives. IEEE Trans. Pattern Anal. Mach. Intell. **35**, 1798–1828 (2013)
2. Najafabadi, M.M., Villanustre, F., Khoshgoftaar, T.M., Seliya, N., Wald, R., Muharemagic, E.: Deep learning applications and challenges in big data analytics. J. Big Data **2**(1), 1–21 (2015). https://doi.org/10.1186/s40537-014-0007-7
3. Dabiri, S., Heaslip, K.: Developing a Twitter-based traffic event detection model using deep learning architectures. Expert Syst. Appl. **118**, 425–439 (2019)
4. Li, P., Abdel-Aty, M., Yuan, J.: Real-time crash risk prediction on arterials based on LSTM-CNN. Accid. Anal. Prev. **135**, 105371 (2020)
5. Chen, C.. Fan, X., Zheng, C., Xiao, L., Cheng, M., Wang, C.: SDCAE: stack denoising con-volutional autoencoder model for accident risk prediction via traffic big data. In: Proceedings of the Sixth International Conference on Advanced Cloud and Big Data (CBD), Lanzhou, China, 15 August 2018, pp. 328–333. IEEE, New York (2018)
6. Bao, J., Liu, P., Ukkusuri, S.V.: A spatiotemporal deep learning approach for citywide short-term crash risk prediction with multi-source data. Accid. Anal. Prev. **122**, 239–254 (2019)
7. Yu, L., Du, B., Hu, X., Sun, L., Han, L., Lv, W.: Deep Spatio-temporal graph convolutional network for traffic accident prediction. Neurocomputing **423**, 135–147 (2020)
8. Kingma, D.P., Ba, J.: Adam: A method for stochastic optimization. arXiv preprint arXiv: 1412.6980 (2014)
9. Hochreiter, S., Schmidhuber, J.: Long short-term memory. Neural Comput. **9**(8), 1735–1780 (1997)
10. Ioffe, S., Szegedy, C.: Batch normalization: accelerating deep network training by reducing internal covariate shift. arXiv preprint arXiv:1502.03167 (2015)
11. Moosavi, S., Samavatian, M. H., Parthasarathy, S., Ramnath, R.: A Countrywide Traffic Accident Dataset. *arXiv*. https://doi.org/10.48550/arXiv.1906.05409 (2019)

# Author Index

Printed in the United States
by Baker & Taylor Publisher Services

Printed in the United States
by Baker & Taylor Publisher Services